Modeling Dynamic Systems

Series Editors

Matthias Ruth
Bruce Hannon

Springer
New York
Berlin
Heidelberg
Hong Kong
London
Milan
Paris
Tokyo

MODELING DYNAMIC SYSTEMS

Modeling Dynamic Biological Systems
Bruce Hannon and Matthias Ruth

Modeling Dynamic Economic Systems
Matthias Ruth and Bruce Hannon

Dynamic Modeling in the Health Sciences
James L. Hargrove

Modeling and Simulation in Science and Mathematics Education
Wallace Feurzeig and Nancy Roberts, Editors

Dynamic Modeling of Environmental Systems
Michael L. Deaton and James J. Winebrake

Dynamic Modeling, Second Edition
Bruce Hannon and Matthias Ruth

Modeling Dynamic Climate Systems
Walter A. Robinson

Dynamic Modeling for Marine Conservation
Matthias Ruth and James Lindholm, Editors

Dynamic Modeling for Business Management: An Introduction
Bernard McGarvey and Bruce Hannon

Landscape Simulation Modeling: A Spatially Explicit, Dynamic Approach
Robert Costanza and Alexey Voinov, Editors

For more information, see:
`www.springer-ny.com/mds`

Bernard McGarvey Bruce Hannon

Dynamic Modeling for Business Management

An Introduction

With 166 Illustrations and a CD-ROM

Springer

Bernard McGarvey
Process Engineering Center
Drop Code 3127
Eli Lilly and Company
Lilly Corporate Center
Indianapolis, IN 46285
USA

Bruce Hannon
Department of Geography
220 Davenport Hall, MC 150
University of Illinois
Urbana, IL 61801
USA

Cover illustration: Top panel—The model with the controls on ORDERING and SELLING.
Bottom panel—Photo by William F. Curtis.

Library of Congress Cataloging-in-Publication Data
Hannon, Bruce M.
Dynamic modeling for business management: an introduction / Bruce Hannon, Bernard McGarvey.
p. cm.
ISBN 0-387-40461-9 (cloth: alk. paper)
1. Management—Mathematical models. 2. Digital computer simulation.
I. McGarvey, Bernard. II. Title.
HD30.25.H348 2003
519.7′03—dc21 2003054794

ISBN 0-387-40461-9 Printed on acid-free paper.

Printed in the United States of America.

9 8 7 6 5 4 3 2 1 SPIN 10938669

www.springer-ny.com

Springer-Verlag New York Berlin Heidelberg
A member of BertelsmannSpringer Science+Business Media GmbH

Series Preface

The world consists of many complex systems, ranging from our own bodies to ecosystems to economic systems. Despite their diversity, complex systems have many structural and functional features in common that can be effectively simulated using powerful, user-friendly software. As a result, virtually anyone can explore the nature of complex systems and their dynamical behavior under a range of assumptions and conditions. This ability to model dynamic systems is already having a powerful influence on teaching and studying complexity.

The books in this series will promote this revolution in "systems thinking" by integrating skills of numeracy and techniques of dynamic modeling into a variety of disciplines. The unifying theme across the series will be the power and simplicity of the model-building process, and all books are designed to engage the reader in developing their own models for exploration of the dynamics of systems that are of interest to them.

Modeling Dynamic Systems does not endorse any particular modeling paradigm or software. Rather, the volumes in the series will emphasize simplicity of learning, expressive power, and the speed of execution as priorities that will facilitate deeper system understanding.

Matthias Ruth and Bruce Hannon

Preface

The problems of understanding complex system behavior and the challenge of developing easy-to-use models are apparent in the field of business management. We are faced with the problem of optimizing economic goals while at the same time managing complicated physical and social systems. In resolving such problems, many parameters must be assessed. This requires tools that enhance the collection and organization of data, interdisciplinary model development, transparency of models, and visualization of the results. Neither purely mathematical nor purely experimental approaches will suffice to help us better understand the world we live in and shape so intensively.

Until recently, we needed significant preparation in mathematics and computer programming to develop, run, and interpret such models. Because of this hurdle, many have failed to give serious consideration to preparing and manipulating computer models of dynamic events in the world around them. Such obstacles produced models whose internal workings generally were known to only one person. Other people were unsure that the experience and insights of the many experts who could contribute to the modeling project were captured accurately. The overall trust in such models was limited and, consequently, so was the utility. The concept of team modeling was not practical when only a few held the high degree of technical skill needed for model construction. And yet everyone agreed that modeling a complex management process should include all those with relevant expertise.

This book, and the methods on which it is built, will empower us to model and analyze the dynamic characteristics of human–production environment interactions. Because the modeling is based on the construction of icon-based diagrams using only four elementary icons, the modeling process can quickly involve all members of an expert group. No special mathematical or programming experience is needed for the participants. All members of the modeling team can contribute, and each of them can tell immediately if the model is capturing his or her special expertise. In this way, the knowledge of all those involved in the question can be captured faithfully and in an agreeable manner. The model produced by such a team is useful, and those who made it will recommend it throughout the organization.

Such a model includes all the appropriate feedback loops, delays, and uncertainties. It provides the organization with a variety of benefits. The modeling effort highlights the gaps in knowledge about the process; it allows the modeling of a variety of scenarios; it reveals normal variation in a system; and, of course, it gives quantitative results. One of the more subtle values of team modeling is the emergence of a way of analogously conceiving the process. The model structure provides a common metaphor or analogous frame for the operation of the process. Such a shared mental analogue greatly facilitates effective communication in the organization.

Our book is aimed at several audiences. The first is the business-school student. Clearly, those being directly prepared for life in the business world need to acquire an understanding of how to model as well as the strengths and limitations of models. Students in industrial engineering often perform modeling exercises, but they often miss the tools and techniques that allow them to do group dynamic modeling. We also believe that students involved in labor and industrial relations should be exposed to this form of business modeling. The importance of the dynamics of management and labor involvement in any business process is difficult to overstate. Yet these students typically are not exposed to such modeling. In short, we want this book to become an important tool in the training of future process and business managers.

Our second general audience is the young M.B.A., industrial engineer, and human-resources manager in their first few years in the workplace. We believe that the skills acquired through dynamic modeling will make them more valued employees, giving them a unique edge on their more conventionally trained colleagues. This book is an introductory text because we want to teach people the basics before they try to apply the techniques to real-world situations. Many times, the first model a person will build is a complex model of an organization. Problems can result if the user is not grounded in the fundamental principles. It is like being asked to do calculus without first doing basic algebra.

Computer modeling has been with us for nearly 40 years. Why then are we so enthusiastic about its use now? The answer comes from innovations in software and powerful, affordable hardware available to every individual. Almost anyone can now begin to simulate real-world phenomena on his or her own, in terms that are easily explainable to others. Computer models are no longer confined to the computer laboratory. They have moved into every classroom, and we believe they can and should move into the personal repertoire of every educated citizen.

The ecologist Garrett Hardin and the physicist Heinz Pagels have noted that an understanding of system function, as a specific skill, must and can become an integral part of general education. It requires recognition that the human mind is not capable of handling very complex dynamic models by itself. Just as we need help in seeing bacteria and distant stars, we need help modeling dynamic systems. For instance, we solve the crucial dynamic modeling problem of ducking stones thrown at us or safely crossing busy streets. We learned to solve these problems by being shown the logical outcome of mistakes or through survivable accidents of judgment. We experiment with the real world as children and get hit by hurled stones; or we let adults play out their mental model of the consequences for us, and we believe them. These actions are the result of experimental and predictive models, and they begin to occur at an early age. These models allow us to develop intuition about system behavior. So long as the system remains reasonably stable, this intuition can serve us well. In our complex social, economic, and ecological world, however, systems rarely remain stable for long. Consequently, we cannot rely on the completely mental model for individual or especially for group action, and often, we cannot afford to experiment with the system in which we live. We must learn to simulate, to experiment, and to predict with complex models.

Many fine books are available on this subject, but they differ from ours in important ways. The early book edited by Edward Roberts, *Managing Applications of System Dynamics* (Productivity Press, 1978), is comprehensive and yet based on Dynamo, a language that requires substantial effort to learn. *Factory Physics,* by Wallace Hopp and Mark Spearman (Irwin/McGraw-Hill, 1996), focuses on the behavior of manufacturing systems. They review the past production paradigms and show how dynamic modeling processes can improve the flow of manufacturing lines. *Business Dynamics,* by John Sterman (Irwin/McGraw-Hill, 2000), is a clear and thorough exposition of the modeling process and the inherent behavior of various if somewhat generic modeling forms.

In a real sense, our book is a blend of all three of these books. We focus on the use of ithink®, with its facility for group modeling, and show how it can be used for very practical problems. We show how these common forms of models apply to a variety of dynamic situations in industry and commerce. The approach we use is to start from the simplest situation and then build up complexity by expanding the scope of the process. After first giving the reader some insight into how to develop ithink models, we begin by presenting our view of why dynamic modeling is important and where it fits. Then we stress the need for system performance measures that must be part of any useful modeling activity. Next we look at single- and multistep workflow processes, followed by models of risk management, of the producer/customer interface, and then supply chains. Next we examine the tradeoffs between quality, production speed, and cost. We close with chapters on the management of strategy and what we call *business learning systems.* By covering a wide variety of topics, we hope to impress on the reader just how easy it is to apply modeling techniques in one situation to another that initially might look different. We want to stress commonality, not difference!

In this book, we have selected the modeling software ithink with its iconographic programming style. Programs such as ithink are changing the way in which we think. They enable each of us to focus and clarify the mental model we have of a particular phenomenon, to augment it, to elaborate it, and then to do something we cannot otherwise do: find the inevitable dynamic consequences hidden in our assumptions and the structure of the model. ithink and the Macintosh, as well as the new, easy-to-use, Windows®-based personal computers, are not the ultimate tools in this process of mind extension. However, the relative ease of use of these tools makes the path to freer and more powerful intellectual inquiry accessible to every student. Whether you are a whiz at math or somewhat of a novice is irrelevant. This is a book on systems thinking and on learning how to translate that thinking into specific, testable models.

Finally, we wish to thank Tina Prow for a thorough edit of this book.

Bernard McGarvey, Indianapolis, Indiana, and
Bruce Hannon, Urbana, Illinois
Summer 2003

Contents

1

Introduction to Dynamic Modeling

> Indeed, from Pythagoras through pyramidology, extreme irrationalities have often been presented in numerical form. Astrology for centuries used the most sophisticated mathematical treatments available—and is now worked out on computers: though there is, or used to be, an English law which provided that "every person pretending or professing to tell Fortunes, or using any subtle Craft, Means or Device . . . shall be deemed a Rogue and Vagabond."
>
> —R. Conquest, *History, Humanity and Truth*

1.1 Introduction

Yet, we hope that our readers will see us neither as vagabonds nor rogues. We do think that modeling *is* a subtle craft, an art form that is intended to help us understand the future. And because of the complexity of dynamic systems, the use of numbers is essential. The use of numbers is needed to dispel complexity, not create it. The use of numbers forces us to be specific. Good dynamic modeling is an art. We cannot teach you THE METHOD for this modeling, for there is none.

Modeling dynamic systems is central to our understanding of real-world phenomena. We all create mental models of the world around us, dissecting our observations into cause and effect. Such mental models enable us, for example, to cross a busy street successfully or hit a baseball. But we are not mentally equipped to go much further. The complexities of social, economic, or ecological systems literally force us to use aids if we want to understand much of anything about them.

With the advent of personal computers and graphical programming, we can all create more complex models of the phenomena in the world around us. As Heinz Pagels noted in *The Dreams of Reason* in 1988, the computer modeling process is to the mind what the telescope and the microscope are to the eye. We can model the macroscopic results of microphenomena, and vice versa. We can simulate the various possible futures of a dynamic process. We can begin to explain and perhaps even to predict.

In order to deal with these phenomena, we abstract from details and attempt to concentrate on the larger picture—a particular set of features of the real world or the structure that underlies the processes that lead to the observed outcomes. Models are such abstractions of reality. Models force us to face the results of the structural and dynamic assumptions we have made in our abstractions.

The process of model construction can be rather involved. However, it is possible to identify a set of general procedures that are followed frequently. These general procedures are shown in simplified circular form in figure 1.1.

Models help us understand the dynamics of real-world processes by mimicking with the computer the actual but simplified forces that are assumed to result in a system's behavior. For example, it may be assumed that the number of people migrating from one country to another is directly proportional to the population living in each country and decreases the farther these countries are apart. In a simple version of this migration model, we may abstract away from a variety of factors that impede or stimulate migration in addition to factors directly related to the different population sizes and distance. Such an abstraction may leave us with a sufficiently good predictor of the known migration rates, or it may not. If it does not, we reexamine the abstractions, reduce the assumptions, and retest the model for its new predictions. Models help us in the organization of our thoughts, data gathering, and evaluation of our knowledge about the mechanisms that lead to the system's change. For example, here is what Daniel Botkin said in 1977 in his *Life and Death in the Forest:*

> One can create a computer model of a forest ecosystem, consisting of a group of assumptions and information in the form of computer language commands and numbers. By operating the model the computer faithfully and faultlessly demonstrates the implications of our assumptions and information. It forces us to see the implications, true or false, wise or foolish, of the assumptions we have made. It is not so much that we want to believe everything that the computer tells us, but that we want a tool to confront us with the implications of what we think we know. (217)

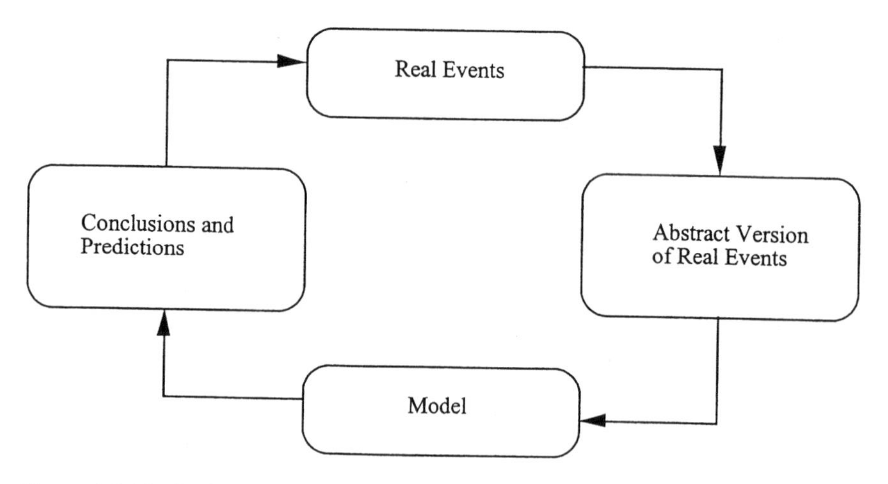

FIGURE 1.1. The basic model configuration

Some people raise philosophical questions as to why one would want to model a system. As pointed out earlier, we all perform mental models of every dynamic system we face. We also learn that in many cases, those mental models are inadequate. We can now specifically address the needs and rewards of modeling.

Throughout this book, we encounter a variety of nonlinear, time-lagged feedback processes, some with random disturbances that give rise to complex system behavior. Such processes can be found in a large range of systems. The variety of models in the companion books naturally span only a small range—but the insights on which these models are based can (and should!) be used to inform the development of models for systems that we do not cover here. The models of this book provide a basis for the formation of analogies.

It is our intention to show you how to model, not how to use models or how to set up a model for someone else's use. The latter two are certainly worthwhile activities, but we believe that the first step is learning the modeling process. In the following section we introduce you to the computer language that is used throughout the book. This computer language will be immensely helpful as you develop an understanding of dynamic systems and skills for analogy formation. Models are developed in nearly every chapter, and we have put most of the developing and final versions of these models on the CD at the back of the book. We refer in some cases to details of the model that are only found in the CD version.

1.2 Static, comparative static, and dynamic models

Most models fit in one of three general classes. The first type consists of *static models* that represent a particular phenomenon at a point of time. For example, a map of the United States may depict the location and size of a city or the rate of infection with a particular disease, each in a given year. The second type, *comparative static models*, compare some phenomena at different points in time. This is like using a series of snapshots to make inferences about the system's path from one point in time to another without modeling that process.

Some models describe and analyze the very processes underlying a particular phenomenon. An example of this would be a mathematical model that describes the demand for and supply of a good as a function of its price. If we choose a static modeling approach, we may want to find the price under which demand and supply are in equilibrium and investigate the properties of this equilibrium: Is the equilibrium price a unique one or are there other prices that balance demand and supply? Is the equilibrium stable, or are small perturbations to the system accompanied by a movement away from the equilibrium? Such equilibrium analysis is widespread in economics.

Alternatively, a third type of model could be developed to show the *changes* in demand and supply *over time*. These are *dynamic models*. Dynamic models try to reflect changes in real or simulated time and take into account that the model components are constantly evolving as a result of previous actions.

With the arrival of easy-to-use computers and software, we can all build on the existing descriptions of a system and carry them further. The world is not a static or comparative static process. The models treating it in that way may be misleading and become obsolete. We can now investigate in great detail and with great precision the system's behavior over time, including its movement toward or away from equilibrium positions, rather than restricting the analysis to the equilibrium itself. "Theoretical Economics will have virtually abandoned classical equilibrium theory in the next decade; the metaphor in the short term future will be evolution, not equilibrium."[1] To this we add our own prediction that the study of economics will evolve over the long run into dynamic computer simulation, based to a significant extent on game theory and experimental approaches to understanding economic processes.

Throughout the social, biological, and physical sciences, researchers examine complex and changing interrelationships among factors in a variety of disciplines. What are the impacts of a change in the El Niño winds, not only on weather patterns but also on the cost of beef in the United States? How does the value of the Mexican peso affect the rate of oil exploration in Alaska? Every day, scientists ask questions like these that involve dissimilar frames of reference and even disciplines. This is why understanding the dynamics and complex interrelationships among diverse systems in our increasingly complicated world is important. A good set of questions is the start—and often the conclusion—of a good model. Such questions help the researcher remain focused on the model without becoming distracted by the myriad of random details that surround it.

Through computer modeling we can study processes in the real world by sketching simplified versions of the forces assumed to underlie them. As an example, you might hypothesize that cities drew workers from farmlands as both became greater users of technology, causing a surplus of jobs in the city and a surplus of labor in the countryside. Another factor could be the feasibility of moving from an agricultural area to the city. A basic version of this model might abstract away from many of the factors that encourage or discourage such migration, in addition to those directly related to job location and the feasibility of relocation. This model could leave behind a sufficiently good predictor of migration rates, or it might not. If the model does not appear to be a good predictor, you can reexamine it. Did your abstractions eliminate any important factors? Were all your assumptions valid? You can revise your model, based on the answers to these questions. Then you can test the revised model for its predictions. You should now have an improved model of the system you are studying. Even better, your understanding of that system will have grown. You can better determine whether you asked the right questions, included all the important factors in that system, and represented those factors properly.

Elementary to modeling is the idea that a model should be kept simple, even simpler than the cause-and-effect relationship it studies. Add complexities to the model only when it does not produce the real effects. Models are sketches of real systems and are not designed to show all of the system's many facets. Models aid us in understanding complicated systems by simplifying them.

---

1. Anderson 1995, 1617.

Models study cause and effect; they are causal. The modeler specifies initial conditions and relations among these elements. The model then describes how each condition will change in response to changes in the others. In the example of farm workers moving to the city, workers moved in response to a lack of jobs in the countryside. But more workers in the city would raise the demand for food in city markets, raising the demand for farm labor. Thus, the demand for labor between the city and the country would shift, leading to migration in both directions and changes in migration over time.

The initial conditions selected by the modeler could be actual measurements (the number of people in a city) or estimates (how many people will be there in four years, given normal birth rate and other specified conditions). Such estimates are designed to reflect the process under study, not to provide precise information about it. Therefore, the estimates could be based on real data or the reasonable guesses of a modeler who has experience with the process. At each step in the modeling process, documentation of the initial conditions, choice of parameters, presumed relationships, and any other assumptions are always necessary, especially when the model is based on the modeler's guesses.

Dynamic models have an interesting interpretation in the world of dynamical statistics. The entire dynamic model in ithink might be considered as a single regression equation, and it can be used that way in a statistical analysis for optimization of a performance measure, such as maximizing profit. It replaces the often arbitrary functional form of the regression equation used in statistical analysis. Thinking of the dynamic model in this way lets one imagine that because more actual system form and information is represented in the dynamic model, that model will produce more accurate results. This is probably so if the same number of parameters are used in both modeling processes. If this is not a constraint, the statistical dynamics process known as cointegration can produce more accurate results. But one can seldom determine in physical terms what aspect of the cointegration model accounted for its accuracy; we are left with a quandary. If we wish modeling to do more than simulate a complex process—that is, if we want a model to help us understand how to change and improve a real-world process—we prefer to use dynamic modeling. Statistical analysis is not involved in optimizing the performance measures to complete the dynamic modeling process; it is key to finding the parameters for the inputs to these models. For example, we need a normal distribution to describe the daily mean temperature to determine the heating fuel demand for an energy wholesale distributor. Statistical analysis is essential in determining the mean and standard deviation of these temperatures from the temperature record. Thus, using statistical analysis, we compress years of daily temperature data into a single equation, one that is sufficient for our modeling objectives.

1.3 Model components

Model building begins, of course, with the properly focused question. Then the modeler must decide on the boundaries of the system that contains the question, and choose the appropriate time step and the level of detail needed. But these are verbal descriptions. Sooner or later the modeler must get down to the business of actually building the model. The first step in that process is the identification of the *state variables,* which will indicate the status of this system through time. These variables carry the knowledge of the system from step to step throughout the run of the model—they are the basis for the calculation of all the rest of the variables in the model.

Generally, the two kinds of state variables are *conserved* and *nonconserved.* Examples of conserved variables are population of an island or the water behind a dam. They have no negative meaning. Nonconserved state variables are temperature or price, for example, and they might take on negative values (temperature) or they might not (price).

Control variables are the ones that directly change the state variables. They can increase or decrease the state variables through time. Examples include birth (per time period) or water inflow (to a reservoir) or heat flow (from a hot body).

Transforming or *converting variables* are sources of information used to change the control variables. Such a variable might be the result of an equation based on still other transforming variables or parameters. The birth rate, the evaporation rate, or the heat loss coefficients are examples of transforming variables.

The components of a model are expected to interact with each other. Such interactions engender feedback processes. *Feedback* describes the process wherein one component of the model initiates changes in other components, and those modifications lead to further changes in the component that set the process in motion.

Feedback is said to be *negative* when the modification in a component leads other components to respond by counteracting that change. As an example, the increase in the need for food in the city caused by workers migrating to the city leads to a demand for more laborers in the farmlands. Negative feedback is often the engine that drives supply-and-demand cycles toward some equilibrium. The word *negative* does not imply a value judgment—it merely indicates that feedback tends to negate initial changes.

In *positive feedback,* the original modification leads to changes that reinforce the component that started the process. For example, if you feel good about yourself and think you are doing well in a course, you will study hard. Because you study hard, you earn a high grade, which makes you feel good about yourself, and that enhances your chances of doing well in the course. By contrast, if you feel bad about yourself and how you are doing in the course, you will not study hard. Because you did not study hard enough, your grade will be low, making you feel bad about yourself. As another example, the migration of farm workers to a city attracts more manufacturers to open plants in that city, which attracts even more workers. Another economic example of positive feedback was noticed by Brian Arthur (1990)—firms that are first to occupy a geographic space will be the first to meet the demand in a region and are the most likely to build additional branch plants or otherwise extend their operations. The same appears to be true with pioneer farmers—the largest farmers today were among the first to cultivate the land.

Negative feedback processes tend to counteract a disturbance and lead systems back toward an equilibrium or steady state. One possible outcome of market mechanisms would be that demand and supply balance each other or fluctuate around an equilibrium point because of lagged adjustments in the productive or consumptive sector. In contrast, positive feedback processes tend to amplify any disturbance, leading systems away from equilibrium. This movement away from equilibrium is apparent in the example of the way grades are affected by how you feel about yourself.

People from different disciplines perceive the role and strength of feedback processes differently. Neoclassical economic theory, for example, is typically preoccupied with market forces that lead to equilibrium in the system. Therefore, the models are dominated by negative feedback mechanisms, such as price increases in response to increased demand. The work of ecologists and biologists, in contrast, is frequently concerned with positive feedback, such as those leading to insect outbreaks or the dominance of hereditary traits in a population.

Most systems contain both positive and negative feedback; these processes are different and vary in strength. For example, as more people are born in a rural area, the population may grow faster (positive feedback). As the limits of available arable land are reached by agriculture, however, the birth rate slows, at first perhaps for psychological reasons but eventually for reasons of starvation (negative feedback).

Nonlinear relationships complicate the study of feedback processes. An example of such a nonlinear relationship would occur when a control variable does not increase in direct proportion to another variable but changes in a nonlinear way. Nonlinear feedback processes can cause systems to exhibit complex—even chaotic—behavior.

A variety of feedback processes engender complex system behavior, and some of these will be covered later in this book. For now, develop the following simple model, which illustrates the concepts of state variables, flows, and feedback processes. The discussion will then return to some principles of modeling that will help you to develop the model building process in a set of steps.

1.4 Modeling in ithink

ithink was chosen as the computer language for this book on modeling dynamic systems because it is a powerful yet easy-to-learn tool. Readers are expected to familiarize themselves with the many features of the program. Some introductory material is provided in the appendix. But you should carefully read the manual that accompanies the program. Experiment and become thoroughly familiar with it.

To explore modeling with ithink, we will develop a basic model of the dynamics of a fish population. Assume you are the sole owner of a pond that is stocked with 200 fish that all reproduce at a fixed rate of 5 percent per year. For simplicity, assume also that none of the fish die. How many fish will you own after 20 years?

In building the model, utilize all four of the graphical tools for programming in ithink. The appendix includes a "Quick Help Guide" to the software, should you

FIGURE 1.2

need one. The appendix also describes how to install the ithink software and models of the book. Follow these instructions before you proceed. Then double-click on the ithink icon to open it.

On opening ithink, you will be faced with the High-Level Mapping Layer, which is not needed now. To get to the Diagram Layer, click on the downward-pointing arrow in the upper left-hand corner of the frame (see figure 1.2).

The Diagram Layer displays the following symbols, "building blocks," for stocks, flows, converters, and connectors (information arrows) (see figure 1.3).

Click on the globe to access the modeling mode (see figure 1.4). In the modeling mode you can specify your model's initial conditions and functional relationships. The following symbol indicates that you are now in the modeling mode (see figure 1.5).

Begin with the first tool, a stock (rectangle). In this example model, the stock will represent the number of fish in your pond. Click on the rectangle with your mouse, drag it to the center of the screen, and click again. Type in FISH. This is what you get (see figure 1.6).
This is the model's first state variable, where you indicate and document a state or condition of the system. In ithink, this stock is known as a *reservoir.* In this model, the stock represents the number of fish of the species that populate the pond. If you assume that the pond is one square kilometer large, the value of the state variable FISH is also its *density,* which will be updated and stored in the

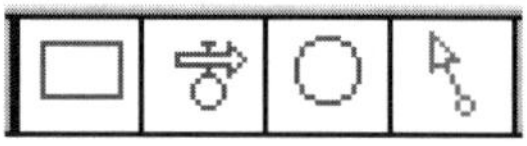

FIGURE 1.3

FIGURE 1.4

FIGURE 1.5

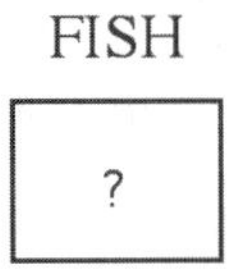

FISH

?

FIGURE 1.6

computer's memory at every step of time (DT) throughout the duration of the model.

> Learning Point: The best guide in determinating the proper modeling time step (DT) is the half rule. Run the model with what appears to be an appropriate time step, then halve the DT and run the model again, comparing the two results of important model variables. (This is step 7 from the modeling steps reminders in chapter 1.) If these results are judged to be sufficiently close, the first DT is adequate. One might try to increase the DT if possible to make the same comparison. The general idea is to set the DT to be significantly smaller than the fastest time constant in the model, but it is often difficult to determine this constant. There are exceptions. Sometimes the DT is fixed at 1 as the phenomena being modeled occur on a periodic basis and data are limited to this time step. For example, certain insects may be born and counted on a given day each year. The DT is then one year and should not be reduced. The phenomenon is not continuous.

The fish population is a stock, something that can be contained and conserved in the reservoir; density is not a stock because it is not conserved. Nonetheless, both of these variables are state variables. So, because you are studying a species of fish in a *specific area* (one square kilometer), the population size and density are represented by the same rectangle. Inside the rectangle is a question mark. This is to remind you that that you need an initial or starting value for all state variables. Double-click on the rectangle. A dialogue box will appear. The box is asking for an initial value. Add the initial value you choose, such as 200, using the keyboard or the mouse and the dialogue keypad. When you have finished, click on OK to close the dialogue box. Note that the question mark has disappeared.

> Learning Point: It is a good practice to name the state variables as nouns (e.g., population) and the direct controls of the states as verbs (e.g., birthing). The parameters are most properly named as nouns. This somewhat subtle distinction keeps the flow and stock definitions foremost in the mind of the beginning modeler.

Decide next what factors control (that is, add to or subtract from) the number of fish in the population. Because an earlier assumption was that the fish in your pond never die, you have one control variable: REPRODUCTION. Use the *flow* tool (the right-pointing arrow, second from the left) to represent the control variable, so named because they control the states (variables). Click on the flow symbol, then click on a point about two inches to the left of the rectangle (stock) and drag the arrow to POPULATION, until the stock becomes dashed, and release. Label the circle REPRODUCTION. This is what you will have (see figure 1.7).

Here, the arrow points only into the stock, which indicates an inflow. But you can get the arrow to point both ways if you want it to. You do this by double-clicking on the circle in the flow symbol and choosing Biflow, which enables you to add to the stock if the flow generates a positive number and to subtract from the stock if the flow is negative. In this model, of course, the flow REPRODUCTION is always positive and newly born fish go only *into* the population. The control variable REPRODUCTION is a uniflow: new fish per annum.

Next you must know how the fish in this population reproduce—not the biological details, just how to accurately estimate the number of new fish per annum. One way to do this is to look up the birth rate for the fish species in your pond. Say that the birth rate = 5 new fish per 100 adults each year, which can be represented as a transforming variable. A transforming variable is expressed as a *converter,* the circle that is second from the right in the ithink toolbox. (So far REPRODUCTION RATE is a constant; later the model will allow the reproduction rate to vary.) The same clicking-and-dragging technique that brought the stock to the screen will bring up the circle. Open the converter and enter the number of 0.05 (5/100). Down the side of the dialogue box is an impressive list of built-in functions that are useful for more complex models.

At the right of the ithink toolbox is the *connector* (information arrow). Use the connector to pass on information (about the state, control, or transforming variable) to a convertor (circle), to the control (the transforming variable). In this case, you want to pass on information about the REPRODUCTION RATE to REPRODUCTION. Once you draw the information arrow from the transforming variable REPRODUCTION RATE to the control and from the stock FISH to the control, open the control by double-clicking on it. Recognize that REPRODUCTION RATE and FISH are two required inputs for the specification of REPRODUCTION. Note also that ithink asks you to specify the control: REPRODUCTION = . . . "Place right-hand side of equation here."

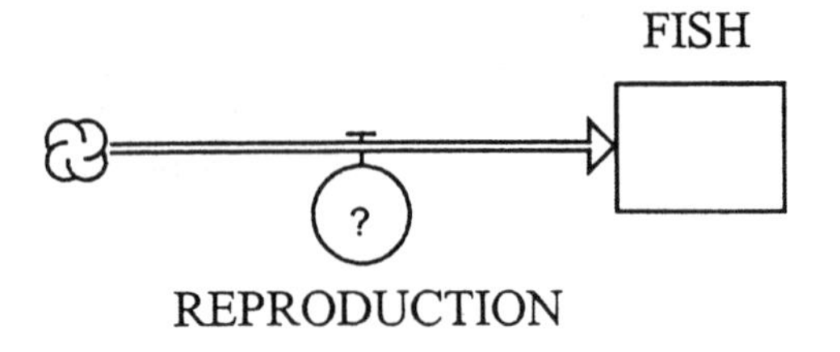

FIGURE 1.7

Click on REPRODUCTION, then on the multiplication sign in the dialogue box, and then on FISH to generate the equation

$$\text{REPRODUCTION} = \text{REPRODUCTION RATE} * \text{FISH} \tag{1.1}$$

Click on OK and the question mark in the control REPRODUCTION disappears. Your ithink II diagram should now look like this (see figure 1.8).

Next, set the temporal (time) parameters of the model. These are DT (the time step over which the stock variables are updated) and the total time length of a model run. Go to the RUN pull-down menu on the menu bar and select Time Specs. A dialogue box will appear in which you can specify, among other things, the length of the simulation, the DT, and the units of time. For this model, choose DT = 1, length of time = 20, and units of time = years.

To display the results of the model, click on the graph icon and drag it to the diagram. If you wanted to, you could display these results in a table by choosing the table icon instead. The ithink icons for graphs and tables can be seen in figure 1.9.

When you create a new graph pad, it will open automatically. To open a pad that had been created previously, just double-click on it to display the list of stocks, flows, and parameters for the model. Each one can be plotted. Select FISH to be plotted and, with the >> arrow, add it to the list of selected items. Then set the scale from 0 to 600 and check OK. You can set the scale by clicking once on the variable whose scale you wish to set and then on the arrow next to it. Now you can select the minimum on the graph, and the maximum value will define the highest point on the graph. Rerunning the model under alternative parameter settings will lead to graphs that are plotted over different ranges. Sometimes these are a bit difficult to compare with previous runs because the scaling changes.

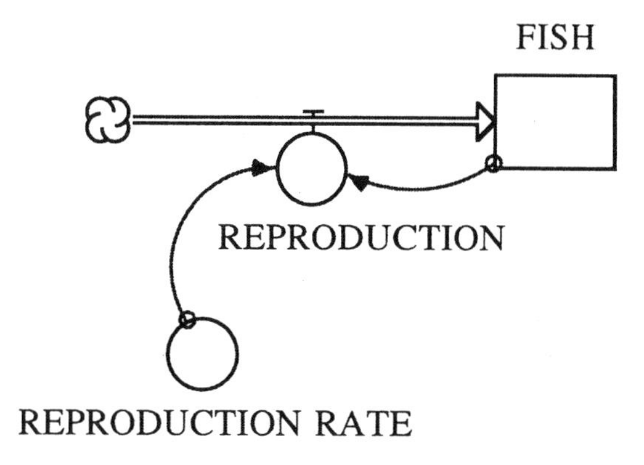

FIGURE 1.8

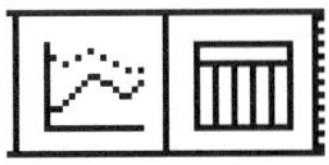

FIGURE 1.9

Would you like to see the results of the model so far? Run the model by selecting RUN from the pull-down menu. Figure 1.10 illustrates what you should see.

The graph shows exponential growth of the fish population in your pond. This is what you should have expected. It is important to state beforehand what results you expect from running a model. Such speculation builds your insight into system behavior and helps you anticipate (and correct) programming errors. When the results do not meet your expectations, something is wrong and you must fix it. The error may be either in your ithink program or your understanding of the system that you wish to model—or both.

What do you really have here? How does ithink determine the time path of the state variable? Actually, it is not difficult. At the beginning of each time period, starting with time = 0 years (the initial period), ithink looks at all the components for the required calculations. The values of the state variables will probably form the basis for these calculations. Only the variable REPRODUCTION depends on the state variable FISH. To estimate the value of REPRODUCTION after the first time period, ithink multiplies 0.05 by the value FISH (@ time = 0) or 200 (provided by the information arrows) to arrive at 10. From time = 1 to time = 2, the next DT, ithink repeats the process and continues through the length of the model. When you plot your model results in a table, you find that, for this simple fish model, ithink calculates fractions of fish from time = 1 onward. This problem is easy to solve; for example, by having ithink round the calculated number of fish—there is a built-in function that can do that—or just by reinterpreting the population size as "thousands of fish."

This process of calculating stocks from flows highlights the important role played by the state variable. The computer carries that information—and only

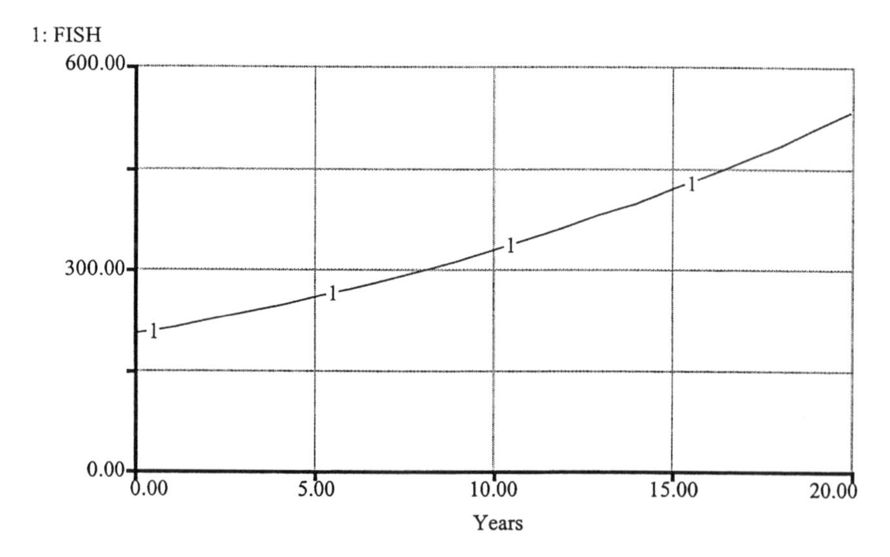

FIGURE 1.10

FIGURE 1.11

that information—from one DT to the next, which is why it is defined as the variable that represents the *condition* of the system.

You can drill down in the ithink model to see the parameters and equations that you have specified and how ithink makes use of them. Click on the downward-pointing arrow at the right of your ithink diagram (see figure 1.11).
The equations and parameters of your models are listed here. Note how the fish population in time period *t* is calculated from the population one small time step, DT, earlier and all the flows that occurred during a DT.

The model of the fish population dynamics is simple. So simple, in fact, that it could be solved with pencil and paper, using analytic or symbolic techniques. The model is also linear and unrealistic. Next, add a dimension of reality—and explore some of ithink's flexibility. This may be justified by the observation that, as populations get large, mechanisms set in that influence the rate of reproduction.

To account for feedback between the size of the fish population and its rate of reproduction, an information arrow is needed to connect FISH with REPRODUCTION RATE. The connection will cause a question mark to appear in the symbol for REPRODUCTION RATE. The previous specification is no longer correct; it now requires FISH as an input (see figure 1.12).

Open REPRODUCTION RATE. Click on the required input FISH. The relationship between REPRODUCTION RATE and FISH must be specified in mathematical terms, or at least, you must make an educated guess about it. An educated guess about the relationship between two variables can be expressed by

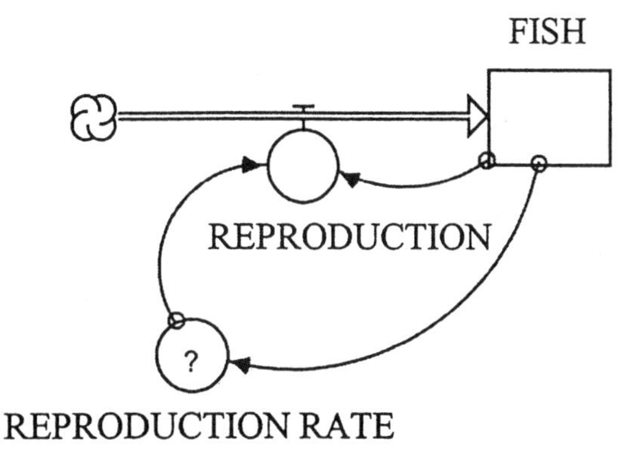

FIGURE 1.12

plotting a graph that reflects the anticipated effect one variable (REPRODUCTION) will have on another (FISH). The feature used for this is called a *graphical function.*

To use a graph to delineate the extended relationship between REPRODUCTION RATE and FISH, click on Become Graph and set the limits on the FISH at 2 and 500. Set the corresponding limits on the REPRODUCTION RATE at 0 and 0.20, to represent a change in the birth rate when the population is between 0 and 500. (These are arbitrary numbers for a made-up model.) Finally, use the mouse arrow to draw a curve from the maximum birth rate and population of 2 to the point of 0 birth rate and population of 500.

Suppose a census were taken at three points in time. The curve you just drew goes through all three points. You can assume that if a census had been taken at other times, it would show a gradual transition through all the points. This sketch is good enough for now. Click on OK (see figure 1.13).

Before you run the model again, consider what the results will be. Think of the graph for FISH through time. Generally, it should rise, but not in a straight line. At first the rise should be steep: the initial population is only 200, so the initial birth rate should be high. Later it will slow down. Then, the population should level off at 500, when the population's density would be so great that new births tend to cease. Run the model. You were right! (See figure 1.14.)

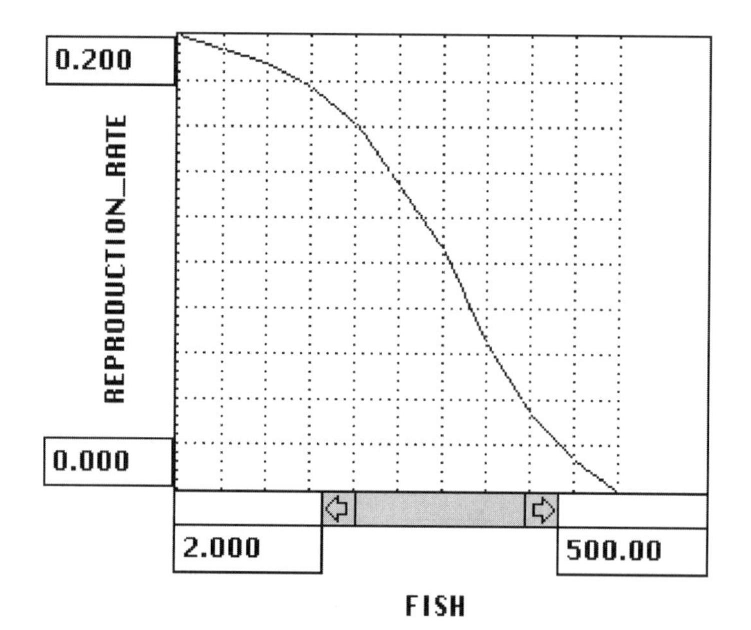

FIGURE 1.13

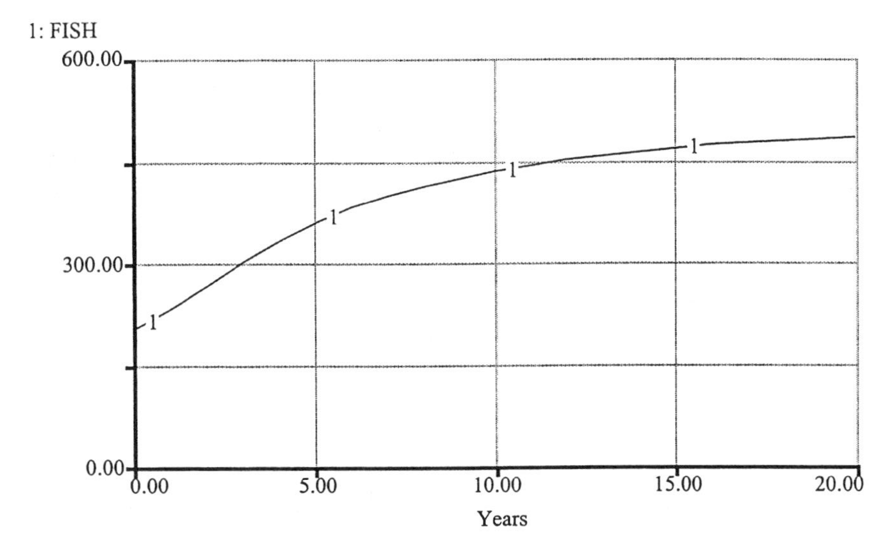

FIGURE 1.14

This problem has no analytic solution, only a numerical one. You can continue to study the sensitivity of the answer to changes in the graph and the size of DT. You are not limited to a DT of 1. Generally speaking, a smaller DT leads to more accurate numerical calculations for updating state variables and, therefore, a more accurate answer. Choose Time Specs from the RUN menu to change the DT. Change the DT to reflect ever-smaller periods until the change in the critical variable is within measuring tolerances. You also may change the numerical technique used to solve the model equations. Euler's method is chosen as a default. Two other methods, Runge–Kutta-2 and Runge–Kutta-4, are available to update state variables in different ways. These methods will be discussed later.

Start with a simple model and keep it simple, especially at first. Whenever possible, compare your results against measured values. Complicate your model only when your results do not predict the available experimental data with sufficient accuracy or when your model does not yet include all the features of the real system that you wish to capture. For example, as the owner of a pond, you may want to extract fish for sale. Assume the price per fish is $5 and constant. What are your revenues each year if you wish to extract fish at a constant rate of 3 percent per year? To find the answer to this question, define an outflow from the stock FISH. Click on the converter, then click onto the stock to have the converter connected to the stock, and then drag the flow from the stock to the right. Now fish disappear from the stock into a "cloud." You are not explicitly modeling where they go. What you should have developed thus far as your ithink model can be seen in figure 1.15.

Next, define a new transforming variable called EXTRACTION RATE and set it to 0.03. Specify the outflow as follows:

$$\text{EXTRACTION} = \text{EXTRACTION RATE} * \text{FISH} \qquad (1.2)$$

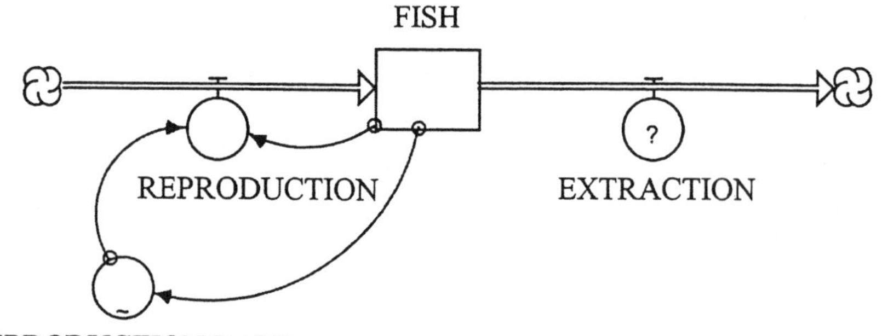

FIGURE 1.15

after making the appropriate connections with information arrows. Then create two more transforming variables, one called PRICE, setting it to 5, and one called REVENUES, specifying it as follows:

$$\text{REVENUES} = \text{EXTRACTION} * \text{PRICE} \tag{1.3}$$

Your model should look like figure 1.16.

Double-click on the graph pad and select REVENUES to plot the fish stock and revenues in the same graph. The time profile of the revenue streams generated by your pond can be seen in figure 1.17.

You can easily expand this model; for example, to capture price changes over time, unforeseen outbreaks of diseases in your pond, or other problems that may occur in a real-world setting. When your model becomes increasingly compli-

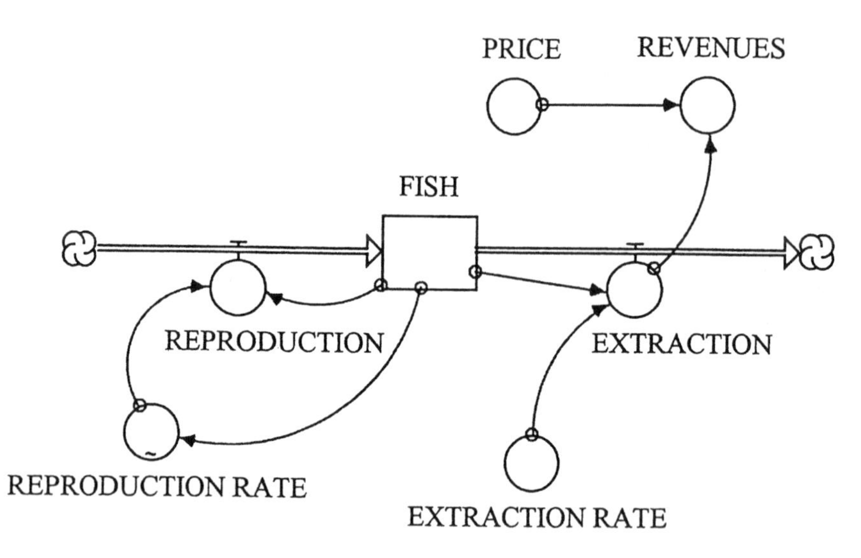

FIGURE 1.16

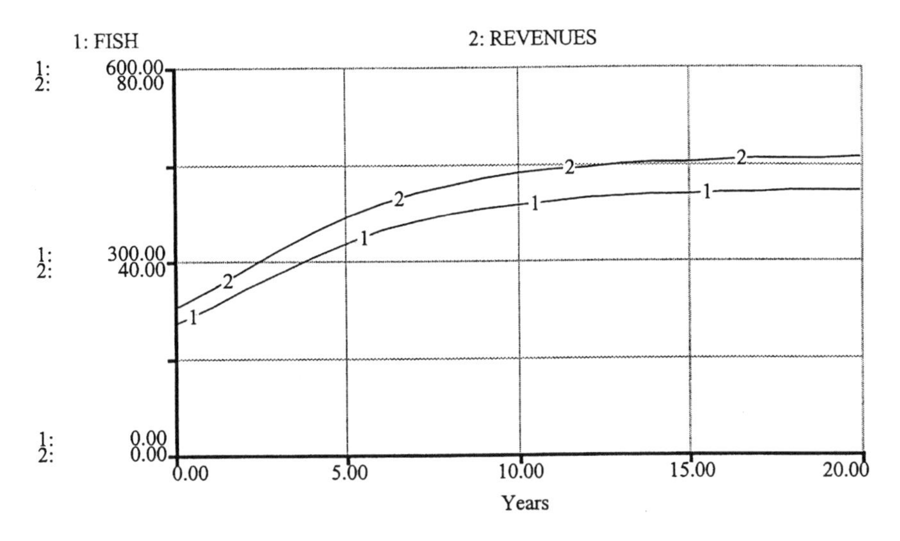

FIGURE 1.17

cated, try to keep your ithink diagram as organized as possible so that it clearly shows the interrelationships among the model parts. A strong point of ithink is its ability to demonstrate complicated models visually. Use the hand symbol to move model parts around the diagram; use the paintbrush symbol to change the color of icons. The dynamite symbol will blast off any unnecessary parts of the model (see figure 1.18).

Be careful when you blast away information arrows. Move the dynamite symbol to the place at which the information arrow is attached and click on that spot. If you click, for example, on the translation variable itself, it will disappear, together with the information arrow, and you may have to recreate it (see figure 1.19).

As the model grows, it will contain an increasing number of submodels, or modules. You may want to protect some of these modules from being changed. To do this, click on the sector symbol (to the left of the A in the next set of pictures) and drag it over the module or modules you want to protect. To run the individual sectors, go to Sector Specs . . . in the RUN pull-down menu and select the ones that you wish to run. The values of the variables in the other sectors remain unaffected.

By annotating the model, you can remind yourself and inform others of the assumptions underlying your model and its submodels. This is important in any model but especially in larger and more complicated models. To do this, click on the text symbol (the letter A) and drag it into the diagram. Then, type in your annotation (see figure 1.20).

FIGURE 1.18

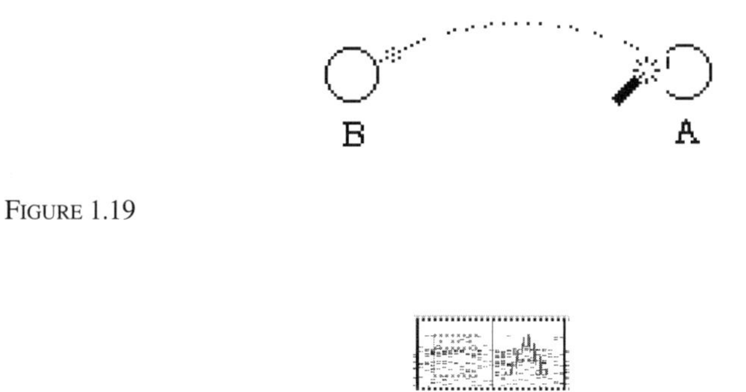

FIGURE 1.19

FIGURE 1.20

The tools mentioned here are likely to prove useful when you develop more complicated models and when you want to share your models and their results with others. ithink contains many helpful tools, which we hope you will use extensively. You will probably want to explore such features as Drill Down (visual hierarchy), Space Compression, High-Level Mapping Layer, and the Authoring features of ithink. The appendix for ithink provides a brief overview of these and other features.

Make thorough use of your model, running it over again and always checking your expectations against its results. Change the initial conditions and try running the model to its extremes. At some point, you will want to perform a formal sensitivity analysis. The excellent sensitivity analysis procedures available in ithink are discussed later.

1.5 The detailed modeling process

Some of the elements that make up the system for which a model is being developed are referred to as state variables. State variables may or may not be conserved. Each conserved state variable represents an accumulation or stock of materials or information. Typical conserved state variables are population, resource endowments, inventories, and heat energy. Nonconserved state variables are pure indicators of some aspects of the system's condition. Typical nonconserved state variables are price and temperature. System elements that represent the action or change in a state variable are called flows or control variables. As a model is run over time, control variables update the state variables at the end of each time step. Examples for control variables are the number of births per period, a variable that changes the state variable population, or the number of barrels of crude oil extracted changing the state variable reserves. These states are changed via controls or control variables. The controls are changed by converters, which translate information coming from other states, controls, or converters. The program ithink is, like many other such programs now available, one that easily allows the graphically based interaction between these three variable forms.

Following is a set of easy-to-follow but detailed modeling steps. These steps are not sacred: they are intended as a guide to get you started in the process. You will find it useful to come back to this list once in a while as you proceed in your modeling efforts.

1. Define the problem and the goals of the model. Frame the questions you want to answer with the model. If the problem is a large one, define subsystems of it and goals for the modeling of these subsystems. Ask yourself: Is my model intended to be explanatory or predictive?

2. Designate the state variables. (These variables will indicate the status of the system.) Keep it simple. Purposely avoid complexity in the beginning. Note the units of the state variables.

3. Select the control variables—the flow controls into and out of the state variables. (The control variables are calculated from the state variable in order to update them at the end of each time step.) Note to yourself which state variables are donors and which are recipients with regard to each of the control variables. Also, note the units of the control variables. Keep it simple at the start. Try to capture only the essential features. Put in one type of control as a representative of a class of similar controls. Add the others in step 10.

4. Select the parameters for the control variables. Note the units of these parameters and control variables. Ask yourself: Of what are these controls and their parameters a function?

5. Examine the resulting model for possible violations of physical, economic, and other laws (for example, any continuity requirements or the conservation of mass, energy, and momentum). Also, check for consistency of units. Look for the possibilities of division by zero, negative volumes or prices, and so forth. Use conditional statements if necessary to avoid these violations.

6. To see how the model is going to work, choose some time horizon over which you intend to examine the dynamic behavior of the model, the length of each time interval for which state variables are being updated, and the numerical computation procedure by which flows are calculated. (For example, choose in the ithink program Time Step = 1, time length = 24.) Set up a graph and guess the variation of the state variable curves before running the model.

7. Run the model. See if the graph of these variables passes a "sanity test": Are the results reasonable? Do they correspond to known data? Choose alternative lengths of each time interval for which state variables are updated. Choose alternative integration techniques. (In the ithink program, for example, reduce the time interval DT by half and simulate the mode again to see if the results are the same.)

8. Vary the parameters to their reasonable extremes and see if the results in the graph still make sense. Revise the model to repair errors and anomalies.

9. Compare the results to experimental data. This may mean shutting off parts of your model to mimic a lab experiment, for example.

10. Revise the parameters, perhaps even the model, to reflect greater complexity and to meet exceptions to the experimental results, repeating steps 1–10. Frame a new set of interesting questions.

Do not worry about applying all of these steps in this order as you develop your models and improve your modeling skills. Do check back to this list now and then to see how useful, inclusive, and reasonable these steps are.

You will find that modeling has three possible general uses. First, you can experiment with models. A good model of a system enables you to change its components and see how these changes affect the rest of the system. This insight helps you explain the workings of the system you are modeling. Second, a good model enables prediction of the future course of a dynamic system. Third, a good model stimulates further questions about the system behavior and the applicability of the principles that are discovered in the modeling process to other systems.

Remember the words of Walter Deming: "All models are wrong. Some are useful." To this we add: "No model is complete."

2

Modeling of Dynamic Business Systems

His driving curiosity was apparent when, in his last media interview, he told the *Boston Globe* last year that his work on the shuttle commission had so aroused his interest in the complexities of managing a large organization like NASA that if he were starting his life over, he might be tempted to study management rather than physics.

—Quotation from the obituary of Richard P. Feynman in the *Boston Globe,* 16 February 1988

2.1 Introduction

The eminent theoretical physicist, Richard P. Feynman, served on the committee that investigated the Challenger disaster in 1986. As a physicist, Feynman was used to the complexities associated with the world of subatomic particles or the motions of stars and galaxies. However, his experience on the committee opened his eyes to the complexities of managing a modern organization, as is shown by the quotation from Feynman's obituary in the *Boston Globe*.

Feynman recognized that managing an organization had become a complex problem and, for a person with his intellectual curiosity, the management of such organizations provided a stimulating area of study.

But where does this complexity come from? An organization is basically a system, which can be defined as "a regularly interacting or interdependent group of items composing a unified whole."

Organizations are composed of a number of interconnected component parts (many of which are people). Like any other system, in order to operate successfully, these component parts must work in a coordinated fashion. The interconnections must be managed. Therefore, the difficulty in operating an organization is directly related to the complexity of individual interconnections and to the number of interconnections that must be managed. Both the number and complexity of the interconnections have changed over time, in part because of the following trends:

- Business size—many organizations have grown in size (through mergers and acquisitions) in order to compete or satisfy the ever-growing expectations of shareholders.[1] Indeed, "merger mania" has been common over recent years. These mergers or acquisitions mean that more interconnections must be managed in order for the new organization to be successful and reap the benefits of the growth in size.

- Globalization—this has added the issues of language, culture, currency, local legal regulations, and so forth, which has made some of the individual interconnections more complex.
- Improving efficiency—the pressure to improve the bottom line ultimately leads to fewer resources being available to buffer components from each other. The typical example here is the impact of removing inventory from supply chains. With lower inventories, an organization has greater difficulty reacting to unexpected production delays, customer demands, and so on. This means that components of the organization that could previously be treated as independent must now be coordinated in order that unexpected situations can be handled successfully.
- Competition/customer expectations—customers expect more and more every day. They have more choice in what is available to them and they can switch suppliers on a whim. In bygone days, a company could survive if communications between research and development, manufacturing, marketing, and sales were poor. But this is clearly no longer the case.
- Technological advances—for example, improvements in the field of communications mean that now it is easy to connect parts of the organization. Use of the Internet in business is a testimony to this. Although these connections may give an organization an advantage initially when they are set up, eventually the organization changes so that it depends on these connections to operate. The Y2K situation was a global-scale example of this dependency not only on individual computers but also on the networks and interconnections they manage.

Whatever the reason for growth in organizational complexity, it is a fact of life and it must be managed. What exactly do we mean by managing a system? The dictionary defines *management* in general as "the judicious use of means (resources) to achieve an end."

This can be translated into a business context as the optimal allocation of resources to achieve the goals of the business.[2] Therefore, the key question that must be asked continuously in order to manage an organization is this: If we allocate our resources in a certain way, what will the impact be on the organization's performance? Given a range of options for allocating resources, we can then choose the option that gives the best results. Put another way, we must be able to predict the performance of an organization under a given set of conditions. Management thus reduces to the ability to predict the future performance of the or-

---

1. Examples are AOL/Time Warner or Pfizer/Warner-Lambert Pharmacian.
2. And as Goldratt and Cox (1992) point out in *The Goal,* the end in mind for any business organization is to make money!

ganization under a given set of conditions.[3] Furthermore, in order to make a prediction, a manager must create a model of the system.

A model of a system is simply a representation of the system that anyone can use to predict the performance of the system—without having to use the actual system. Such models can range from simple mental models to sophisticated computer simulations. The value of a model is not measured by its sophistication but by its ability to predict the real system performance. In this book, we provide readers with insights and tools to model and then predict the performance of an organization so that they can improve their capability of managing the organization. We approach this objective with a definite strategy, which can be outlined as follows:

1. Break up the organization into smaller, more manageable parts.
2. Choose an appropriate modeling technique.
3. Build the model.
4. Validate the model by predicting known historical behavior.
5. Make new predictions.
6. Propose and implement changes to the organization using the new predictions.

Steps 1 and 2 will be discussed in this chapter. In particular, we will look at the role of dynamic modeling in business.

2.2 Making the organization more manageable: Systems and processes

Given the complex nature of the entire organization when viewed as a single system, it is natural to try to break the system into smaller, more manageable units using organizing principles. For example, organizing by skill set leads to an organization in which activities emphasize functional abilities. The business has engineering activities, financial activities, and so forth. A business that is supply chain-focused will organize by product or product group.

In general, no single organizing principle will suffice for the whole business—the business is just too complex. More often, multiple organizing principles are used, which leads to the concept of the matrix organization. People working in such a matrix organization will see their roles from multiple perspectives. For example, a person might be an engineer from a functional perspective but a member of a product team from a supply chain perspective. This can present a difficult work environment for people because it may appear that their loyalties are divided. Am I an engineer first or a supply-chain person first? Hammer (1996, 128) discusses the issues around the matrix organization, where he refers to management with this organizing principle as "notorious matrix management."

---

3. Of course, these conditions may involve assumptions about the world outside the organization.

In more recent years, the strategy of using process as the key organizing principle has received more attention. In fact, Hammer (1996, xii) has said that in his definition of reengineering—"the radical re-design of business processes for dramatic improvement"—the original emphasis was on the word *radical* but it really should have been on the word *process*. On top of this, most if not all of the major quality improvement approaches focus on process, defined in the dictionary as "a series of actions, activities or operations conducing to an end," as the key organizing principle.[4]

This process focus is also the organizing principle favored in this book. To apply this approach, we must break up the system into its component processes. For example, consider the system consisting of an automobile, the driver, the road, other drivers and pedestrians, and environmental conditions. Within this system, the driver wants to drive from point A to point B. The driver can identify a number of processes to help accomplish this overall goal: starting the car, accelerating the car, braking, keeping the car in the lane, avoiding other traffic, and so on. Similarly, looking at a business system or organization, we can identify the major processes within the organization and study these processes individually.

There are many ways to break an organization into its major processes. The one favored by the authors is to start by looking at the value chain of the organization. This allows us to identify the main value adding processes, which add value to the products or services that the organization sells to customers to make money. A typical value chain is shown in figure 2.1.

Therefore, we have major processes for market research, product development, and so on. These processes are still large. Typically, these major processes will be subdivided into smaller processes. For example, process development might be broken into processes such as manufacturing route selection, process optimization, validation, and so forth.

Next, we can consider the business value adding processes. These are processes that must be executed in order for the value chain to operate successfully but do not add any value to the products or services of the organization. For example, in manufacturing, production must be planned, people must be trained, safety must be managed, and so on.

Finally, we must identify the non-value adding processes. These are processes that add no value but consume resources. (Consequently, the organization would prefer not to have to do them). This group includes rework processes, inspection processes, and so on.

Be aware that breaking down the organization into its component processes does not solve the interdependency problem. It is still there, and it shows up in the way the processes are interconnected. Outputs of one process become inputs to other processes. In analyzing a particular process, we must know of any other connected processes so that we can take account of them in our analyses. In ana-

4. For example, the six-sigma approach developed by Motorola and popularized by GE.

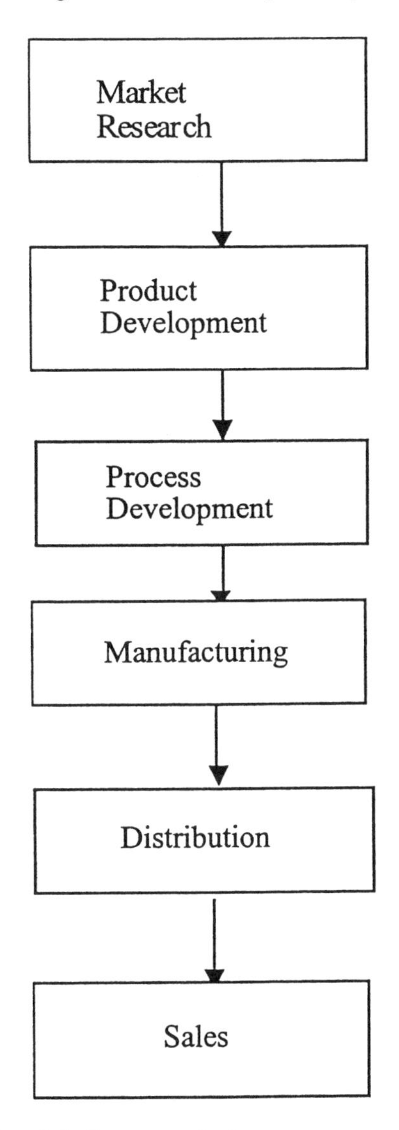

FIGURE 2.1. Typical value chain for an organization

lyzing a process, it is helpful to have a common description of a business process. A simple but useful general description of a process is the supplier input process output key stakeholder (SIPOKS) description, shown in figure 2.2.

In this model, the flow of the process is from suppliers who give us the required inputs to our process. These inputs are used in the process. The process then produces a set of outputs that are used by the key stakeholders. This model is discussed fully by Scholtes (1998). In his description, he uses "customer" instead of "key stakeholder," giving the acronym is SIPOC. However, using the notion of a

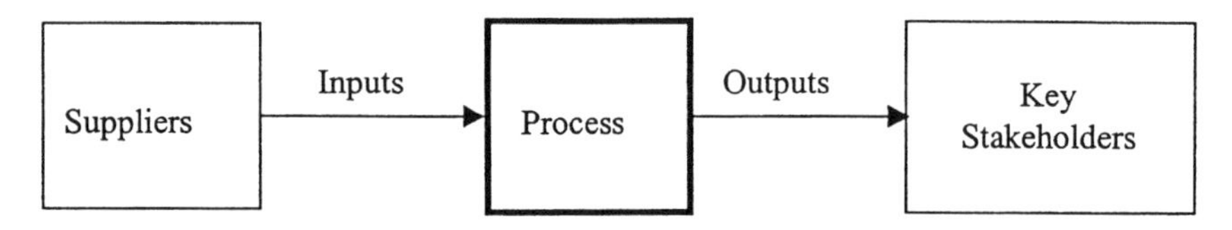

FIGURE 2.2. The SIPOKS description of a general process

key stakeholder is more general than that of a customer. A stakeholder is defined as any group that is impacted or interested in the performance of the process and the word key denotes the important stakeholders. Thus, a customer is an obvious key stakeholder. Also, the public may be a key stakeholder. If the organization potentially can pollute the environment, then the public will be keenly interested in the performance of the process from an environmental perspective. The key stakeholders are the ultimate evaluators of organizational and/or process performance. In general, we can divide the key stakeholders into three groups:

1. Compliance Key Stakeholders—these groups regulate the operation of an organization. They represent and protect the public. Examples are the Environmental Protection Agency (EPA), the Food and Drug Administration (FDA), and the Occupational Health and Safety Administration (OSHA). Being compliant means meeting all key stakeholder requirements.
2. Effectiveness Key Stakeholders—these are the customers of the product and/or services provided by the organization. Being effective means meeting all customer requirements.
3. Efficiency Key Stakeholders—this group includes management and shareholders who are interested in the financial performance of the organization. Being efficient means using resources efficiently. Of course, this group realizes that the financial performance depends on being compliant and effective as well and will be interested in them, too. We will return to the stakeholders again when we discuss performance measures in chapter 3.

Once we have defined a process according to a description like SIPOKS, we can create a model of the process.

2.3 Creating and using a model

It would be nice if people could create system models and make predictions from them using their mental capabilities alone. However, in general, the human mind does not have the capability to make predictions from a model using pure mental reasoning alone. Processes have inherent computational complexities that prevent the human mind from being able to take a model and infer behavior from it. For this reason, people need help. Traditionally, this help has come in the form of mathematical modeling methods. More recently, computer-based modeling techniques have become more popular, and we use the latter approach in this book. The advantage of using computer-based techniques is that people can create sophisticated models without having advanced mathematical knowledge. In this book, we purport that there are four different types of complexity in a real system as shown in figure 2.3.

The following discussion will focus on each of the four types of computational complexity: size of solution space, physics, uncertainty, and structure.

Size of the Solution Space

Here the number of possible configurations of the resources available to the manager is immense, but only one of these configurations leads to an optimal system. The system manager is faced with the task of identifying the one optimal configuration from the immense number of possibilities. One of the classic problems in this area is the Traveling Salesperson Problem (TSP). In the TSP, a salesperson must make a trip regularly to a number of cities to visit customers. Starting from a home city, the salesperson visits each customer city just once, ending up back at the home city again. In order to achieve the goal of minimizing time on the road, the salesperson must find the most time-efficient route. The salesperson

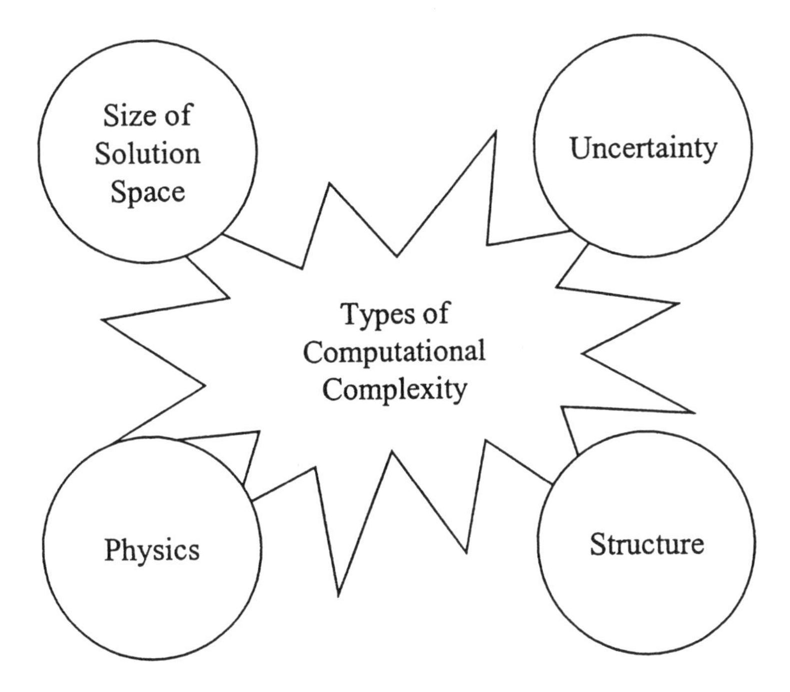

FIGURE 2.3. Types of computational complexity

knows the time it takes to travel between each city. Suppose that 19 customer cities are on the route. There are 19! (factorial 19)[5] possible ways to visit all the cities, ending back at the starting point. How is the salesperson going to choose the optimal route? One could try to enumerate all the possibilities. However, even if the salesperson could evaluate one million tours a second, it would take 3,857 years to evaluate all routes. Clearly, entire enumeration is not a practical option. Models that incorporate this type of complexity are referred to as normative models. These models allow us to answer this question: "If we want the system to give a certain output, how should the system be set up; that is, how should the inputs be specified?" This is an optimization problem and arises in many areas in business—from scheduling a factory to optimizing an entire supply chain. Among the tools available to help manage this complexity are linear and nonlinear programming and integer programming. For more details, see Winston (1994).

Physics

Here the basic relationships in the system are complex. This usually means that we must be able to represent complex physical relationships in order to predict the system behavior. An example would be an engineer trying to understand the behavior of a complicated series of chemical reactions. The engineer may have to use complicated relationships from chemistry and thermodynamics to specify the relationships and, hence, predict the system behavior. Generally the models used to represent these relationships are referred to as predictive models.

Uncertainty

Here some aspects of the inputs are not fixed or known. Uncertainty in an input implies that there is a probabilistic aspect to input. Uncertainty can show up in two distinct ways.

An input may be uncertain if it varies every time an activity or task is repeated, because it is not possible to repeat an activity exactly the same every time. For example, the time it takes typically to complete a task would have some uncertainty associated with it. The best we can do is to say that the task time will be within some range with some probability. So we might be able to say that there is a 90 percent chance that a task time is in the range of 5 to 10 minutes. This type of uncertainty is referred to as random variation: we say that the input has random variability associated with it.[6]

By contrast, an input can also have uncertainty associated with it if it has a definite value but that value is unknown to us. This lack of knowledge may be because we do not have enough information about the input or because the input is based on an event that will happen in the future. An example of the former could

---

5. $19! = 1 \times 2 \times 3 \times \ldots \times 17 \times 18 \times 19 = 1.216451004088 \times 10^{17}$.

6. In fact, the impact of random variation is so important that Hopp and Spearman (1996) include a chapter titled "The Corrupting Influence of Variability."

be the average salary of people in a certain country. Unless we get the data for each person and calculate the average, we must take a sample of people and estimate the average. An example of the latter case is the price of a stock at some future date. We can guess what it might be and make a decision based on that guess, and when the date actually occurs, the stock will have a definite value.

Irrespective of the type of uncertainty involved, models that incorporate this type of complexity are called *stochastic models.* In this book, the models that we will discuss will involve only random variation. In order to create such models, we must be able to describe the uncertainty in the input. This involves the use of probability distributions. Although readers can use this book without getting involved in the details of these probability distributions, it is beneficial to understand these distributions and where they tend to occur. This will ensure that incorrect probability distributions are not used when readers create their own models. Appendix C contains a discussion of some of the more common probability distributions, where they occur, and how they can be implemented in ithink.

Note that Hopp et al. (1996) make the important point that any source of variability, whether it is random or deterministic, will have an impact on the performance of a process. Such deterministic sources of variation can be related to nonuniform staffing levels, production levels, and so on. Dynamic models make it easy to separate the impact of these two sources of variation. One can simply turn off all sources of random variation and run the model to see the best performance possible if all random variation could be removed (remembering that this is an interesting ideality only). In this book, we consider the impact of both random and deterministic sources of variation.

Structural Complexity

In this case, the structure of the system makes it difficult to see what the output of the system might look like. Structure refers to the relationships between the various parts of system. The outputs of the system depend on these relationships, and the evolution of the outputs in time becomes complex because of these relationships. However, it is known that people cannot look at the structure of a system and reliably predict its evolution in time.[7] Three major structural relationships that can exist in a system make it difficult to predict the behavior of a system over time. These are feedback, time delays, and nonlinearity.

Following are definitions and some brief discussion of how each of these elements can affect a system.

Feedback

Feedback occurs when a variable within a process is used to modify the value of another variable in the process. The variable is "fed back" from one part of a process to another. This feedback can be of two types: *reinforcing feedback*, which

7. See Senge 1990.

causes a variable to increase or decrease in a sustained fashion, or *balancing feedback,* which causes a variable to return to a target value if a change has moved it away from target. Suppose that a company's profits go up. Then it has more money to invest in research that leads to new products, new sales, and more profits. Of course the opposite could also happen. A decrease in profits leads to less investment in research, fewer new products, fewer new sales, and hence reduced profits. Now, suppose a manager controls costs closely. If costs rise above plan, the gap is noted and activities are put in place to reduce the costs. If the costs go below plan, the system might look for ways to use the extra cash for unfunded activities. Balancing feedback is critical to the notion of management controls. Such controls are used by management to keep performance at a target (expenses on budget, projects on track, and so forth).

Time Delays

In real systems, a time delay always occurs between taking an action and seeing the result of this action. Many people consider time delays to be one of the primary difficulties in managing systems. When managers make changes to a system to create improvement, normal business pressures force them to look for instant results. When this does not happen, they feel compelled to keep doing other things that may undermine the original action. The original action may have been the right thing to do; they just did not give it enough time to see the results. In effect, such managers end up tampering with the system. Also, they may erroneously conclude that the original action was incorrect and reverse their original decision.

Nonlinearity

A *linear relationship* is one where the sensitivity of an output to an input is constant over the entire range of possible values of the input. Suppose a change of one unit of input changes the output by 2 units: then the sensitivity would be 2/1 = 2. If this value of 2 does not change as the input value is changed, then the relationship is linear. A linear relationship is an idealization. It is a useful approximation to the real world. In general, relationships will be *nonlinear* (although the error introduced by assuming linearity may be negligible). Relationships involving people are notoriously nonlinear. For example, if the gap between actual performance and target is small, a manager may well ignore the gap. This may continue until a threshold value is reached. Then the manager takes a huge action.

These three elements of system structure make it difficult to predict how a system will behave over time. That is, it is difficult to predict the dynamics of the system. Models used to predict the behavior of a system over time are referred to as descriptive models. In particular, models used to predict the dynamics of a system are referred to as dynamic models or simulations. We set up the model structure, give it a set of initial conditions, and then run the model (simulation) to see the predicted behavior.

While all four types of computational complexity (size of solution space, physics, uncertainty, and system structure) appear in business organizations, the main focus of this book is on system structure. We focus on business organizations as comprising of processes whose dynamics must be modeled so that performance can be predicted. Random variation is included because it can have a significant impact on the dynamics. In general, any relationships between variables are described with simple mathematical formulas to avoid any computational complexity because of physics. An example of optimization is included in appendix D, which is based on the models used in chapter 8, but the reader can omit this appendix if desired.

To justify our claim that structure and random variation introduces computational complexity, we now discuss two examples illustrating these complexities.

2.4 Structural complexity: A market share model

In order to show how structural complexity can make it difficult to predict the behavior of a system, consider a marketing department that is getting ready to launch a new product. The department wants to study the process by which the product is taken up by the market. The total marketplace for the product is given the value of 1 so that uptake in the marketplace is represented by a number from 0 (no uptake) to 1 (maximum uptake). The market uptake, or market share, is represented by the variable M. The time horizon of the model is divided into equally spaced segments, and the model should predict the value of the market share at a period T+1, M_{T+1}, from the market share at period T, M_T. Initially, the company intends to invest in creating a small seed market, so that $M = M_0$ at time 0.

Once the product is launched, two opposing mechanisms will change the market share from its initial value of M_0:

- As the product gets used, satisfied users of the product will spread the word so that the value of M_{T+1} will be proportional to M_T.
- As the market uptake grows, less of the marketplace remains. It becomes harder to gain more share. This means that the value of M_{T+1} will be proportional to $(1 - M_T)$.

We can summarize this model with the following equation:

$$M_{T+1} = RM_T (1 - M_T) = RM_T - RM_T2 \qquad 0 \leq M_T \leq 1 \qquad (2.1)$$

The first term is the growth term for the market share and the second term is the decay term. R is a constant that will vary from system to system. It is a combination of how much effort the company puts into marketing the product and how much resistance there is to growth as the market share gets closer to 1. If the company invests in marketing the product or if the resistance to growth is large, then R will be large. Equation 2.1 is nonlinear because of the M_T squared term. Equation 2.1 is now cast into a model as shown in figure 2.4.

The market share at time T is represented as a stock, MT. This stock is set up so that it cannot go negative and its initial value is set to the value of the converter M_0. Every time period, an amount DELTA M is added to this stock. The equation for this flow connector follows:

$$R*MT*(1 - MT) - MT$$

This is equation 2.1 with M_T subtracted from both sides. Note that the flow connector is configured to be a biflow because the population can decrease as well as increase. The constant R is represented by a converter with a constant value. Figure 2.4 illustrates the feedback in the market share model. The current value of the market share is fed back to determine the market share at the next time period. The parameter R is now seen to represent the strength of the feedback. That is, the larger the value of R, the greater the amount of feedback. Figure 2.5 shows the profile for the population M over 500 time periods with M_0 set to 0.1 and R set to 0.9 (profile 1) and 2.5 (profile 2).

This chart shows that in both cases the market share quickly reaches a steady state value. In fact, it is easy to predict this value directly from equation 2.5 by setting $M_{T+1} = M_T = M_\infty$. This gives the following equation:

$$M_\infty = RM_\infty(1 - M_\infty) \quad \text{which simplifies to} \quad M_\infty = (R - 1)/R \qquad (2.2)$$

From this it can be seen that if $R <= 1$ then the steady state value M_∞ is 0,[8] which is in agreement with profile 1 in figure 2.5. This means that if the effort that the company puts into the product is low, then the market share will decay to

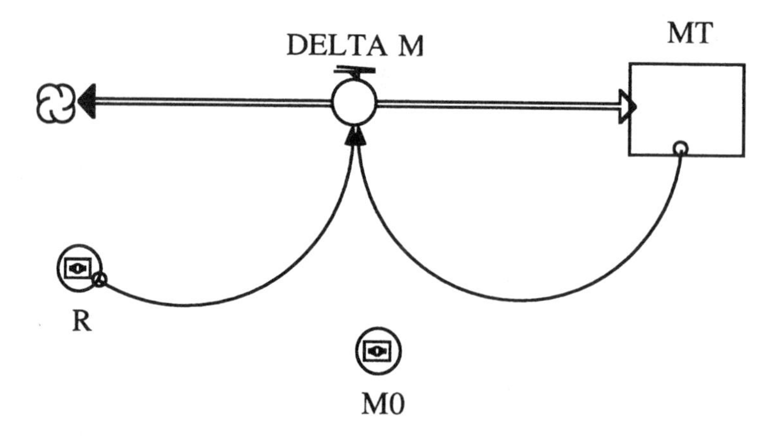

FIGURE 2.4. Model for market share equation 2.1

8. The equation for the steady-state market share gives a value < 0; but, because the market share cannot be negative, the model gives a steady state value of 0.

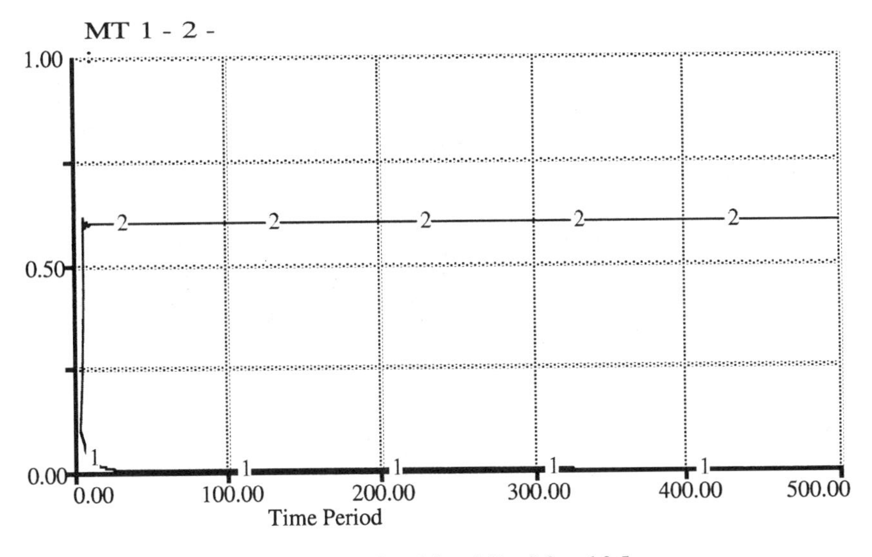

FIGURE 2.5. Profiles of market share $M_0 = 0.1$ and $R = 0.9$ and 2.5

the point that the product loses all market share. This certainly makes sense. Setting R to value of 2.5 gives $X_\infty = 0.6$ as seen in figure 2.5, in agreement with profile 2 in figure 2.5. This means that if the marketing effort is large enough, eventually a constant market share is obtained where the resistance to growth in the marketplace balances the growth effort. Again, this appears to be a reasonable result.

At this stage, it appears that the system as defined by equation 2.1 or the model in figure 2.4 is simple. So where is the complexity? If our understanding of the role of feedback is correct, then increasing the amount of feedback may introduce complex behavior in the system. To increase the amount of feedback, we must increase the value of R. Figure 2.6 shows the profile for the market share with R set to 3.9.

Now the complexity is revealed. The steady-state value for the population with this value of R should be $(3.9–1)/3.9 = 0.744$. However, there is no evidence from figure 2.6 that a steady state will ever be reached. The reader can increase the amount of time periods in the model to verify this. The profile in figure 2.6 is reminiscent of random variation but there is no random variation in the model. In fact, the market share profile shows evidence of chaotic behavior.[9] It is difficult to imagine how a profile as complex as that shown in figure 2.6 could be predicted by looking at the simple structure in figure 2.4. Essentially, the high value of R means that the market share keeps bouncing quickly between high and low values. It is as if the high growth and resistance means that the market is never still

9. For more information on chaos, see Gleick 1998.

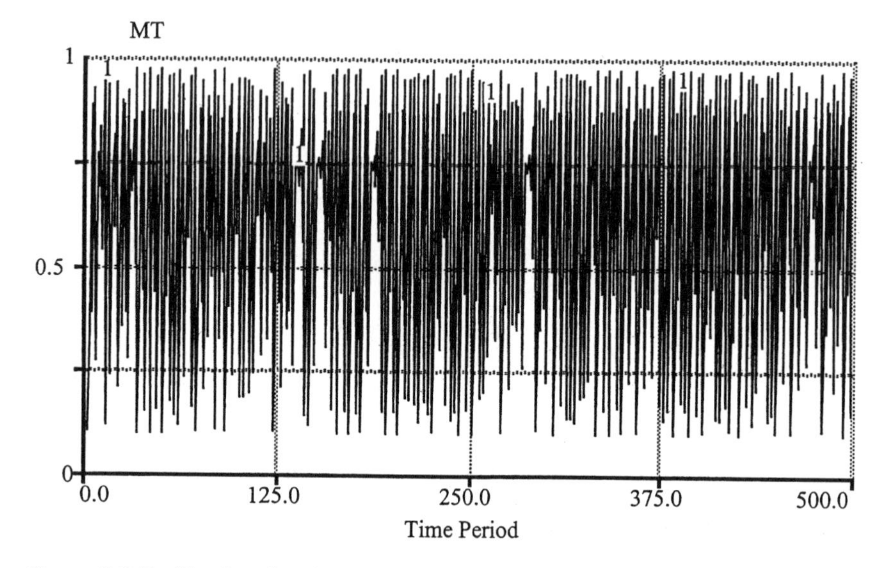

FIGURE 2.6. Profile of market share with R = 3.9 and $X_0 = 0.1$

long enough for a steady state to be reached. Detailed analysis of the behavior of the model shows that the reason for the complex behavior at the higher values of R is the nonlinear term in equation 2.1.[10] So it is the combination of the nonlinearity and feedback in the model that leads to the complex behavior.

It must be pointed out that the profile in figure 2.6 would never be seen in a real marketing situation. No sane organization would ever have a product marketed with this profile. What the model does point out, however, is that if the marketing department uses an aggressive marketing plan, or if the product takes off quickly, the initial fast growth rate may turn to decay rapidly. Perhaps we have seen some of this behavior in the rapid growth and demise of the dot-com companies.[11]

We can add further complexity to the system in figure 2.4 by adding time delays. These time delays can occur because the market may not react immediately to changes. The time delay for the growth term and the decay terms may be different. Consequently, separate time delays are added to the model. The modified model for this system is shown in figure 2.7.

Converters have been added between the stock and the flow to allow for the time delay. Following are equations in the delay converters:

DELAY(MT,GROWTH_DELAY_TIME)
DELAY(MT,DECAY_DELAY_TIME)

---

10. Bequette 1998.
11. Senge (1990) discusses the case of the rapid growth and subsequent demise of the People's Express airline company, which also may be an example of this behavior.

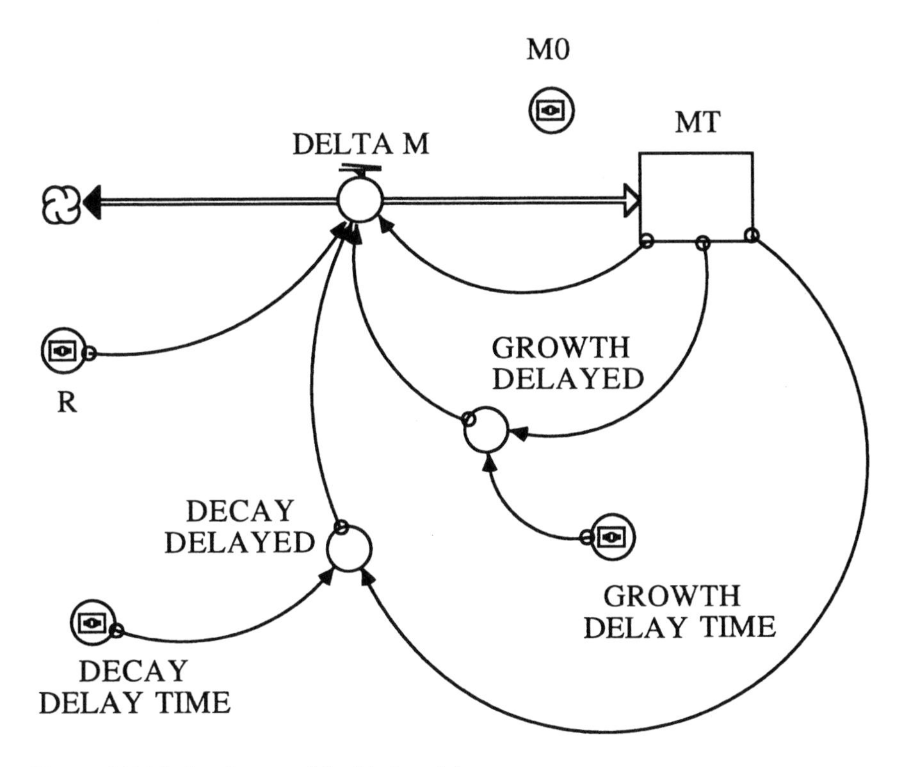

FIGURE 2.7. Market share model with time delay

By increasing the value of the converters—GROWTH_DELAY_TIME and DECAY_DELAY_TIME—we can increase the amount of time delay in the model. The equation in the inflow is

R*GROWTH_DELAYED-R*DECAY_DELAYED*DECAY_DELAYED-MT[12]

We will now look at the behavior of the model for the three cases, R = 0.9, 2.5, and 3.9. Figure 2.8 shows the profile with R = 0.9 and with delays of 10 in both growth and decay.

Comparing figure 2.8 to profile 1 in figure 2.5 shows that the time delays have not changed the profile by much. The steady state value is still 0. The only impact of time delays appears to be that the decay to 0 is slower when time delays are present. In effect, the time delays slowed the response of the market. This makes sense. By running the model with different combinations of the time delay values, readers can also verify that having different values for the time delays does not change the general behavior of the market share profile.

Figure 2.9 shows the profile for the case of R = 2.5 and both time delays set to 10.

---

12. Readers should satisfy themselves that the last term, MT, should not be delayed.

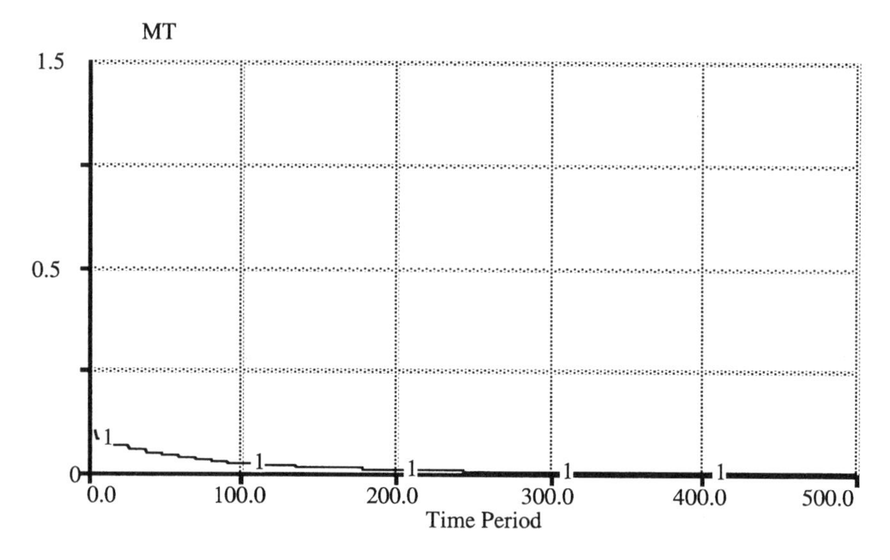

FIGURE 2.8. Market share model with R = 0.9 and delays of 10 time periods

Again, we see that the behavior in figure 2.9 is the same as for profile 2 in figure 2.5, except that it takes longer to reach the steady-state value. Further investigation shows, however, that this situation only holds when the time delays in both growth and decay are the same. If the time delays are different from each other (even by 1 time period only) then the profiles become unstable and the market

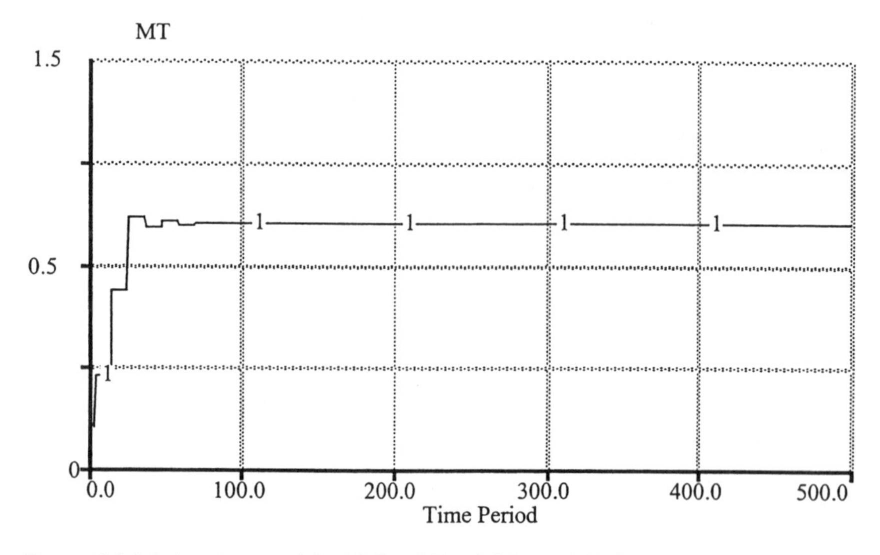

FIGURE 2.9. Market share model with R = 2.5 and delays of 10 time periods

share profile oscillates in ever-increasing cycles. Readers should verify this by running the model with various combinations of the time delays. It would appear that when a time delay is present, it introduces instability into the model. When both time delays are the same, they effectively cancel each other out. It should not surprise us that the time delays have a different impact on the behavior of the model. After all, the growth time delay enters through a linear term and the decay time delay enters through a nonlinear term.

Figure 2.10 shows the market share profile for R = 3.9 with both time delays set to 10. As in the previous two cases, if the growth and decay time delays are the same, the basic profile is the same as when there was no time delay. The only impact of the time delay is to slow down the response of the market. In figure 2.10, this slowing down is seen as a stretching out of the oscillations. If the time delays are not the same, then the market share profiles depend on the relative values of the time delays. It is left as an exercise to the readers to investigate the impact of the time delays by running the model.

This analysis can be summarized with the following learning point.

> Learning Point: The structure of a system (feedbacks, time delays, nonlinearities) adds computational complexity to a system and makes it difficult, if not impossible, for a person to infer the dynamic behavior of a system by mental modeling alone.

It should be pointed out that the earlier model for market share (figure 2.7) is not presented as an accurate model for market share dynamics. As we have already said, all models are only approximations to the real world. The ultimate jus-

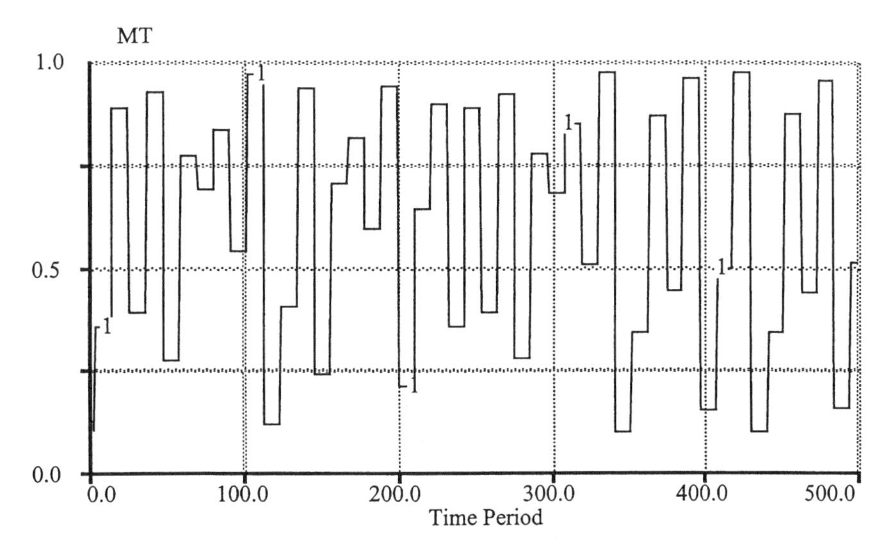

FIGURE 2.10. Market share model with R = 3.9 and delays of 10 time periods

tification for any model will be its usefulness to the user. Various modifications to the model presented here could be proposed and investigated. In particular, consider a model where the growth rate term and decay rate term have different constants, R_1 and R_2, instead of a single constant, R. This modification could be made in two ways, as shown by equations 2.3 and 2.4:

$$M_{T+1} = R_1 M_T \left(1 - \frac{M_T}{R_2}\right) = R_1 M_T - \frac{R_1}{R_2} M_T^2 \qquad 0 \le M_T \le 1 \qquad (2.3)$$

$$M_{T+1} = R_1 M_T \left(1 - \frac{R_2}{R_1} M_T\right) = R_1 M_T - R_2 M_T^2 \qquad 0 \le M_T \le 1 \qquad (2.4)$$

These two models are basically the same. These models are not analyzed here. However, the second model is included as model 2.7a to enable readers to investigate their properties as a function of R_1, R_2 and the time delays.

2.5 Complexity due to random variation: An order control process

In order to understand how random variation can make it difficult to predict the behavior of a system, consider the following example. You are employed in an order entry process in which orders arrive from customers and are processed. The downstream part of the process can only handle one order per minute, but orders do not necessarily enter the process at this rate. It is your job to adjust the upstream part of the process so that the rate of flow of orders moving downstream is one/minute. Based on the input order flow rate, you are able to adjust the rate to get it back to 1. A model for this system is shown in figure 2.11.

The base order input rate is modeled using the stock BASE INPUT RATE. It might seem strange that a rate is modeled as a stock. In theory, this model could be set up using only flows and converters. However, within the ithink architecture, this would show up as problem with a circular reference.[13] The converter INPUT RATE VARIATION allows us to add a change to the input rate, so that the converter ORDER INPUT RATE adds the variation to the BASE INPUT RATE. The actual input rate that the process sees is the converter ORDER INPUT RATE. It is this number that must be kept at the target. Once a change occurs because of the value in INPUT RATE VARIATION, we must decide on an adjustment. This adjustment should depend on how far the ORDER INPUT RATE is from target. The equation in the INPUT RATE ADJUSTMENT converter is

IF TIME < = 500 THEN 0 ELSE (TARGET - ORDER_INPUT_RATE)

---

13. See MODEL 2.11A.STM for an example of trying to make this model work without a stock.

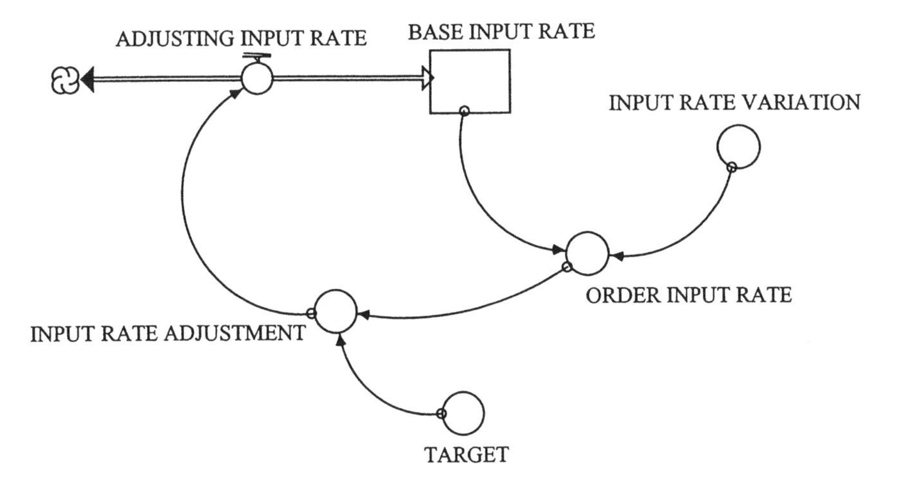

FIGURE 2.11. Model for the order rate adjustment process

The reason for the IF statement with TIME < = 500 is that we will not turn on any adjustment strategy until time = 500 minutes. This will allow us to see the impact of the adjustments on ORDER INPUT RATE. Once the order input rate is off target, an adjustment will be made. Note that the measurement and control of ORDER INPUT RATE is considered to be part of the process because you cannot understand the behavior of the process without understanding how the process is being adjusted when it is off target.

To show how the model reacts to a change, we use the equation STEP(0.5,200) in INPUT RATE VARIATION. This will cause the order input rate to jump from 1.0 to 1.5 orders/minute at the 200th minute. The profile for ORDER INPUT RATE is shown in figure 2.12.

The input order rate stays at 1.5 until the 500th minute, when the adjustment strategy is engaged. The input order rate is immediately adjusted back to the target of 1.0. The reader is invited to try different functions for the input rate variation to see the impact of the adjustment strategy. The behavior of the order input rate is easy to understand, and it behaves as expected.

Now we modify the input variation using the function RANDOM(-0.5,0.5). This function adds random variation that is uniform in the range –0.5 to +0.5. The resultant profile for the order input rate is shown in figure 2.13.

Until the 500th minute, the order input rate varies randomly about the target of 1.0. Because the additional variation is in the range –0.5 to +0.5, the input order rate varies in the range 0.5 to 1.5. It remains in this range until the 500th minute, when the adjustment strategy is engaged. We would now expect the variation in order input rate to be reduced and eliminated. The graph shows a completely different behavior. Not only is the variation not reduced, it actually increases! The data in figure 2.13 are also contained in table 2.13 in the model. If these data are copied into a spreadsheet, the standard deviation of the data can be calculated for

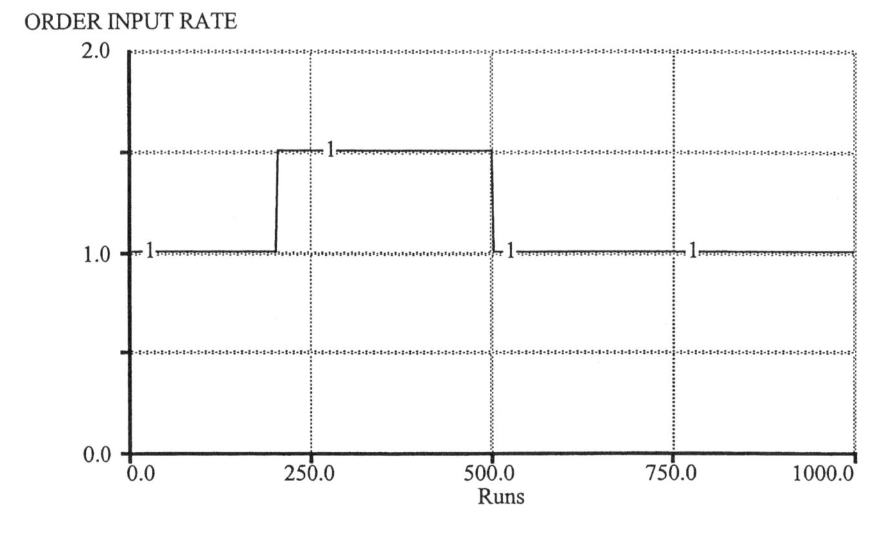

FIGURE 2.12. Profile for order input rate with a step change at t = 200 minutes

the time periods before and after 500 minutes. The values thus obtained are .29 before and 0.40 after. This represents an increase of 40 percent in the standard deviation. Thus, the addition of an adjustment strategy that was designed to eliminate the variation actually increased it. This leads to another learning point.

> Learning Point: When a measure is subject to random sources of variation, it is not possible to reduce the variation in the parameter by making adjustments based solely on the last value of the parameter. Such adjustment strategies actually increase the variation!

In fact, this principle is a key part of the process adjustment strategies used in statistical process control (SPC).[14] It is not difficult to see how a lack of understanding of this principle can impact a manager. When random variation exists in the process measure, the measure will never be on target. If the manager responds to every gap by taking a control action, the manager only ends up making things worse! Suppose the measure represents the average unit cost over a month and is reported on a monthly basis. The value this month is $12, whereas last month it was $11.50. The manager is asked to explain why costs have risen, and therefore tries to find some reason for the increase—in effect, valuable people resources must be used to track down the problem. However, any actions the manager might take based on just this one measurement could actually make things worse.

14. For example, see McConnell 1997 or Deming 1986.

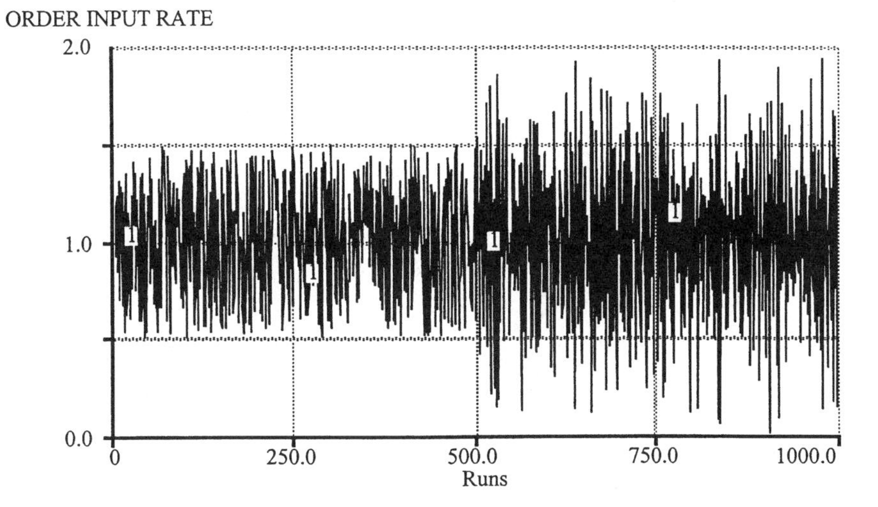

FIGURE 2.13. Profile for order input rate with random variation

Suppose next month the unit cost drops, as it might because the measurement is affected by random variation. Then the manager might attribute this drop to interventions that were made during the month. This convinces the manager to take further action because the interventions appear to be working. The following month, the manager is expecting the unit costs to drop again, but instead the value increases again. Now the manager is confused. More interventions must be identified—and quickly. What the manager is failing to realize is that the control system being used is reacting to pure random variation. The only change that will result is that the variation in the measure will increase.

Of course, the manager cannot simply ignore the values of the unit cost altogether. Some of the changes in the unit cost might represent a genuine trend in the unit cost. For example, a drop in productivity may result in an upward trend in the unit cost. This is a genuine signal to which the manager should react as soon as possible. Therefore, the manager needs a way to separate the real upward trend from the "normal" background random variation and must be able to do this as soon as possible after the trend has started, or the business will suffer a loss in profitability for a longer period of time. Again, the ability to differentiate between trends and random variation is the subject of statistical process control.

> Learning Point: In order to operate a business as successfully as possible, a manager must be able to differentiate between changes in a measure due to genuine trends in the measure and random variation in the measure.

Figure 2.11 shows the feedback that is involved in the control strategy used for the input order rate. The value of the order input rate is fed back to calculate the

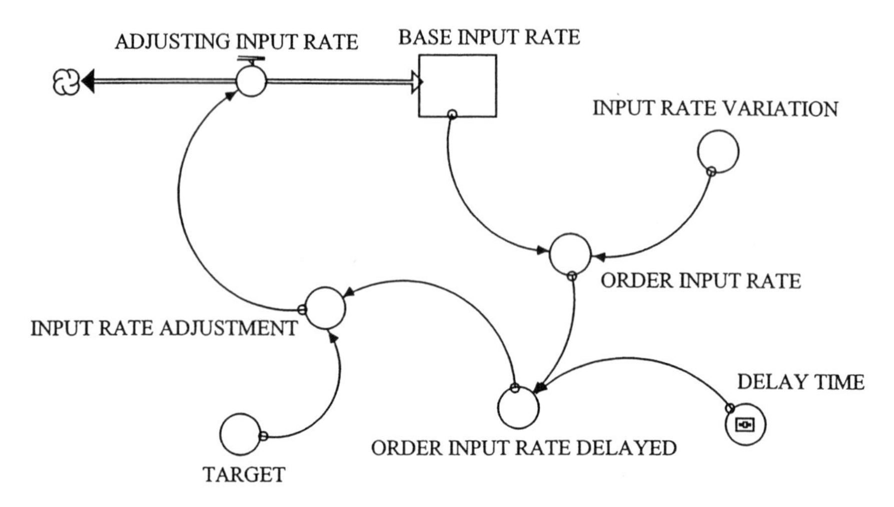

FIGURE 2.14. Model for the order rate adjustment process with time delay

adjustment. There is no time delay involved. The model in figure 2.14 is similar to figure 2.11 except that a time delay has been added.[15]

The model in figure 2.14 contains the time delay explicitly. However, the presence of random variation is equivalent to a time delay when the monitoring of a process for control is part of the process. Consider figure 2.11 again, in which the system changed at t = 500. The change in the profile was immediate and there would be no discussion as to where the change occurred. However, what about figure 2.13, which has random variation included? How quickly could we decide that an increase in variation had occurred? Figure 2.15 shows the same profile as figure 2.13, except that it shows only the range t = 490 to t = 510.

Now it is more difficult to answer this question: "Where did the change occur?" Perhaps by time 503 or 504, one might be prepared to say that a change in the amount of random variation has occurred. This delay in our ability to detect a change will delay our ability to react to that change. Thus in effect, the random variation has introduced a time delay into the process.

Learning Point: Random variation in a process introduces a time delay in our ability to monitor change in a process.

2.6 Further benefits of dynamic modeling

There are other benefits that can be gained from the dynamic modeling of business processes. These include the following:

---

15. This model will not be discussed here and is provided to readers so that they can assess the impact of the time delay.

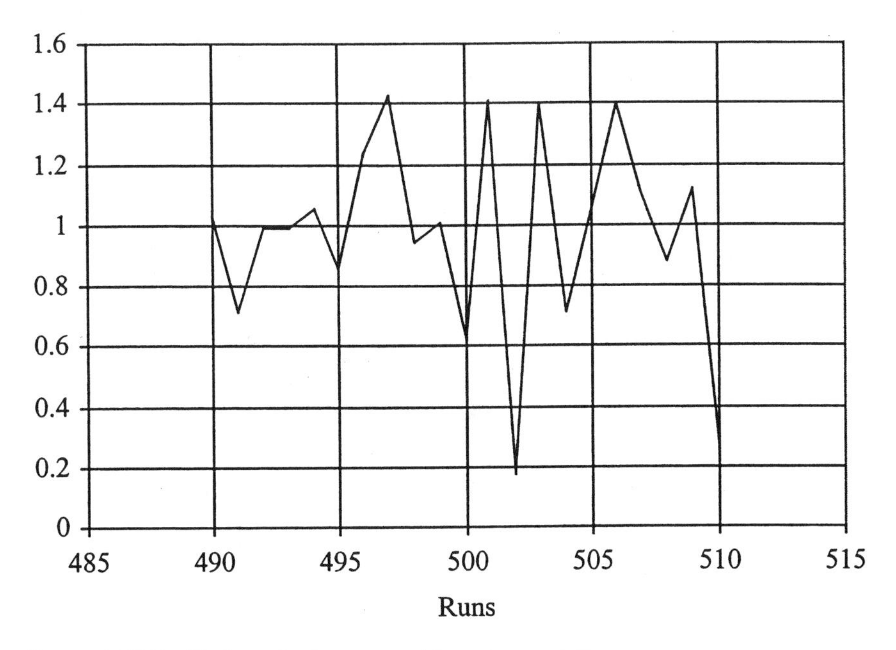

FIGURE 2.15. Profile for order input rate with random variation

• Find possible emergent properties of a system. Even though managers have been working with particular systems for a considerable time, they may not have seen all the possible ranges of behaviors that a system can exhibit. This emergent behavior may take a long time to occur in the actual system, that is, in real time. Dynamic models provide a way to explore such emergent behavior because such models can cover these real-time spans in the short time it takes to run the model. We can model five years of behavior in 60 seconds of real time!

• Provide quantitative assessments of qualitative ideas. Generally, most systems are only understood at a qualitative level. A model can allow the user to turn qualitative understanding into quantitative understanding. For example, we all qualitatively understand that compounding interest can cause an investment to grow rapidly over time. However, without a model, we cannot quantitatively understand just how large it will grow.

• Allow the identification of the most important parameters in a system. In any real system or process, many inputs are typical. In theory, all of the inputs can have an impact on the output. However, if we have to monitor all of the inputs to manage the outputs, we are in big trouble. We hope that we can use the Pareto Principle. This principle states that 80 percent of the output is affected by 20 percent of the inputs. Said another way, all inputs are not created equally. In general, there will be a subset of these inputs that have the greatest impact on the output. If we know what inputs are in this subset, we can focus on these inputs. But how are we to identify these inputs? If we have a model available for the process, we can perturb each input and see how it affects the output. Then we can rank the inputs in terms of their impact on the output and pick those that have the highest impact. This is called a *sensitivity analysis.* We will say more about this in chapter 3.

• Allow testing of "what-if?" scenarios. When planning how we can improve a process, we generally propose some specific improvement actions and then set up some tests to see which action leads to the biggest improvement. In effect, we set up some experiments and carry out some "what-if" scenarios. Traditionally, such experiments have been physical in nature. We set up a physical model of the system or process and test the changes on the physical model. This can be an expensive way to test, however. If we have good computer model of the system or process, the computer model becomes the equivalent of the physical model. The main expense for the computer model is typically the initial set up. After that, testing these "what-if" scenarios is relatively cheap.

• Facilitate communication through both model results and the model structure. After "what-if" scenario tests have been carried out, it is important that results of the testing can be communicated easily to those who are interested but have not been involved in the testing directly. Models can be useful to communicate the structure of a system and the results obtained from testing. Also, and perhaps just as important, models help show people why the actions taken lead to the results obtained.

• Provide a system memory. When we learn something about how a system works, we may be able to use this immediately to create improvement. As well as the immediate use of learning, it is also important that we provide some way to archive the learning so that sometime in the future, the information is easily available. It is not unusual in organizations to find situations where an issue arose because something that was once learned was forgotten. Models provide a way to create system memory. A model can be set up to contain current knowledge about a system or process. Any changes to the system or process must be tested in the model, and if the real system or process is modified, then the model is modified as well.

• Use as a training tool. It would be great if we could train a person on a process and then be sure that this person would always work on the process. However, in reality, people move around an organization and new people must be trained. A model can be invaluable in helping train to new people, especially if the model is also being used as an important part of the overall learning strategy as described in the previous paragraph. In fact, a popular idea today is the management flight simulator. A model of a system is created and then set up so that a user can interact with the model to make decisions and see the impact of these decisions. The model is set up to give feedback when the user takes actions that are not optimal.[16] By running the model multiple times, the user can try different approaches to managing the system. Of course, if wrong decisions are made, only a virtual

---

16. High Performance Systems, the company that produces the ithink software, also produces what they refer to as learning environments—these are essentially the same as management flight simulators

business is affected. It is better to make these mistakes on this virtual system than on the real system!

• Facilitates the formation of a dynamic hierarchy of exploration within a group. In many situations, getting management to commit to giving resources to create a detailed model of a system is difficult. They want to see some benefit before they make the commitment. A compromise can be reached by starting with a high-level model that has just enough detail to make the model useful. Then as management sees the usefulness of the model, they might commit resources to add more detail to the model. In fact, reluctance on the part of management has some positive benefit. By forcing us to start off at a high level, we get a feel for where greater detail will be most useful. Sometimes, addition of detail to a certain part of a model may involve a lot of work but add little or nothing to our understanding. Knowing how much detail to add to a model in order to make it useful is a skill that can be obtained only through experience.

2.7 Organizing principle of this book

The remainder of this book focuses on the application of dynamic modeling to business systems. Chapter 3 is devoted to the modeling of performance measures. In keeping with the approach of starting simple and then expanding to more complicated models, chapter 4 discusses the simplest process possible, a single-step process. Then, between chapters 5 and 10, we expand this to cover multistep processes, supplier and customer interfacing, multiple business objectives, and supply chains. Think of this as expanding the simple process in a single dimension. In chapter 11, we show how to expand into a second dimension by considering the relationship between strategy and tactics. Finally, in chapter 12 we consider how organizations improve their process performance. Improvement is a process in itself; it can be modeled like any other process.

Validation of models will not be covered in this book. Readers will do this as they apply models to their own organizations. Instead, we will use the models presented throughout this book to explain the general behavior of organizations. The principles that we uncover can be applied to the many different types of organizations found in business today.

Model Equations

```
MODEL 2.4.ITM
{ INITIALIZATION EQUATIONS }
M0 = 0.1
INIT MT = M0
R = 0.9
DELTA_M = R*MT*(1-MT)—MT

{ RUNTIME EQUATIONS }
```

```
MT(t) = MT(t—dt) + (DELTA_M) * dt
DELTA_M = R*MT*(1-MT)—MT
```

MODEL 2.7.ITM

```
{ INITIALIZATION EQUATIONS }
M0 = 0.1
INIT MT = M0
R = 0.5
GROWTH_DELAY_TIME = 0
GROWTH_DELAYED = DELAY(MT,GROWTH_DELAY_TIME)
DECAY_DELAY_TIME = 0
DECAY_DELAYED = DELAY(MT,DECAY_DELAY_TIME)
DELTA_M = R*GROWTH_DELAYED-R*DECAY_DELAYED*DECAY_
DELAYED-MT

{ RUNTIME EQUATIONS }
MT(t) = MT(t - dt) + (DELTA_M) * dt
GROWTH_DELAYED = DELAY(MT,GROWTH_DELAY_TIME)
DECAY_DELAYED = DELAY(MT,DECAY_DELAY_TIME)
DELTA_M = R*GROWTH_DELAYED-
R*DECAY_DELAYED*DECAY_DELAYED-MT
```

MODEL 2.11.ITM

```
{ INITIALIZATION EQUATIONS }
TARGET = 1.0
INIT BASE_INPUT_RATE = TARGET
INPUT_RATE_VARIATION = RANDOM(-0.5,+0.5)
ORDER_INPUT_RATE = BASE_INPUT_RATE +
INPUT_RATE_VARIATION
INPUT_RATE_ADJUSTMENT = IF TIME < = 500 THEN 0 ELSE
(TARGET- ORDER_INPUT_RATE)
ADJUSTING_INPUT_RATE = INPUT_RATE_ADJUSTMENT

{ RUNTIME EQUATIONS }
BASE_INPUT_RATE(t) = BASE_INPUT_RATE(t—dt) +
(ADJUSTING_INPUT_RATE) * dt
INPUT_RATE_VARIATION = RANDOM(-0.5,+0.5)
ORDER_INPUT_RATE = BASE_INPUT_RATE +
INPUT_RATE_VARIATION
INPUT_RATE_ADJUSTMENT = IF TIME < = 500 THEN 0 ELSE
(TARGET- ORDER_INPUT_RATE)
ADJUSTING_INPUT_RATE = INPUT_RATE_ADJUSTMENT
```

MODEL 2.14.ITM

```
{ INITIALIZATION EQUATIONS }
TARGET = 1.0
```

```
INIT BASE_INPUT_RATE = TARGET
INPUT_RATE_VARIATION = RANDOM(-0.5,+0.5)
ORDER_INPUT_RATE = BASE_INPUT_RATE +
INPUT_RATE_VARIATION
DELAY_TIME = 0
ORDER_INPUT_RATE_DELAYED =
DELAY(ORDER_INPUT_RATE,DELAY_TIME)
INPUT_RATE_ADJUSTMENT = IF TIME < = 500 THEN 0 ELSE (TARGET-
ORDER_INPUT_RATE_DELAYED)
ADJUSTING_INPUT_RATE = INPUT_RATE_ADJUSTMENT/DT

{ RUNTIME EQUATIONS }
BASE_INPUT_RATE(t) = BASE_INPUT_RATE(t—dt) +
(ADJUSTING_INPUT_RATE) * dt
INPUT_RATE_VARIATION = RANDOM(-0.5,+0.5)
ORDER_INPUT_RATE = BASE_INPUT_RATE +
INPUT_RATE_VARIATION
ORDER_INPUT_RATE_DELAYED =
DELAY(ORDER_INPUT_RATE,DELAY_TIME)
INPUT_RATE_ADJUSTMENT = IF TIME < = 500 THEN 0 ELSE (TARGET-
ORDER_INPUT_RATE_DELAYED)
ADJUSTING_INPUT_RATE = INPUT_RATE_ADJUSTMENT/DT
```

3

Measuring Process Performance

> I often say that when you can measure what you are speaking about and express it in numbers you know something about it; but when you cannot express it in numbers your knowledge is a meager and unsatisfactory kind; it may be the beginning of knowledge but you have scarcely, in your thoughts, advanced to the stage of science, whatever the matter may be.
>
> —William Thompson (Lord Kelvin)

3.1 Introduction

Given that we want to predict process performance under various conditions, it is important to be able to measure what we mean by process performance. If we cannot measure performance or if we measure it incorrectly, then we cannot decide if the performance of the organization is acceptable, and we cannot decide if the operating conditions of the organization must be changed and resources redeployed. Also, we cannot know if improvement activities are yielding anticipated results. Furthermore, if we are going to model a business process in an organization, we must know what we want to measure, and we must set up the model to include those measurements. The model should reflect how the measurements would be made in the real process. We must model the measurement process itself.

The question that we must ask is this: Is there is a single, overarching measure of performance that incorporates all aspects of process performance? If such a measure exists and we can actually measure it, then we would have an unambiguous way of deciding the impact of any decision on the performance of the process.

When a person invests in an organization, they are tying up resources with the objective of making a return on the investment of these resources.[1] If they knew they could get greater return elsewhere, they would certainly want to take these resources and put them in the alternative investment. The purpose of an organization is not survival, because many times the best way to create wealth for an owner is to sell the business, nor is the purpose of an organization to satisfy its customers. Selling a Porsche for $100 might create many happy customers, but it

1. This is true whether the organization is privately or publicly owned.

would not create wealth for the business owners. So, it would be difficult to find people willing to invest in such an organization. Deming (1986) recognized this when he defined quality as a "predictable level of variation suited to the market" and not as perfection. This is not meant to imply that satisfying customers is not important. You cannot create wealth if you do not sell to some set of customers. And these customers will not buy if they are not satisfied with the organization's products or services. To see what such a measure might be, we should first ask: What is the overall purpose of an organization?

We contend that the purpose of an organization is to create wealth for its owners greater than what could be created by other means over some time horizon. That is why people will invest in an organization that is losing money today but promises the rewards of the money it will make in the future. Goldratt and Cox (1992) take a similar view by stating that the purpose of an organization is simply to make money.

> Learning Point: The purpose of an organization is to create more wealth for its owners (over some time horizon) than they could get by investing elsewhere.

This means that the most fundamental measure of an organization's performance is the money it will make over the given time horizon. Section 3.2 is devoted to the modeling of financial performance. Then we will look at the issues with monitoring performance based solely on financial measures. We will illustrate a process model that provides a framework for developing nonfinancial measures of performance. These measures will supplement the financial measures and help overcome the limitations of using just financial measures.

3.2 Financial measures of performance

A financial measure of process performance will involve some measurement of the overall profitability of the organization. Such measures are based on the difference between the cost of operations and the revenues generated by the sales of the organization. We start by looking at a basic profit model and then build our model up from there. In developing these models, we will show how these measures can be used to make decisions about how to run the business.

3.3 The basic profit model

Consider the most basic dynamic description of production. We have a PRODUCTION RATE, measured in thousands of widgets produced per week. Widgets are assumed to be sold immediately (no inventory). In this case, we can equate production and sales. Our firm has investigated the demand for widgets and found that if we lower the price, we can sell more. That equation is: PRICE = 10—2.5*PRODUCTION RATE. This is the demand curve for our product. We really want to know which price to set for our widgets, so we make PRICE a parameter and use the demand curve to solve for the PRODUCTION RATE. Then REVENUES are simply PRICE times PRODUCTION RATE. We still need more information to solve for the optimum PRICE. We need either a production function, which relates production rate to the use rates of inputs, or we need a cost function, which relates the inputs to the cost of production, COSTS. The following example model uses the cost function approach.

Suppose company production data show that COSTS equal 0.1*PRODUCTION RATE. Knowing both the REVENUES and COSTS, we can easily calculate PROFIT as simply their difference. Our objective should now be clear. We want to find the maximum profit by adjusting the selling PRICE. Figure 3.1 is the ithink model of this process.

We now use the Sensitivity Analysis feature of ithink to home in on the PRICE that leads to the maximization of PROFIT rate (e.g., $ per time period). Repeatedly narrowing the range of the PRICE leads to a profit maximum of $9.80 at a PRICE of $5.04 per widget. Note that here we use an explicit trial and error procedure to find the optimal PRICE.

Realize that this solution is for a monopoly because our firm is setting its own price and finding its optimal amount of production, the amount that maximizes its PROFIT. A competitive firm in a mix of many identical competitive firms would set its own price equal to the market price. It would need no information about the demand–production relationship. It would then proceed to find the level of output that would maximize its profit. The link between PRICE and PRODUCTION RATE would be broken; PRICE would become a given constant and PRODUC-

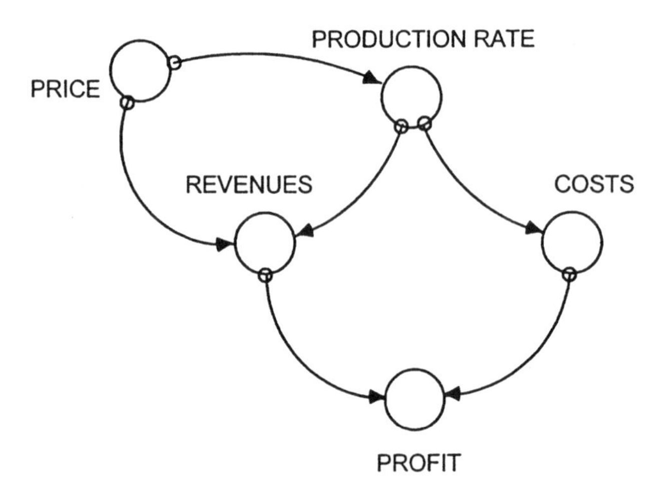

FIGURE 3.1. A model of our widget firm, relating price to profit through production rate and then revenues and costs

TION RATE would become the parameter. For example, chose a PRICE of $3 and assume that the COSTS are now the total costs of all the identical firms in this market. We find the profit maximizing PRODUCTION RATE to be 15 widgets per week: lower the price; increase the production rate (but share those profits). These competitive firms will unwillingly, inadvertently share profits and invite more firms into the market, until they fill the market demand at the equilibrium price. At this point there is no profit in the industry and the firms are said to be in equilibrium.

Learning Point: The tools and procedure used to solve simple problems can be used to solve more complex problems as well.

To improve your modeling skill, recast the monopoly problem in terms of a production function. Use the same demand and COST equation. The optimal PRICE and PROFIT should be the same as we found here. In doing so, you can prove that either form of the solution is independent of the unit cost.

Learning Point: A great number of financial numbers are involved in production. The most common idea is not to minimize cost nor maximize revenues but to maximize the difference between revenues and costs.

3.4 The role of time, borrowing, and lending

In the world of profit maximizing, revenues can be thought of as benefits, and the general concept of benefit-cost analysis can be applied to the firm. Profit is the difference between revenues and costs. To gain the maximum profit, we must find that level of output (correlated with the size of the firm) that makes the variation in profit zero. The variation in profit with respect to size is also equal to the variation in revenues (Marginal Revenue, MR) minus the variation in cost (Marginal Cost, MC), each with respect to firm size. Therefore, at the profit maximizing size, the MR is equal to the MC for the firm. The firm size at which the slope of the revenue line is equal to the slope of the cost line is where the largest difference between revenues and costs is found—this is the size that produces the maximum profit. This is benefit-cost analysis at its simplest.

The benefit-cost issue becomes more complicated when we realize that for most firms some of the costs occur well before the revenues appear. How can we possibly do a benefit-cost analysis on such a case? To answer this question, we must digress into the world of the Time Value of money to consider the concept of interest.

We begin with a bank account. Suppose that right now you have $100 dollars in a savings account at the local bank and that the savings annual interest rate is a constant 6 percent compounded annually. How much do you have after 10 years under these conditions (figure 3.2)?

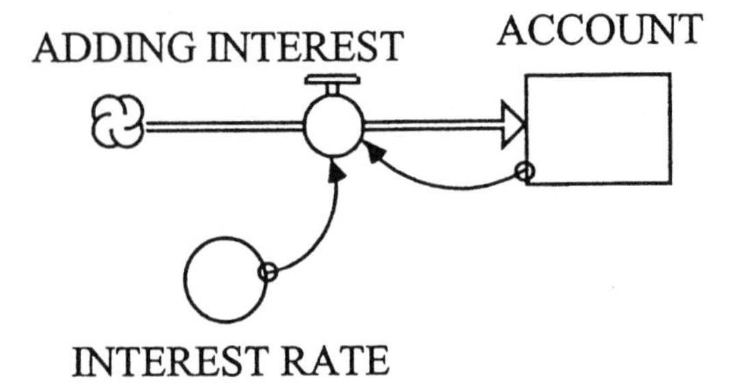

FIGURE 3.2. A model of the bank account accruing compound interest

We set the DT equal to one year to cause the compounding to be done every year. The \$100 bank account rises exponentially to \$179.08 at the end of year 10. Suppose that we compound the interest every six months. Halve the interest rate, double the number of periods, run the model, and we find the account at the end of year 10 to be \$180.61. Or, we could have just changed the DT to 0.5 years and run for 10 years. The same result occurs. We can see that this is so by flipping to the equations in the ithink model and examining the equation for the ACCOUNT:

ACCOUNT(t) = ACCOUNT(t – dt) + (ADDING_INTEREST) * dt
ADDING_INTEREST = INTEREST_RATE*ACCOUNT

We can quickly see that the second term on the right in the first equation is the key. It is here that the interest rate is halved. The first term on the right is the ACCOUNT value every DT or every six months. This is exactly the way a bank would compound the account.

If we continue this process, until the DT is small, we can see that nearly instantaneous or continuous compounding approaches a fixed value. When the DT = 1/1024, for example, the ACCOUNT at 10 years is \$182.21. If we resort to basic mathematics here and let the DT $\rightarrow$ 0, the interest formula becomes EXP(0.06*time) = EXP(0.06*10). This represents true continuous compounding and gives a 10-year annual compounding account value of \$182.21. We were close enough with our small DT.

Note that we can turn this problem around. We can ask what is today's value of \$182.21 in 10 years? Suppose someone says that they will repay your loan to them of \$182.21 in 10 years and suppose that your interest rate (now playing the role of a discount rate) is again 6 percent. Today's value of this future repayment is \$100; the previous problem is simply worked in reverse. This direction of the analysis is called Present Value Analysis.

The exponential interest rate formula comes in handy when there is a routine addition to the account. Suppose that our account begins with a \$100 deposit at time = 0 and we continue to deposit \$100 at the beginning of each year. What is the value of the account at the end of 10 years (ten deposits) of continuous interest accumulation? We set the ACCOUNT initial value at 100 to represent the very first deposit. We know that this deposit has a continuously compounded 10-year value of $182.21. So one instant after the end of the first year, the ACCOUNT value is going to be $282.21. We continue in this way, using the ADDING INTEREST formula of: EXP(INTEREST_RATE*(10—TIME))*100. Thus each new deposit at the end of its respective year is compounded less each year, representing the time that each one was actually deposited. The ACCOUNT value at the end of year 10 is therefore $1,511.71. We could not accurately represent the process with a small DT (ADDING INTEREST = INTEREST RATE*ACCOUNT + 100); if we did, the deposit would be represented by tiny amounts spread across the year rather than periodic deposits. The control name in this case is a bit of a misnomer, because this formula adds both interest and deposit.

A second way to model this process is one that more accurately represents the actual process used by a bank. The interest formula becomes: ADDING INTEREST = (EXP(INTEREST_RATE*1) – 1)*ACCOUNT. It is based on the diagram shown in figure 3.3.

This ADDING INTEREST formula computes compounded interest on the current account (including the current ACCOUNT value), then removes the current ACCOUNT value. The control, DEPOSITING, Pulse(100,0,1), adds the first-of-the-year $100 deposits and another $100 every year. The final account value (DT = 1) is $1,511.71, as before.

Now let us look at borrowing. Suppose that we borrow $10,000 for a car. The annual borrowing rate is 8 percent. The question is this: What is our monthly payment for a 36-month period if the interest rate is compounded monthly? The model, shown in figure 3.4, is similar to the last one.

The INTEREST RATE is 0.08/12 per month. ADDING INTEREST is just INTEREST RATE*ACCOUNT to calculate the annual compounding the remaining account. The initial ACCOUNT value is $10,000 from which PAYING OFF (= PAYMENTS) is extracted monthly. The DT is 1 (month). The solution is one of trial and error. Be sure to uncheck the nonnegative box in ACCOUNT. Now, experiment with the payment size such that the ACCOUNT is 0 at the month 36. We find the monthly PAYMENT to be $313.36. If the compounding were continuous (see the previous problem), the monthly PAYMENT would be $313.49, hardly different from the monthly compounding value.

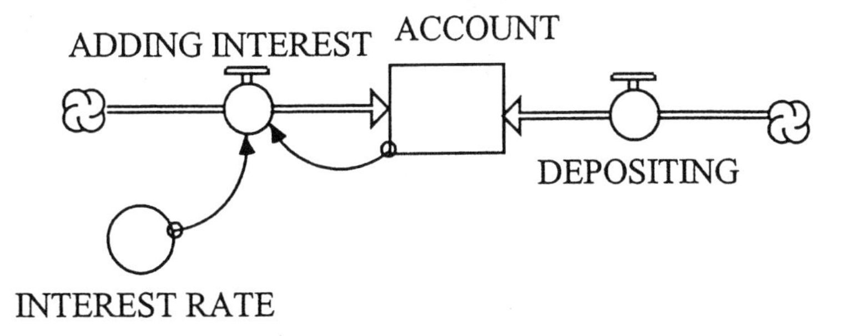

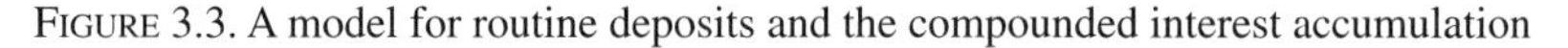

FIGURE 3.3. A model for routine deposits and the compounded interest accumulation

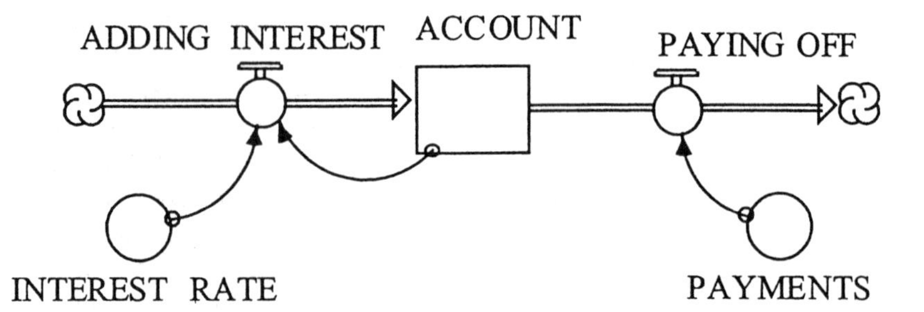

FIGURE 3.4. A model for amortizing a loan

Now that we can do everything that a bank can do, we are prepared to look at the world of benefit-cost analysis. Here the issue is one of collapsing a string of expected (future) benefits and of costs so that their ratios can be compared. In the next chapter we consider full-sized projects, but for now consider just a stream of expected benefits, B, transformed into dollars: Let B = sinwave(1,5) + 1: B is periodic, rising to 2 and dropping to 0, every five years. What we want is a present value of this entire stream of benefits extending infinitely far into the future. So we use the interest rate in a discount function, exp(-INTEREST RATE*TIME)*B and sum up the discounted value of each future year's B. This is easily done in ithink (see figures 3.5 and 3.6).

The PRESENT VALUE reaches a steady value after 100 or so years and a peak at 17.81. It will not grow beyond this level even as $t \rightarrow \infty$. We have transformed this infinitely long cycling into a single number. We recognize that most financial agents *do* value benefits less in the future than they value them now. We assume away the challenge of a variable discount rate. We also assume that the benefits do not rise exponentially at or faster than the interest rate. Were benefits to occur in this manner, the present value would not reach a steady value. Next we will examine the present value principle in use by a real firm.

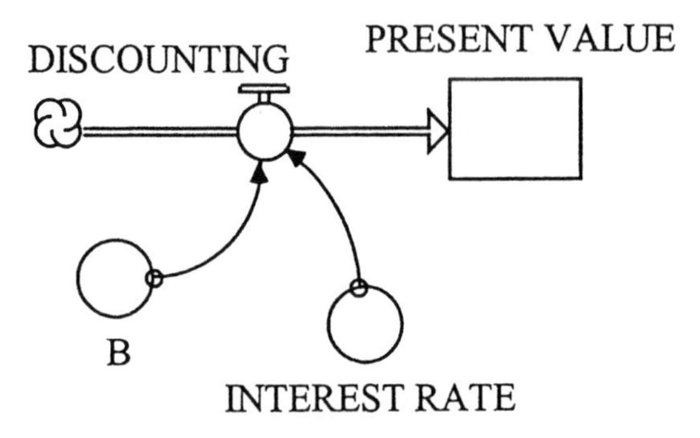

FIGURE 3.5. A model for finding the present value of the periodic benefits B with a 6 percent interest rate

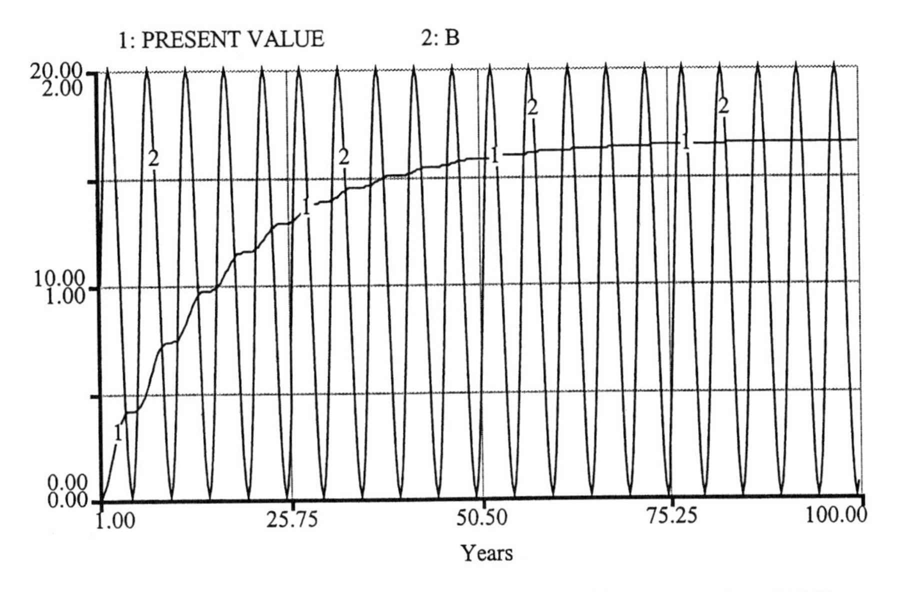

FIGURE 3.6. The graph of the periodic benefit function B and its present value of 16.51

> Learning Point: Time is money, in the world of business.

3.5 Choosing among alternatives

Now that we have learned how to calculate the present value of future payments, let us employ the method of discounting revenues and cost to decide among alternative investment options. Such comparisons are regularly made in policy decision making when a society needs to decide, for example, whether to build a dam for flood control or a new highway to improve the accessibility of a region. Let us use this method of comparing discounted cost and benefits of a project for an individual firm.

Suppose that a company is ready to commit \$5 million in profits from last year's operations to a possible combination of the available company projects or to a bank account. The market interest rate for money in the bank account is 10 percent. We will abstract away from the question of future taxes and inflation. Call the three projects A, B, and C. Distinct construction, maintenance, and revenue streams characterize each project. For project A, these streams are specified in figures 3.7 through 3.9.

The model and the construction, maintenance, and revenue streams for projects A, B, and C are in MODEL 3.1.STM on the CD that accompanies this book. (They are labeled Data for Projects 3B and 3C.) Which of the investment options

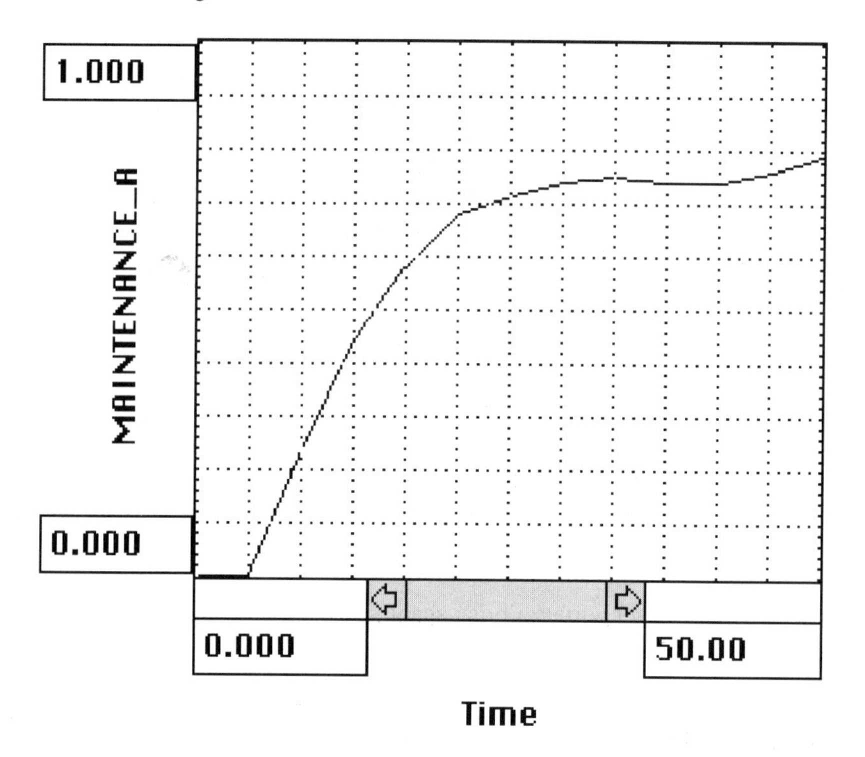

FIGURE 3.7. Construction cost schedule

should your firm choose to invest in first? Second? Third? At what point would any remaining profit be placed in the money market?

To answer these questions, form a discounted net present value, PVP, of each investment; then, accumulate it as CPVP, for an arbitrary choice of the return rate. Run the model first with a low rate of return; note the result. Then, increase the rate of return for subsequent runs and compare the model results. Find that return rate that causes the cumulative discounted net present value, CPVP, of the project to be exactly 0. This is defined as the rate of return for the project. Compare this rate to those of the other investment options. We have chosen a 50-year lifetime for each project, after which time we assume it is abandoned. However, as you will see from the proper result, the CPVP curve is nearly flat after 50 years, so an infinite project horizon would not significantly change the result (figure 3.10).

The preceding graph shows the cumulative net present value of project A over the 50-year time horizon for rates of returns set at 5, 6, 7, 8, and 9 percent, respectively. We can see from the graph that the fourth choice (8 percent) is close to the desired value. Run the results in a table to find the rate of return that comes closest to CPVP = 0 in year 50. The analysis of project A should give a rate of return of approximately 7.89 percent.

Now go to the data on the CD and find the rates of return for the other two proposed projects and determine which of the three possible investments should

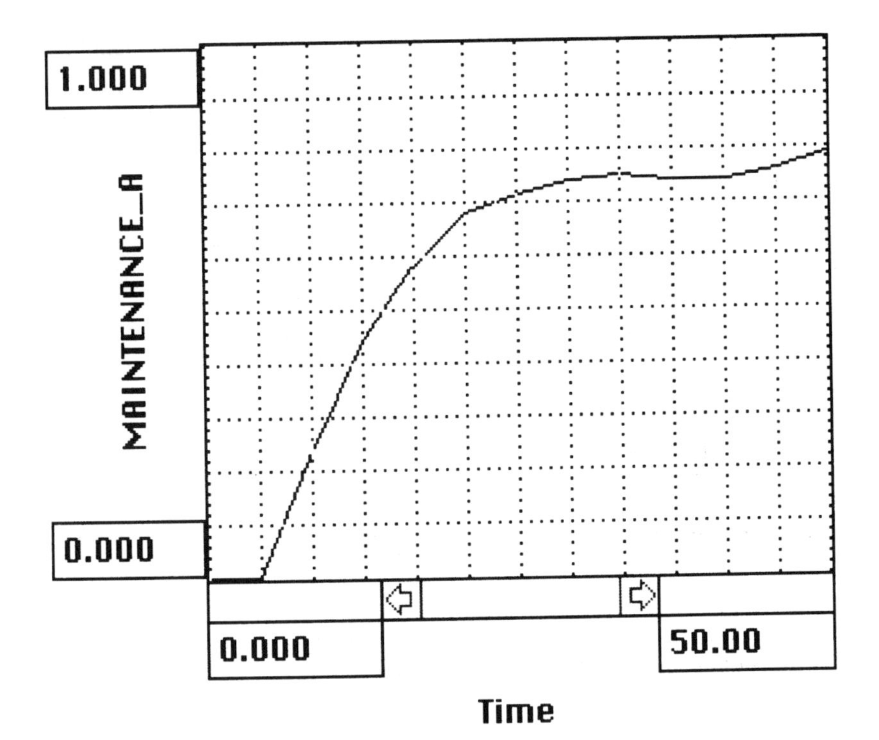

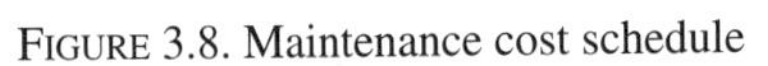

FIGURE 3.8. Maintenance cost schedule

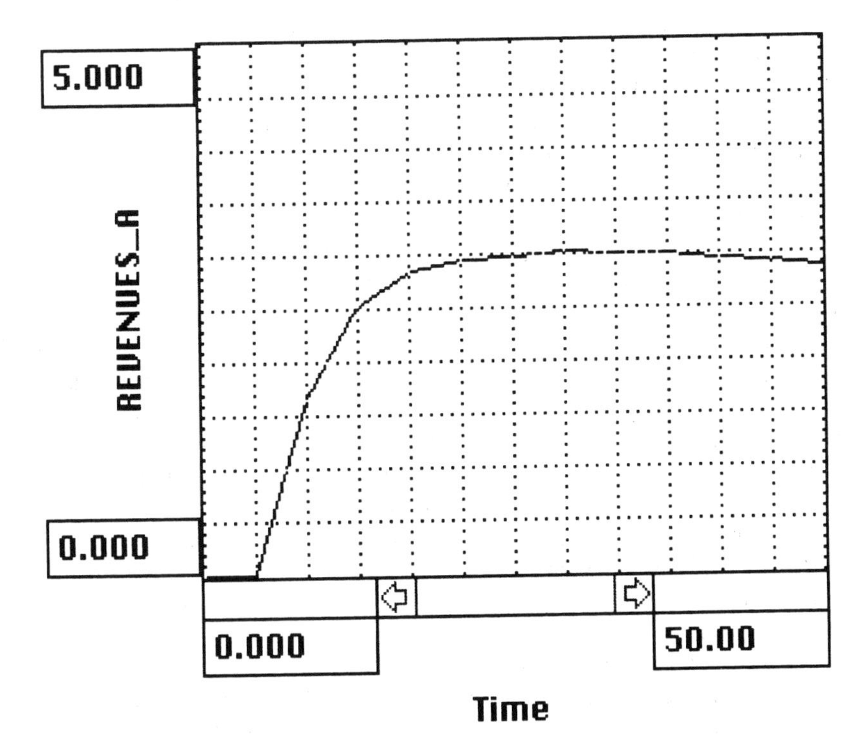

FIGURE 3.9. The schedule of expected revenues

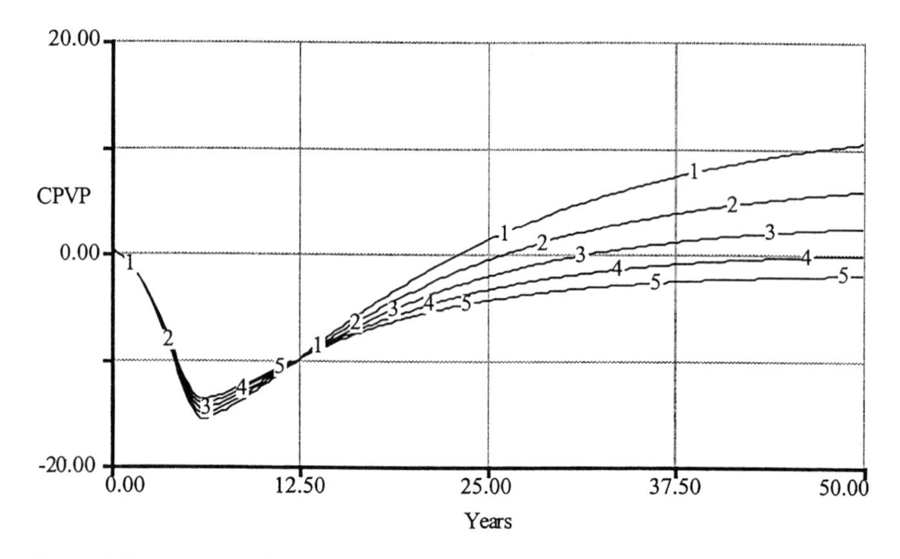

FIGURE 3.10. The sensitivity of cumulative present value of the profit, CPVP, to various return rates, for Project A over the 50-year project life

draw first on the profit funds. Then invest the money in the market while drawing it down over the construction years to finance the winning project(s). What is the present value of the invested profit over the 50-year time span?

Earlier we noted that cost–benefit analysis is a method frequently employed both at the level of an individual firm and for policy decision making on projects such as dams for flood control or construction of highways. Assume that either an individual firm or a government agency may carry out such projects. Under each scenario, the construction, maintenance, and revenue streams are the same. What may compel an individual company to reject the project even though it may be beneficial from a societal perspective and perhaps be carried out by the government agency?

Of course, the discount rate is a crucial determinant for the choice of a project, and we may expect that individual firms have shorter planning horizons than society as a whole. The firm may go out of business if it does not generate sufficient profits. Its decisions should be made in the interest of the owners of the firm, all of who have a finite life expectancy and wish to see their investments pay off soon. Government, in contrast, may be around for a long time to come, and is meant to represent the interests not just of the current generation but also of that generation's children and grandchildren (via the Social Discount Rate). As a consequence, the firm's discount rate may be higher than the government's, and therefore, its choice of an interest rate with which to compare the rate of return of the projects may be different from the government choice.

Then there is the issue of what scale to construct the favored project. This becomes a trial and error process. The analysis has to be repeated for many different sizes of each project and the optimal size—the project size that maximizes the gap between revenues and costs. In cases in this chapter, the project sizes were assumed to be fixed by other operations within the firm.

Which discount or interest rate to apply in cost–benefit analysis is a hotly debated issue,[2] and the model (project A) presented here should help you investigate the implications of alternative assumptions about the discount rate. For models of the optimal use of renewable and nonrenewable resources, the choice of discount rates is most important.[3]

We choose to discount at a rate that large investors can expect to receive, EXPECTED RETURN, usually thought to be about 15 percent.

Learning Point: In any investment process, one of the alternatives is the market itself. Internal investing is not always the most profitable path.

Learning Point: When costs and income necessarily vary over time, one must use a discounting process to properly compare proposed projects. The choice of the discount rate to be used in the process is crucial.

3.6 Optimizing at the level of the firm

We turn to the final example of our short lesson on financial analysis. Here we have a firm with only one input of constant unit cost and one output with a constant price. This firm starts small and grows until its profit begins to turn down. The model is presented in figure 3.11.

Begin with the stock INPUT. This is the use level of the single input. INPUT is used in a production function equation to yield OUTPUT. The coefficients of this production function are ALPHA and A. PRICE times OUTPUT give revenues. INPUT times UNIT COST gives total cost. Revenues less total costs give PROFIT. Δ PROFIT causes an increase in Δ INPUT. When Δ PROFIT turns negative, Δ INPUT becomes 0 and the firm has reached its maximum size.

We can see from the equations that the only trick here is to monitor the change in PROFIT for a downturn. When that downturn occurs, stop the expansion of the firm by stopping the growth of the INPUT, Δ_INPUTS:

```
INPUT(t) = INPUT(t—dt) + (Δ_INPUTS) * dt
    INIT INPUT = 1
Δ_INPUTS = IF Δ_PROFIT > 0 THEN 3 ELSE 0
OUTPUT = A*INPUT^ALPHA
PROFIT = PRICE*OUTPUT—TOTAL_COST
```

2. For a detailed discussion, see R.C. Lind et al. 1982.
3. Ruth and Hannon 1997.

TOTAL_COST = UNIT_COST*INPUT
Δ_PROFIT = PROFIT - DELAY(PROFIT,DT,.1)
A = 50
ALPHA = .6
UNIT_COST = 16
PRICE = 5

We see also that we must resort to a production function that gives OUTPUT (widgets) from the INPUT (e.g., labor). This is a standard means of mathematically describing the transformation of one thing into another in both economics and biology (see figure 3.12). The constant A carries the units of the output and the input is normalized so that it is dimensionless. The fact that ALPHA is less than 1 means we are experiencing diminishing returns from added input increments.

Starting small, the firm expands its output until its marginal TOTAL COST is equal to the PRICE of its output. In the competitive case, the price is constant and marginal revenue equals price. Taking the derivative of revenue with respect to price can easily show this.

Form this marginal cost by using the derivative function in ithink on both TOTAL COST and OUTPUT and forming a ratio of the two. Plot this ratio and note how it becomes equal to PRICE just as PROFIT reaches the maximum. Discount PROFIT at 15 percent using EXP(–0.15*time)*PROFIT and accumulate

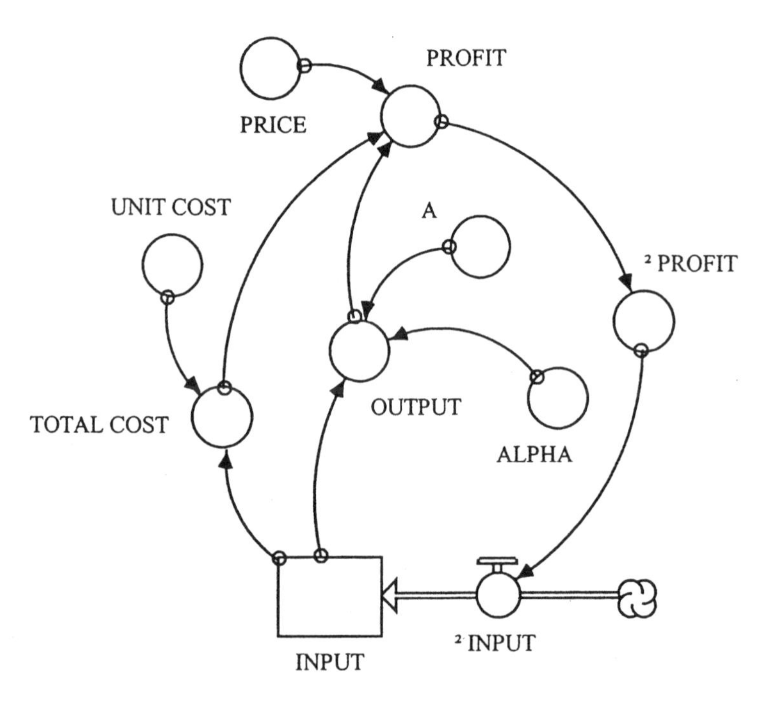

FIGURE 3.11. The model of a single input, single output profit-maximizing firm

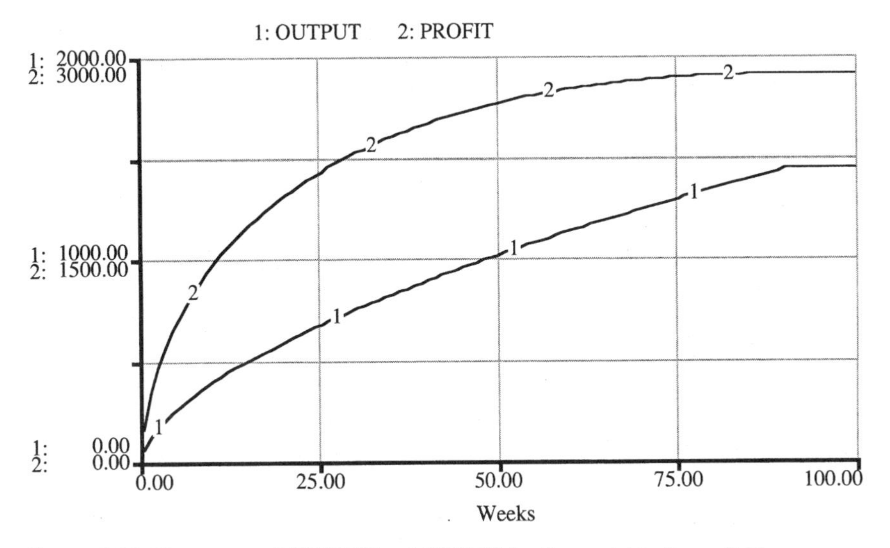

FIGURE 3.12. The curves of OUTPUT and PROFIT for the model in figure 3.11

this figure. Change the model such that you maximize this cumulative present-valued profit. Compare the production level to the one found here and explain any differences.

What you should realize is that the way we have posed the original maximum profit problem precludes it from being a dynamic problem. We were simply looking for the output level that would maximize profits. This is a kind of set point problem: what is the optimal output level? The path to that level made no difference in our original problem. Time was just a parameter to allow us find the optimal firm size. But in the real world this is not so. The profit stream and expectations of that stream allow the firm to borrow the money and to finance its expansion. In each period, the firm assesses its profits and invests in itself, borrows, and perhaps sells a portion of its stock to get financing for expansion. This puts the firm on a critical path to its optimal size. It is for this (and similar but alternate paths) that the discounted cumulative net profit calculation is useful and essential. Shall I choose a path that accelerates my growth immediately, moderately or slowly? Each path choice has its problems. The firm cannot expand to its optimal size instantly because it has neither the profits stream nor the confidence of lenders when it is small. Yet if it expands too slowly, the time penalty of the discount rate weighs heavily against it. Consider the nature of these two optimal profit algorithms. Note how the first strategy involves finding a peak while the second requires that you find the largest asymptote. Make this model a true dynamic one by investing all of its own profits into the firm. Do this simply by making the purchase rate of the input proportional to the profit rate, or in a more realistic way by naming the input steel and building a capital stock that is needed with steel in the production function.

Of course there are many other considerations that should be included in this path choice, such as product quality, the emergence of new and better ways of making the widgets, the emergence of systems that do not need widgets at all, and the strategies of competitive widget makers.

Learning Point: Even the dynamics of a single firm can become complex quickly.

Profit and Economic Value Added (EVA) are closely linked. EVA is a more sophisticated statement of profits (see Appendix B). This measure takes into account the amount of capital that must be used to create profits. It is not enough to just generate profits—we must generate these profits with some control of the capital it took to generate them. A model that penalizes profits based on the amount of investment it takes to generate these profits will support our premise that the only way an organization will survive is if it provides a better investment for its owners/shareholders than they could get from other investments. It is left as an exercise for readers to generate a profit model using EVA.

3.7 Issues with financial measures

A financial measure of performance is a measure of how well the system is meeting one of its stakeholder requirements. In fact, financial performance is one of the requirements of the efficiency stakeholders (see section 2.2). Thus, when it comes to managing systems or processes, we must pay close attention to this financial measure. In addition, we must manage other output measures that reflect the compliance and effectiveness stakeholders. We can then manage a process by monitoring these output measures and taking actions when the values of these output measures do not meet our expectations. (Sometimes the expectation is to keep a measure on target and sometimes the expectation will be to improve the measure. At any point in time, an organization may choose to focus on improving some of its measures while keeping others at their current levels. Which measures are put in each category will depend on the current business issues and opportunities and on resource availability.) While managing an organization or process by monitoring output measures can work in theory, in practice it is not a good strategy because such output measures are lagging indicators of process performance. First, they only tell us we have an issue once the issue has already occurred. Second, if we implement an improvement activity, it may be a long time before we are able to decide if the anticipated results were obtained. This problem is further compounded because all performance measures exhibit random variation. As we pointed out in section 2.5, this adds a time delay to our ability to detect change. Also, once we have detected the issue, in general we will not be able to implement a countermeasure immediately, adding a further time delay to our ability to react to an issue. This is the reason that these measures are referred to as lagging measures. To give a simple illustration of what can happen when such time delays are present, consider the model shown in figure 3.13.

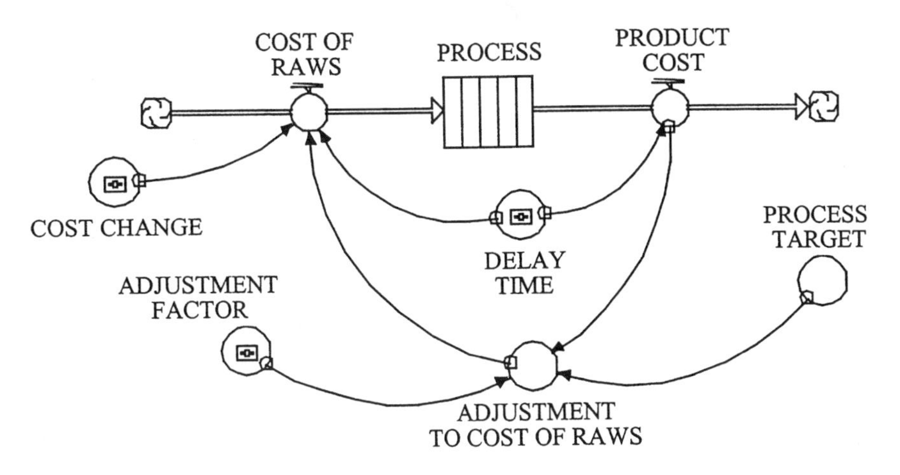

FIGURE 3.13. A simple cost control model

This shows a simple cost control model. A process converts raw materials into a product. The process is represented by a conveyor with a time delay of 10 time units. The target cost is defined in PROCESS TARGET as 10. When the actual cost is measured in the outflow PRODUCT COST, it is compared with the target. If there is a difference between actual and target, an adjustment is calculated to the inflow representing the cost of raws. In effect, we are monitoring an output and then adjusting an input to compensate for change to the outputs. The equation for this adjustment is

ADJUSTMENT_FACTOR*(PROCESS_TARGET-PRODUCT_COST)

The converter ADJUSTMENT FACTOR is used to control the size of the adjustment. The converter COST CHANGE represents a change to the cost of raws. The change is made at time period 50 in the inflow, COST OF RAWS. The equation for this inflow is

IF TIME < (DELAY_TIME+10) THEN 10 ELSE
10 + ADJUSTMENT_TO_COST_OF_RAWS + STEP(COST_CHANGE,50)

The reason for the IF statement is to make sure that the process has time to reach steady state before the adjustment or cost change is introduced. Figure 3.14 shows the profile for PRODUCT COST for the conditions TIME DELAY = 10 time periods, COST CHANGE = 5 units, and ADJUSTMENT FACTOR = 1.0.

The value of PRODUCT COST cycles continuously.[4] The change is introduced at time 50. However, it takes 10 time periods before the change is seen. The value of PRODUCT COST jumps to 15 at time period 60. Now we begin to react by

4. Readers can run the model with different values of ADJUSTMENT FACTOR to see the impact this parameter has on the profile for PRODUCT COST. Values less than 1 will still lead to cycles but the cycles will decay over time.

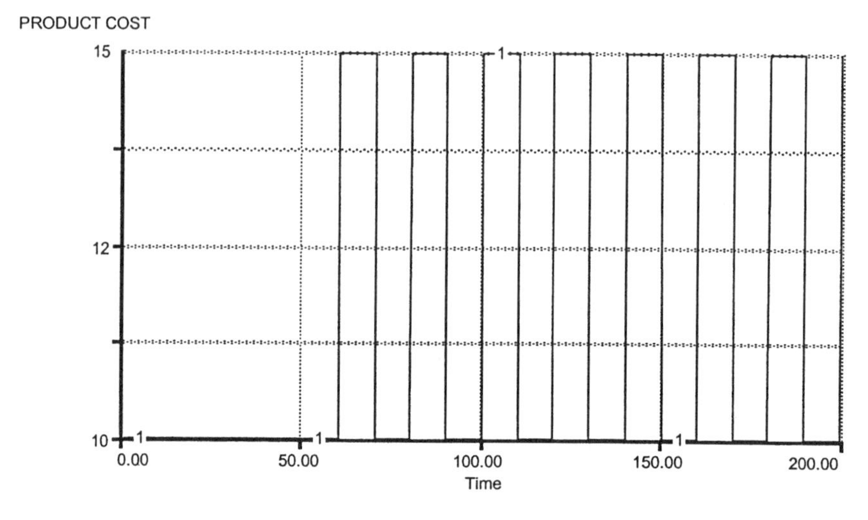

FIGURE 3.14. Profile for PRODUCT COST

making the appropriate adjustment. Between time periods 60 and 70, PRODUCT COST is off target. Consequently, the product in the conveyor during this time period will be adjusted. This product comes out in time periods 70 and 80 and is on target, so we stop adjusting. Now the material in the conveyor from period 70 to 80 is off target, and the cycle repeats itself. Thus we can see that because we had a time delay of 10 time periods, we have introduced a cycle of 20 time units. This leads to a general result when we try to manage a process that has time delays. Such a process will tend to cycle with a period related to the time delays present in the process. Again, the reason for this is the combination of the time delays and the fact that we manage solely on the output measures of the process. Because all real processes have time delays present, we can state the following:

> Learning Point: Processes that are managed solely using output performance measures may show cyclical behavior in the performance of these measures.

Cyclical behavior of the output measures is not desirable. Consequently, we must look for ways to better manage the process. Note that if we were measuring COST OF RAWS directly, then we would be able to react as soon as a change occurred. In this case, the cyclical behavior would not occur.

3.8 Beyond process output measures

It should be obvious that, although output measures are the proper measurement of process performance, they are not the best management measures for a process. We need other measures. But what should such measures look like? The key to progress is to remember that we labeled the output measures as lagging measures of performance. What we need, therefore, are some leading measures of performance, which will show change faster than the output measures and are at the same time predictive of the future performance of the output measures. By monitoring these leading measures, we hope to see change quickly and react to it much faster than if we had to depend solely on the output measures. So how do we generate such leading measures? One popular way to generate such measures is the balanced scorecard approach developed by Kaplan and Norton (1996). This is a useful approach when we are looking at an organization in total. The basic idea is that measures are broken down into four layers that are all linked:

- The financial layer—these measures reflect measures of financial performance such as profitability or EVA. The balanced scorecard approach recognizes that the ultimate goal of the organization is to make money.
- The customer layer—these measures reflect how well the organization is satisfying customer requirements in existing market segments or perhaps in new segments in which the organization wants to develop market share.
- The internal business process layer—these measures reflect the performance of key business processes in which the organization must excel to be successful.
- The learning and growth layer—these measures reflect the state of the infrastructure that the organization must build to create long-term growth and improvement. The infrastructure is generally considered to be composed of three parts: people, systems, and organizational procedures.

In the balanced scorecard model, the first two layers contain the lagging measures. The leading measures come from the second two layers. The theory is that by improving elements of the organizational infrastructure, we then improve our internal business processes, which in turn leads to improved customer satisfaction, which finally leads to improved financial performance. The specific components of each layer on which we focus depends on the organization's current understanding of what they need to do to improve. For example, in order to improve overall financial performance, the organization may decide that it must develop a new customer segment. Successful penetration of this market will depend on two key business processes working well. And, finally, the successful operation of these two key business processes will depend on improvements in information technology and certain people skills. Then the organization would have lagging measures relating to overall financial performance and the amount of penetration into the new market segment. It also would have some leading measures relating to the performance of the two key business processes and the status of information technology and people skills. In essence, the leading and lagging measures must be related, one leading to the other in a causal way. In effect, what the organization is saying is that improving the measures in the learning and growth layer will improve the measures in the internal business process layer, which in turn will improve the measures in the customer layer, which then will lead to improvement in the financial layer. An interesting way to view the balanced scorecard approach is to think of a focus on the lagging measures as managing by results, whereas a focus on the leading measures is managing by means.[5] The infrastructure and business processes are the means by which the customer and financial objectives are realized. If we understand how the organization works and know its objectives, then we can get the results by managing the means.

3.9 The process model approach

Although the balanced scorecard approach is appealing at the level of the overall organization, it does not provide a good framework at the level of the individual process. What is preferred is an approach that ties in much more directly with the SIPOKS model of a process that we outlined in section 2.2 but that keeps the essential idea of related measures that can be grouped into layers. This can be accomplished by using an expanded view of the SIPOKS model as shown in figure 3.15.

We refer to this model simply as the process model. It views the process as an interaction of a set of inputs where the result of this interaction is a set of outputs (results). In this model, we see measures as being organized into three layers:

- Output measures—these are the overall performance measures of the process and include all three stakeholder groups, compliance, effectiveness, and efficiency. These are the lagging indicators of process performance.
- Process measures—these are measures that reflect the interaction of the inputs. For example, the throughput of product from a certain machine may be the result of the interaction of equipment, people and raw materials.
- Input measures—these measures reflect the "quality" of the inputs. Input measures reflect attributes of the inputs that will be important for process performance. For example, the quality of a raw material could certainly impact the overall performance of a process. We generally organize inputs into six groups as shown in figure 3.15. The only group that the reader may not be familiar with is the environment group. This group represents inputs which can impact

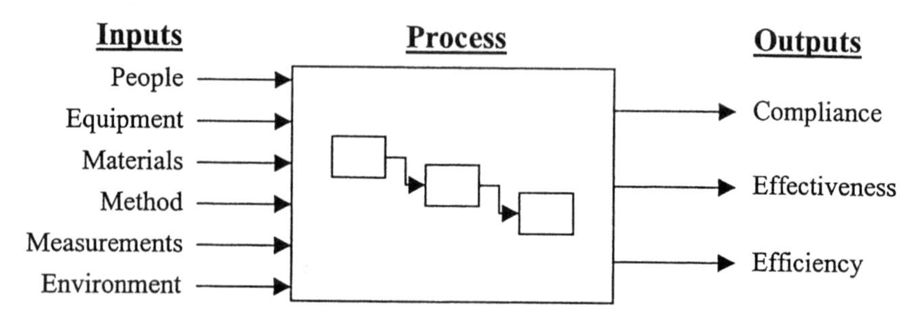

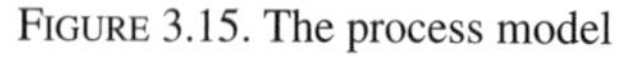
FIGURE 3.15. The process model

5. See Johnson 2001.

the process but over which we have no direct control, either because we cannot have control or we have chosen not to have control. For example, we cannot control whether an earthquake will occur. And while we may be able to control humidity, we may choose not to because of cost constraints. We must ensure that the process is robust to the expected variation in any environment variables.

It is worth pointing out that when we want to modify a process, the only place we can intervene is at the inputs. We modify inputs, which modifies the input measures, which in turn modifies the process measures, which then modifies the output measures. (The process model has the same causal alignment as the balanced scorecard.) For example, the quality of a plastic molding might depend on the temperature at which the mold is formed. The quality is an output measure and the temperature is a process measure. We cannot modify either of these directly. The molding temperature may be set by a controller set point on the molding machine, and it is this set point that is the input. This differentiation is important because changing the set point does not guarantee that the actual temperature will change exactly from the old set point to the new set point. Anyone who has worked in control applications will know that good control at one set point is no guarantee of good control at a different set point.

In the process model, the output measures are the lagging measures and the process and input measures are the leading measures. In fact, input measures are better leading measures than process measures. Given a choice, we will take an input measure over a process measure, all other things being equal. As in the balanced scorecard approach, we propose that by focusing on the process and input measures, we can anticipate the performance of the outputs and take actions long before we see any impact on the outputs. To illustrate this point, consider the model shown in figure 3.16.

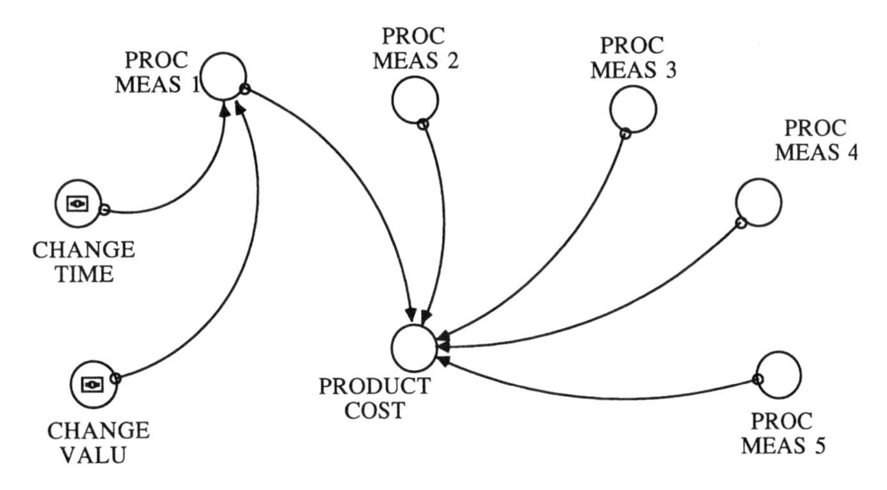

FIGURE 3.16. Simple product cost model

In this model, we suppose that the cost of a product is driven by five key process variables. For simplicity, the value of PRODUCT COST is assumed to be the sum of the values of the five process measures so that the equation for PRODUCT COST is this:

PROC_MEAS_1+PROC_MEAS_2+PROC_MEAS_3
+PROC_MEAS_4+PROC_MEAS_5

Each of the process measures exhibits random variation. Again for simplicity, we assume that each measure has the same amount of random variation, given by the function NORMAL(1.0,0.05). At time CHANGE TIME, the value of PROC MEAS 1 is changed by the value CHANGE VALUE so that the equation for PROC MEAS 1 is this:

IF TIME< = CHANGE_TIME THEN NORMAL(1.0,0.05)
ELSE NORMAL(1.0,0.05) + CHANGE_VALUE

We know that the random variation impairs our ability to see the change. The question is this: Will we see the change faster in PRODUCT COST or PROC MEAS 1? Figure 3.17 shows the profile for PRODUCT COST with a change value of 0.05.

Before proceeding, readers should try to decide when they think the change occurred. It is clear that the time of the change is not easy to see. But now look at figure 3.18, which shows the profile for PROC MEAS 1 that generated the PRODUCT COST values in figure 3.17.

Obviously, the change occurred around time 50. If we were just monitoring the PRODUCT COST measure, it would be a long time before we detected a change.

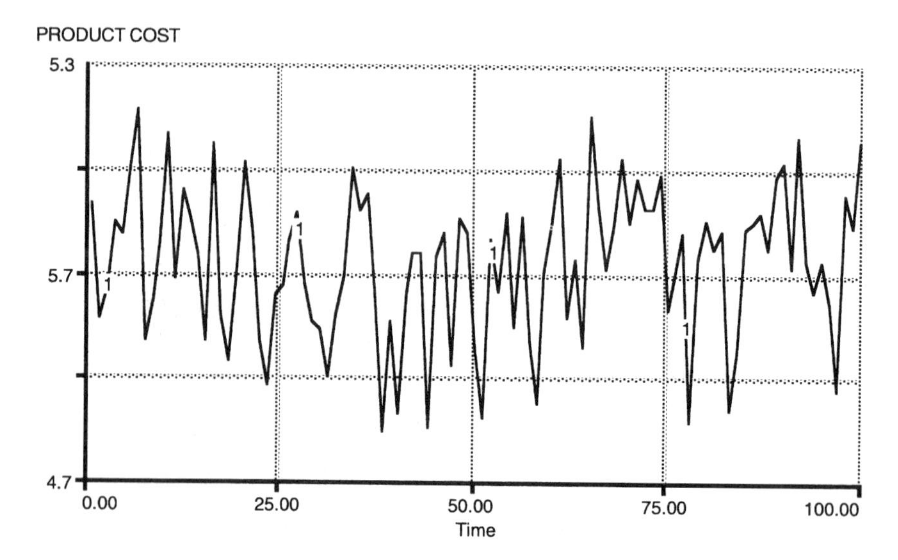

FIGURE 3.17. Profile of PRODUCT COST

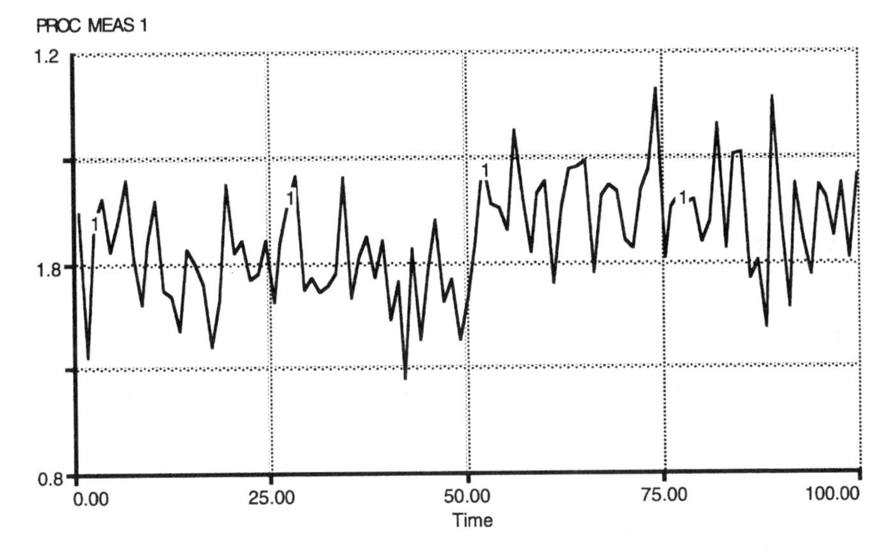

FIGURE 3.18. Profile for PROC MEAS 1

But by monitoring a key process measure, the change is detected much sooner. This then is the value of a leading measure. Readers should satisfy themselves that the same observation could be made if we had pure time delays in the system due to processing times, delay times, and so forth. For example, consider a customer service measure. This is an output measure. A key process measure might be whether we start production when we are supposed to (referred to as conformance to schedule). A key input measure might be availability of raw materials, which we might measure as the inventory level of raw materials. Then if we have an issue with raw material inventories, this may lead to delays in starting production, which will eventually lead to customer service problems.

> Learning Point: To effectively manage a process, it is important to measure key process and input measures as well as the output measures.

The phrase "*as well as*" and the word "*key*" are two important ideas in this learning point. As in the case of the balanced scorecard, we do not propose that output measures be abandoned in favor of process and input measures. Rather, the process and input measures are used to support the output measures. The reason we cannot abandon the output measures is that they are still the ultimate measure of process performance as judged by the only people who can judge our performance, namely, our key stakeholders. There are many process and input measures that we could measure, but it will be impossible to monitor all of them. We must understand the relationship between the process/input measures and the output measures. Otherwise, we may focus on improving a measure that has little effect on the outputs. Although we will do our best to choose only the best process and input measures, there is always a risk that we will make an incorrect choice. If we change an input or process measure and it does not have the desired impact on the output measures, then we must go back and re-evaluate the usefulness of the process/input measure. Therefore we need to look for *key* input and process measures, that is, look for those input/process measures that provide the most benefit in predicting the behavior of the process output measures.

Choosing key input and process measures will involve determining the sensitivity of an output measure to the process or input measure. Many of the models that we will discuss in the rest of this book will allow the user to calculate such sensitivities. We can illustrate the approach here with a simple model (figure 3.19).[6]

Figure 3.19 represents a process that takes input and processes it through three steps to produce a product. Each process step takes 10 time periods to be completed. Each input item produces an output item. FEED RATE represents the number of items input to the process per unit time period. FEED RATE represents the capacity of the process, which we assume is running as fast as it can. The PRODUCT flow rate is a key output for the process because the company can sell all of the widgets it can make. At each step there are losses as expressed by the yield at each step, the yield being a number between 0 and 1 and representing the fraction of items at each step that are deemed good at the end of the step. The loss fractions, which are set in the losses outflows, are simply (1-yield). The current yield for step 1 has a base case value of 0.85, and we want to understand the impact of changing the value to 0.95. The current FEED RATE base case value is 85 items per unit time period, and we want to understand the impact of changing the

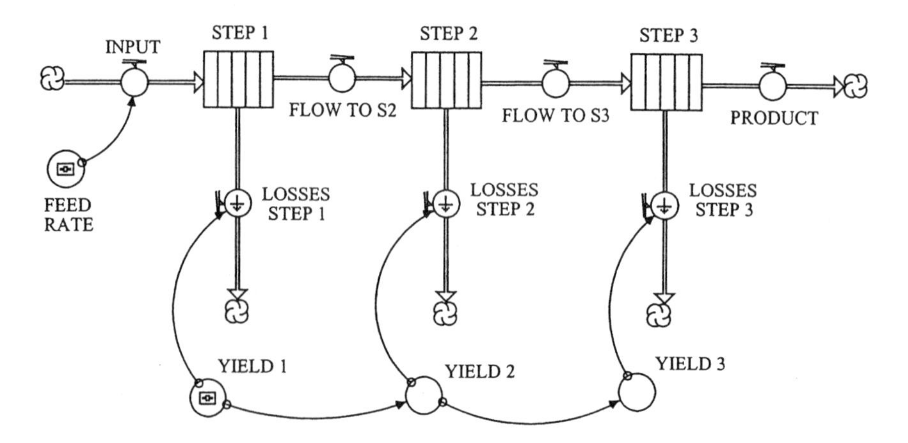

FIGURE 3.19. Model of three-step process with losses at each step

6. In fact, this model is simple enough that it can be modeled in a simple spreadsheet. Readers are invited to do this to compare results if so desired.

value to 95 items per unit time period. The yields for steps 2 and 3 are related to the yield of step 1, with the yield for step 2 given by (YIELD 1)^2 and the yield for step 3 given by (YIELD 2)^2. FEED RATE is an input measure and YIELD 1 is a process measure. We would like to adopt one of these measures as a leading indicator for PRODUCT. Consequently, the focus is on improving that measure to improve PRODUCT. So the question we want to answer is this: Is the output more sensitive to FEED RATE or YIELD 1?

In order to calculate the sensitivities, we first run the model at the base case values. The resultant profile for PRODUCT is shown in figure 3.20.

We get no product for the first 30 time periods because each step has a cycle time of 10 time periods. The steady state value is 27.2 items per unit time period. We now run the model for value of FEED RATE and YIELD 1 between 85/95 and 0.85/0.95 respectively. The results are shown in table 3.1.

The change column represents the changes in PRODUCT for the changes in FEED RATE and YIELD 1. The data show that as we change the value of FEED RATE from 85 to 95, the value of PRODUCT changes by (3.3+6.9)/2 = 5.1 on average.[7] For the change in YIELD 1, the change is (32.2+25.8)/2 = 29.0. Clearly PRODUCT is more sensitive to YIELD 1 than it is to FEED RATE; therefore, we focus first on YIELD 1 as a process measure so that we can implement improvements and

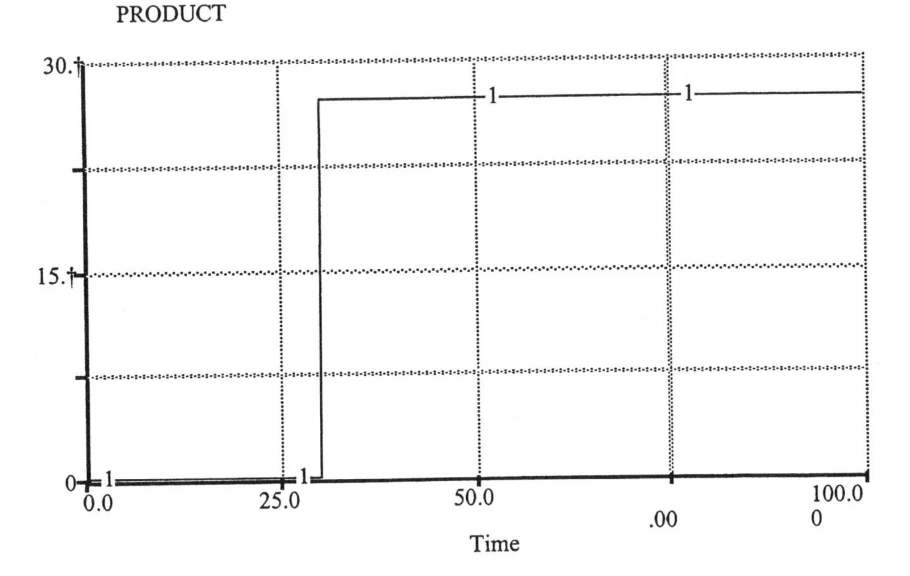

FIGURE 3.20. Product rate with base case values of INPUT and YIELD 1

7. Note that the change in PRODUCT with FEED RATE depends on the value of YIELD 1 and vice versa. This is because of the nonlinear relationships in the model. In fact, this type of behavior defines what we mean by a nonlinear model.

TABLE 3.1. PRODUCT sensitivity results for model in figure 3.19.

		Feed Rate		
		85.0	95.0	Change
Yield 1	0.85	27.2	30.5	3.3
	0.95	59.4	66.3	6.9
	Change	32.2	25.8	

see the impact on YIELD 1. It is important to point out that this conclusion is not an inherent property of the process but, rather, a property of the *process under current operating conditions and potential change in performance.* For example, we know that even with all yields at their best-case value of 1, the maximum value of PRODUCT is equal to FEED RATE. Therefore, if the current value of YIELD 1 were close to 1, then the sensitivity to FEED RATE would be much greater than the sensitivity to YIELD 1. In this instance, we would create an input measure for FEED RATE and look at how we could increase the inherent capacity of the process.

Note that even though FEED RATE is an input (and in general we prefer input measures to process measures), in this case, the sensitivity of the proposed process measure is so much greater than the sensitivity of the proposed input measure that we would choose the process measure over the input measure. Having chosen YIELD 1, however, we would now begin to look at this measure in more detail. (Recall that you cannot change the value of YIELD 1 directly). We would try to understand what inputs impact the value of YIELD 1. We could then build a model relating these inputs to YIELD 1. In effect, we would want to take the converter YIELD 1 in figure 3.19 and explode it to reveal greater detail about what factors impact YIELD 1. This would lead to a model with greater detail than the model in figure 3.19. This illustrates a general principle of model building. We tend to start with a model with a low level of detail. Then we analyze this model and use the analysis to decide where greater detail is required, if it is required at all.

In the remainder of this book, we will encounter more complex models in which we will be able to use this type of analysis to determine what process and/or input measures might be appropriate for a particular process. It is important to remember that modeling the process performance measures is an important part of the overall modeling activity.

Model Equations

```
MODEL 3.11
INPUT(t) = INPUT(t—dt) + (Δ_INPUTS) * dt
INIT INPUT = 1
Δ_INPUTS = IF Δ_PROFIT > 0 THEN 3 ELSE 0
OUTPUT = A*INPUT^ALPHA
PROFIT = PRICE*OUTPUT—TOTAL_COST
TOTAL_COST = UNIT_COST*INPUT
Δ_PROFIT = PROFIT—DELAY(PROFIT,DT,.1)
```

```
A = 50
ALPHA = .6
UNIT_COST = 16
PRICE = 5
```

```
MODEL 3.13.ITM
{ INITIALIZATION EQUATIONS }
DELAY_TIME = 10
INIT PROCESS = 100
TRANSIT TIME = varies
INFLOW LIMIT = INF
CAPACITY = INF
ADJUSTMENT_FACTOR = 1.0
PROCESS_TARGET = 10
PRODUCT_COST = CONVEYOR OUTFLOW
TRANSIT TIME = MAX(DELAY_TIME,DT)
ADJUSTMENT_TO_COST_OF_RAWS = ADJUSTMENT_FACTOR*(PRO-
CESS_TARGET-PRODUCT_COST)
COST_CHANGE = 1
COST_OF_RAWS = IF TIME < (DELAY_TIME+10) THEN 10 ELSE 10 +
ADJUSTMENT_TO_COST_OF_RAWS + STEP(COST_CHANGE,50)

{ RUNTIME EQUATIONS }
PROCESS(t) = PROCESS(t—dt) + (COST_OF_RAWS—PRODUCT_COST) *
dt
PRODUCT_COST = CONVEYOR OUTFLOW
TRANSIT TIME = MAX(DELAY_TIME,DT)
ADJUSTMENT_TO_COST_OF_RAWS = ADJUSTMENT_FACTOR*(PRO-
CESS_TARGET-PRODUCT_COST)
COST_OF_RAWS = IF TIME < (DELAY_TIME+10) THEN 10 ELSE 10 +
ADJUSTMENT_TO_COST_OF_RAWS + STEP(COST_CHANGE,50)
```

```
MODEL 3.16.ITM
{ INITIALIZATION EQUATIONS }
CHANGE_TIME = 50.0
CHANGE_VALUE = 0.1
PROC_MEAS_1 = IF TIME< = CHANGE_TIME THEN NORMAL(1.0,0.05)
ELSE NORMAL(1.0,0.05) + CHANGE_VALUE
PROC_MEAS_2 = NORMAL(1.0,0.05)
PROC_MEAS_3 = NORMAL(1.0,0.05)
PROC_MEAS_4 = NORMAL(1.0,0.05)
PROC_MEAS_5 = NORMAL(1.0,0.05)
PRODUCT_COST =
PROC_MEAS_1+PROC_MEAS_2+PROC_MEAS_3+PROC_MEAS_4
+PROC_MEAS_5
```

```
{ RUNTIME EQUATIONS }
PROC_MEAS_1 = IF TIME< = CHANGE_TIME THEN NORMAL(1.0,0.05)
ELSE NORMAL(1.0,0.05) + CHANGE_VALUE
PROC_MEAS_2 = NORMAL(1.0,0.05)
PROC_MEAS_3 = NORMAL(1.0,0.05)
PROC_MEAS_4 = NORMAL(1.0,0.05)
PROC_MEAS_5 = NORMAL(1.0,0.05)
PRODUCT_COST =
PROC_MEAS_1+PROC_MEAS_2+PROC_MEAS_3+PROC_MEAS_4
+PROC_MEAS_5

MODEL 3.19.ITM
{ INITIALIZATION EQUATIONS }
INIT STEP_1 = 0
TRANSIT TIME = 10
INFLOW LIMIT = INF
CAPACITY = INF
INIT STEP_2 = 0
TRANSIT TIME = 10
INFLOW LIMIT = INF
CAPACITY = INF
INIT STEP_3 = 0
TRANSIT TIME = 10
INFLOW LIMIT = INF
CAPACITY = INF
FEED_RATE = 90.0
INPUT = FEED_RATE
YIELD_1 = 0.9
LOSSES_STEP_1 = LEAKAGE OUTFLOW
LEAKAGE FRACTION = 1-YIELD_1
NO-LEAK ZONE = 0
FLOW_TO_S2 = CONVEYOR OUTFLOW
YIELD_2 = YIELD_1^2
LOSSES_STEP_2 = LEAKAGE OUTFLOW
LEAKAGE FRACTION = 1-YIELD_2
NO-LEAK ZONE = 0
FLOW_TO_S3 = CONVEYOR OUTFLOW
YIELD_3 = YIELD_2^2
LOSSES_STEP_3 = LEAKAGE OUTFLOW
LEAKAGE FRACTION = 1-YIELD_3
NO-LEAK ZONE = 0
PRODUCT = CONVEYOR OUTFLOW

{ RUNTIME EQUATIONS }
STEP_1(t) = STEP_1(t—dt) + (INPUT—FLOW_TO_S2—LOSSES_STEP_1)
* dt
```

STEP_2(t) = STEP_2(t—dt)
+ (FLOW_TO_S2—FLOW_TO_S3—LOSSES_STEP_2) * dt
STEP_3(t) = STEP_3(t—dt) + (FLOW_TO_S3—PRODUCT—
LOSSES_STEP_3) * dt
INPUT = FEED_RATE
LOSSES_STEP_1 = LEAKAGE OUTFLOW
LEAKAGE FRACTION = 1-YIELD_1
FLOW_TO_S2 = CONVEYOR OUTFLOW
YIELD_2 = YIELD_1^2
LOSSES_STEP_2 = LEAKAGE OUTFLOW
LEAKAGE FRACTION = 1-YIELD_2
FLOW_TO_S3 = CONVEYOR OUTFLOW
YIELD_3 = YIELD_2^2
LOSSES_STEP_3 = LEAKAGE OUTFLOW
LEAKAGE FRACTION = 1-YIELD_3
PRODUCT = CONVEYOR OUTFLOW

4

Single-Step Processes

That's one small step for [a] man, one giant leap for mankind.

—Neil Armstrong, 20 July 1969

4.1 Introduction

As we said in chapter 2, the most basic activity in any business is the management of a workflow process, or simply, a process. In this and subsequent chapters, we are going to explore the operation of such processes. Processes can be complicated, with many individual tasks in series and parallel. In order to understand the behaviors that can be exhibited by such processes, we will begin with the simplest possible representation of such a workflow process. Surprisingly, we will be able to learn a lot about the general behavior of real processes from this simple representation. In chapters 5 and 6, we will expand the single-step process to multistep processes.

The single-step workflow process is the simplest representation of a process. It is modeled using a single conveyor. Units[1] or items are fed into the process on the input side and emerge on the output side some time later. At any time, such a simple process only has three basic attributes:

- Thruput rate (or simply thruput) of units through the process—this is rate at which units exit on the output side of the process.
- Cycle time of the process—this is the time a unit spends in the process.
- Amount of units in the process—this is referred to as the work in progress (WIP) for the process.

We will begin by looking at the relationships that exist between these variables under steady state conditions with no constraints on the capability of the process. We will demonstrate a fundamental law relating thruput, WIP, and cycle time and investigate the circumstances under which this law can be applied. Finally, we

1. For continuous processes, a unit can be thought of as a unit volume of material rather than a discrete unit or item.

will add capability constraints to see the impact of these constraints on performance. This will lead directly to simple single-step, single-routing[2] queuing applications. Then, in chapters 5 and 6, we will expand these simple queuing models to more complex queuing networks with multiple processing steps and multiple routings.

Note: We will devote a significant part of this chapter to the setup and analysis of the simplest single-step process in sections 4.2 and 4.3 because this is an introductory text. It is important to develop a good understanding of the basics before tackling the more complex models.

4.2 The basic process model and Little's Law

Figure 4.1 represents a single-step process. The time basis for this model is hours. The process is represented in ithink as a single conveyor, PROCESS. The conveyor has an inflow going into it, WORK FLOWING INTO PROCESS, which represents the flow of work into the process. The input rate of the process is specified in this flow. In general, the input rate of items is not expressed directly as a rate, but rather by specifying the time interval between arriving items. We refer to this as the interarrival time, IIT. This is because the IIT is directly observable, whereas the thruput is not. Of course, the average thruput is simply the inverse of the average IIT. To generate items with an interarrival time of IIT, we send items from the stock SUPPLY OF WORK through a conveyor with capacity 1 and conveyor time IIT. The equation in FLOW TO WORK CONTROL is simply 1/DT. A little thought will show that this construct gives a stream of items that arrive at PROCESS with the correct interarrival time IIT.

Each unit of work that enters the conveyor remains in the conveyor for some period of time and then exits the conveyor through the outflow, WORK COMPLETING. This time period is the cycle time of the process. In ithink, this cycle time is specified in the WORK COMPLETING outflow. For this simple model,

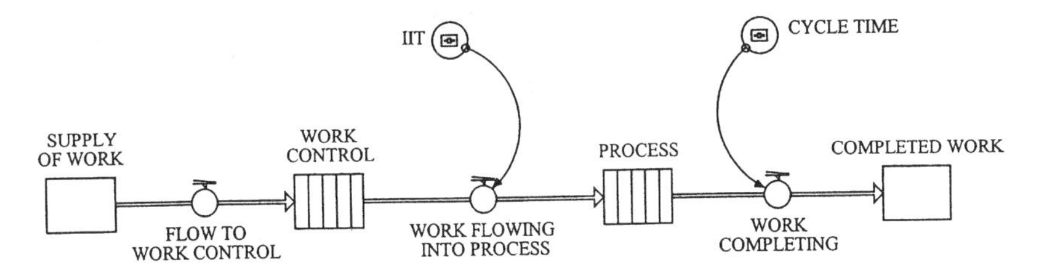

FIGURE 4.1. Single-step process model

2. A routing is a path that an item can follow in the process. The possibility of multiple pathways in a model obviously complicates the model.

the process cycle time is an input to the model. It is important to realize that in real processes, the process cycle time is not an input but, rather, arises because of the detailed way in which the units flow through the process. If the model contained more detail of the process, we would be able to predict the actual process cycle time of the process. In other words, the cycle time would be an output of the model. This is a typical situation in modeling. The less detailed the model, the more fundamental the data that must be entered as inputs to the model. Later in this chapter and in chapters 5 and 6, when we model processes in more detail, we will predict the process cycle time.

> Learning Point: Adding more detail to a model generally involves adding causation to input variables so that they become predictions of the model rather than inputs.

As units move through the process, they stay in the process until their cycle time is reached, whereupon they exit the process. At any time there may be multiple units in the conveyor moving through the process. These units represent the work in progress or WIP for the process. Once a unit enters the conveyor, we lose our ability to "see" and control that unit. This places a limitation on the detail of our modeling. For example, we cannot control how units interact with each other while in the conveyor. In fact, for this simple model, we must assume that the units do not interact with each other. Such an interaction can only be taken into account with a more detailed model.

On both sides of the process we have placed two stocks. The stock SUPPLY OF WORK is the source of the units for the process. The initial value of this stock must be set high enough so that it will not run out of units over the simulation. The stock COMPLETED WORK accumulates all the completed units that have gone through the process. It is easy to see that the number of units that come from SUPPLY OF WORK must equal the number of units in PROCESS plus the number of units in COMPLETED WORK. This helps us keep track of the units during a simulation. While keeping track of units may not appear necessary in the simple model we are developing here, it is a good practice to develop because when you develop more complex models, it will aid in debugging when the model does not appear to be working correctly. Note that in this simple model we do not worry about where the units come from prior to PROCESS or where they go to after PROCESS. We assume that units are always available when needed and always have a place to go once they have been processed in the conveyor. In other words, we have a large number of items initially in SUPPLY OF WORK and no restriction on how large COMPLETED WORK can grow. These conditions will be relaxed in chapters 7 and 8, in which we look at the source and destination of units in more detail.

The model in figure 4.1 also has two converters added to specify the interarrival time and the conveyor cycle time. These converters have been added as a convenience. They facilitate making model runs with different values of the thruput and conveyor cycle time. This will be useful when we want to conduct sensitivity runs with the model.

Base case run: For a base case run of the model, the following parameters will be used:

Model time units	hours
DT	0.0625
End Time	1,000 hours
IIT	2 hours
CYCLE TIME	100 hours
SUPPLY OF WORK	initial value 2,000
process	initial value 0
completed work	initial value 0

Because both the interarrival time and the cycle time have been specified as inputs in the model, the only variables that are free and that we need to look at are the value of the conveyor, PROCESS, and the value of the two stocks, SUPPLY OF WORK and COMPLETED WORK (as we said earlier, the thruput is simply the inverse of IIT). The results of the base case run are shown in figure 4.2.

As would be expected, the value of the stock SUPPLY OF WORK drops immediately and keeps dropping in a linear fashion as time proceeds. The stock COMPLETED WORK remains at 0 for the first 100 hours of the simulation. (It takes 100 hours for a unit to get through the process.) During this first 100 hours, the amount of units in PROCESS grows from 0 to 50 units. This amount is growing because new units are entering every 2 hours but no units are exiting in the first 100 hours. After 100 hours, the first completed unit exits the conveyor. Now

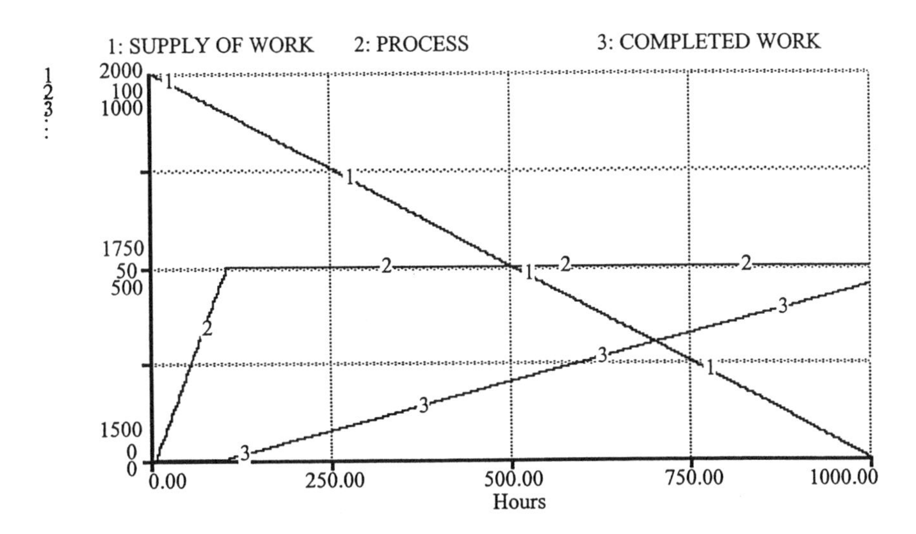

FIGURE 4.2. Base case run of model 4.1

the amount exiting the conveyor exactly balances the amount entering the conveyor. The value of PROCESS remains constant.

Steady state behavior: The process has reached a steady state after 100 hours. Although the notion that the model reaches a steady state may not seem such a significant discovery for our simple model, it can be important to know whether or not a steady state will ever be reached when working with more complex models. For example, if the model fails to reach a steady state but we know in practice that the real process does, then there is a problem with the model that must be fixed if the model is to be useful. Even if the model reaches a steady state, it may not reach the steady state as quickly as we think it should. This also may be an indication of a problem with the model.

When the model has just reached steady state, we can see that the number of units in PROCESS is exactly equal to the rate at which units are entering PROCESS, multiplied by the cycle time that each unit spends in PROCESS. Because the model remains in steady state after this point, this relationship holds for the rest of the simulation. Thus we can write this equation:

$$\text{PROCESS} = \text{THRUPUT*CYCLE TIME} \tag{4.1}$$

This relationship can be written for any process as this equation:

$$\text{Work in Process (WIP)} = \text{Thruput*Cycle Time} \tag{4.2}$$

The next question to ask is this: Does this relationship hold for all combinations of thruput and cycle time? It is easy to verify by running the model several times with various combinations of interarrival and process cycle times that equation 4.2 holds every time. In fact, equation 4.2 is the well-known Little's Law.

> Learning Point: Little's Law, equation 4.2, is a fundamental relationship for a process.

This law is considered to be the most fundamental law in workflow analysis. This law is useful because it says that if we know two out of the three unknowns in equation 4.2, we can easily calculate the remaining unknown. It is the fundamental relationship for a single-step process and represents all that we really need to know about the behavior of the simple process we have just modeled. It is a simple law and looks easy to use.[3] However, before we rush to use it on real systems, it is extremely important to know whether or not some assumptions must be satisfied before it can be applied.

Testing Little's Law: There are two potential limitations to Little's Law that we would like to test:

- What happens if there is a change to the interarrival times and/or process cycle times at some point during the simulation? (That is, the process is not run in steady state).

3. A rigorous proof of Little's Law is not trivial; see Ross 1970.

- What happens if there is random variation in the interarrival times and process cycle times?

Before we can investigate potential limitations behind this law, we first modify the model to allow us to directly see whether Little's Law is valid or not. In order to do this, we will create a variable in the model called LLV, which stand for Little's Law Variable. We can rewrite equation 4.1 as this:

$$\frac{THRUPUT * CYCLETIME}{PROCESS} = 1 \tag{4.3}$$

Then we define LLV as the left-hand side of equation 4.3. In this way, if we measure LLV in a model according to equation 4.3 and it has the value 1, then we know Little's Law holds. In equation 4.3, THRUPUT and CYCLETIME represent the measured values of thruput and cycle time, so we must also modify the model to include the measurements. By doing this, we are now considering the measured values of the thruput and cycle time as outputs of the model. We know that for our simple model the input values of thruput and cycle time are the same except for the time delay because of the cycle time. We also must add functionality to the model so that changes to the IIT and process cycle times can be made. The complete model is shown in figure 4.3.

The clock typefaces on the flows indicate that cycle time functionality is being referenced by these flows. The flow stream WORK FLOWING INTO PROCESS is modified by checking the cycle time box in the flow dialog box. This marks

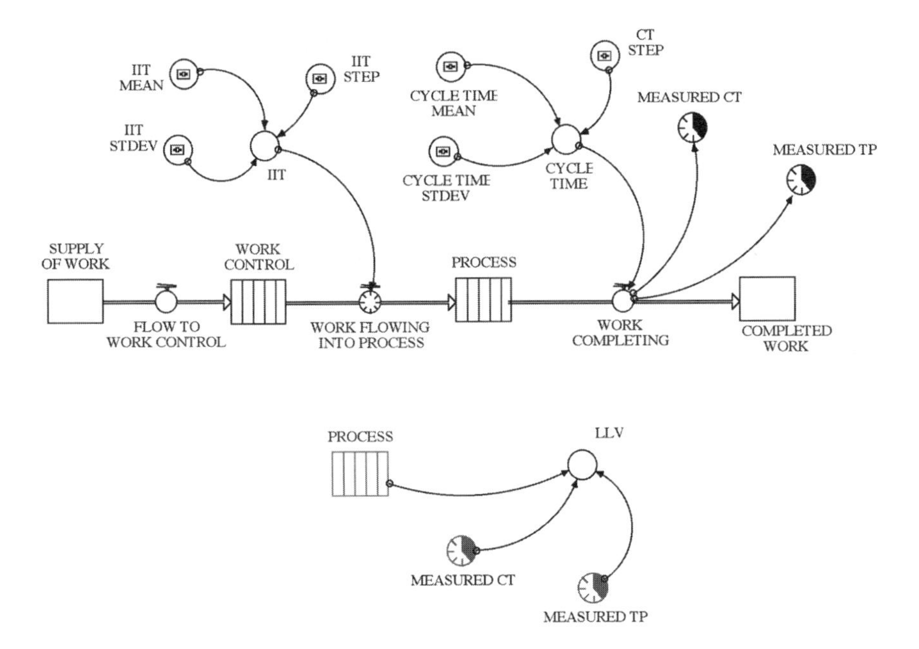

FIGURE 4.3. Basic process model with Little's Law calculations

each unit flowing into PROCESS with a time stamp. At any point later, we can then use the cycle time functionality to measure the cycle time of units. This is the purpose of the MEASURED CT converter. It contains the function CT-MEAN(WORK COMPLETING). This function measures the average cycle time of all units that pass through the flow WORK COMPLETING. Of course, in the first 100 hours, no units will reach the WORK COMPLETING flow. For this part of the simulation, the MEASURED CT value will be 0. Similarly, the THROUGHPUT function is used in MEASURED TP to calculate the average thruput. Like the average cycle time, it has a value 0 until a unit exits the process. Finally, the Little's Law Variable is calculated in the converter LLV using equation 4.3. The actual equation in the LLV converter is

IF PROCESS > 0 THEN
(MEASURED_TP*MEASURED_CT)/PROCESS ELSE 0

This means that for the first 100 hours, we expect LLV to be 0 because the values of MEASURE TP and MEASURED CT will be 0 until time 100. The ability to model random variation and changes in the interarrival times and process cycle times is reflected in the equations for IIT and CYCLE TIME. These equations follow:

IIT:

IF IIT_STDEV < = 0.0 THEN IIT_MEAN + STEP(IIT_STEP,300) ELSE
NORMAL(IIT_MEAN,IIT_STDEV) + STEP(IIT_STEP,300)

CYCLE TIME:

IF CYCLE_TIME_STDEV < = 0.0 THEN CYCLE_TIME_MEAN
+ STEP(CT_STEP,300) ELSE
NORMAL(CYCLE_TIME_MEAN,CYCLE_TIME_STDEV)
+STEP(CT_STEP,300)

We use a normal distribution to insert random variation and a step function to add changes at time 300 hours. With this model, many combinations are possible and readers are encouraged to try some to ensure they have a good understanding of how the model works. Here we just give an example—what happens if we decrease the interarrival time (IIT) by 0.5 hours at time 300 hours? The resultant profile is shown in figure 4.4.

As is expected, the value of LITTLE'S LAW VARIABLE is 0 up to 100 hours. Then it rises and eventually settles down to the expected value of 1.0. Then at time 300 the value of LLV drops. This is because when the IIT value drops, the input rate to PROCESS increases but it takes time for this change to reach the output side of the process. During this time, the value of PROCESS is increasing without a similar change in thruput. Thus the value of LLV drops. Because the value in MEASURED TP covers the entire simulation period, it never gets to the new average thruput. Consequently, the value of LLV never recovers to 1. Thus, once we have a change in IIT, Little's Law no longer holds.

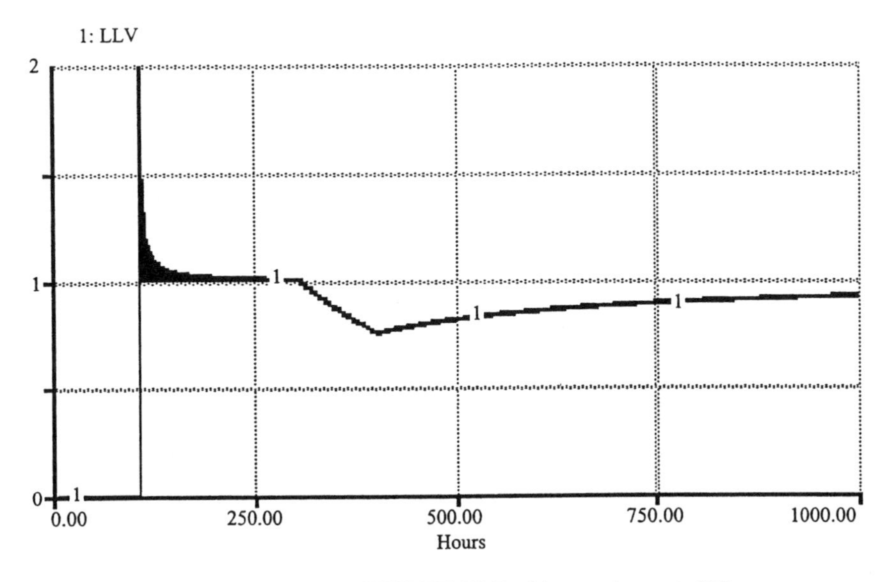

FIGURE 4.4. Graph of LITTLE'S LAW VARIABLE with step change in IIT

Note that initially at time 100 hours oscillation occurs in the LLV value. Observation of the model shows that this can be attributed to oscillation in the value of MEASURED TP. This is because of the way the average thruput is calculated by ithink. When the first unit arrives, the thruput is calculated as 1 unit in a DT so that it will be 1/0.0625 = 16 units per hour. After another DT, the thruput will be 1/(2*DT) = 8 units per hour. Just before the second unit arrives, the thruput will have the expected value of 0.5 units per hour. When the second unit arrives, the calculated thruput jumps again and the cycle repeats. This phenomenon decays as the number of units that have passed though the outflow increase. It would not occur if we were modeling a continuous flow process rather than a discrete flow process. Thus the oscillation occurs because we are trying to update the thruput more frequently than the time interval between units. Any software package that updates at this frequency will have this problem. It is a problem with measuring thruput for discrete item processes but not an issue with the modeling package per se.

> Learning Point: Any aspect of behavior in a model may be because of a real phenomenon or it may be simply because of some attribute of the modeling environment. We should never assume that the behavior is a result of real process behavior without first trying to see if it is related to the modeling environment.

Cautionary note: Models that include random variation are generally more difficult to debug than models that do not. This is particularly true when the models are complex and include many variables that have random variation functions associated with them. The reason for this is that, by definition, every time we run the model we will get a different result. So if we have a problem with the model and we make a change to try to fix the model, any change in the model output may be because of the impact of the change or to the random variation. Most simulation tools, ithink included, that provide random functions have the ability to set a seed number when setting up the function. When such a seed number is used, the resultant stream of random numbers is reproducible. So even though the random number function will give different numbers every time the function is called within a single run of the simulation, it will generate the same stream of random numbers from one simulation to the next. Readers are advised to use seed numbers when developing their models, particularly for their first models or for complex models. Later, when the model is working, seed numbers can be removed.

> Learning Point: When developing models that include random variation, it is recommended that all random number functions be specified with seed numbers included. When the model is working to the user's satisfaction, the seed numbers can be removed so that the entire impact of the random variation can be seen.

So what do we learn by running the model for various combinations of step changes in IIT and CYCLE TIME? It turns out that random variation and unsteady state behaviors limit the application of Little's Law. These limitations can be summarized in the following learning point.

> Learning Point: Little's Law holds only on the average over a time period when the process is in steady state.

Just how long a time period we need to use so that Little's Law is accurate is not obvious. However, a model such as figure 4.3 can give us a lot of guidance on how to choose such a time period.

Using Little's Law: Many people use Little's Law without even knowing that they are using such a fundamental law of process behavior. For example, suppose that we are managing a department with an average staffing level of 100 people and that on average each person remains in the department for five years before being reassigned to another department. We can imagine that each employee is going through an "employment" process. The average cycle time of the process is five years and the average WIP of the process is 100. By Little's Law, the average thruput is 100/5 = 20 employees per year. Thus as a department manager, we must cope with 20 new employees every year (on average). Most people would accept this calculation without ever having heard of Little's Law.

One powerful use of Little's Law is to get an organization focused on cycle time reduction as a main lever for overall improvement. If we look at the three components of Little's Law, we usually set average thruput to equal the average demand rate. The WIP then takes the value it must take to satisfy Little's Law. The measure we can change to change the overall thruput, WIP, and cycle time relationship is the average cycle time. Furthermore, in the next section of this chapter as well as in chapters 5 and 6, we will show that reducing the amount of random variation in the process will lead to reduction in the average cycle time. Thus, Little's Law can help justify a focus on variation reduction activities and hence support the application of the variation reduction philosophy of Deming (1986). Of course, in any real situation when there are more opportunities to reduce variation than resources to effect such reductions, some prioritization must occur. This prioritization should ultimately be done based on the financial impact of such variation reduction efforts. Any model can be expanded to take such financial considerations into account, as we indicated in chapter 3.

Most organizations are interested in understanding how their processes are performing from an inventory perspective. Generally speaking, reducing inventory levels is seen as an improvement in performance.[4] Another way that we can use Little's Law is in normalizing the performance of a process with respect to inventory. As an example, suppose we are running a process that has a WIP or inventory level of 10 units. In a certain time period, the average WIP level increases from 10 to 15 units. At first, we might say that the performance has deteriorated. However, it also turns out that in the same time period, the average thruput increased from 10 units/day to 20.0 units/day. Now we get a different picture of performance. The average cycle time of the process has been reduced from 10/10 = 1 day to 15/20 = 0.75 days. So in fact, the organization has improved its cycle time from 1 to 0.75 days while increasing its thruput. Said another way, the inventory level has increased but at a rate slower than the increase in thruput. Any reasonable management would certainly classify this as an improvement in performance. What Little's Law has allowed us to do is to normalize inventory performance by looking at the ratio of inventory to thruput as a more fundamental measure of performance than inventory level alone.

The normalized inventory metric is usually referred to as Days of Stock (DOS). For a simple single-step process, it simply represents the cycle time of the process step. It can also be interpreted as how long the process could satisfy demand if we stopped entering new units into the process, assuming that the thruput is set at the demand rate. If we define DOS as inventory divided by the anticipated future demand rate, it can be used as a measure for contingency planning. With this interpretation, the larger the value of DOS, the more contingency the process has to satisfy the market should there be a problem with the process. Thus, while we want to reduce DOS to reduce inventory carrying costs, we do not want to reduce it so far as to remove contingency. These two uses of a DOS measure should not be confused with each other. The first definition can be used as a measure of performance, whereas the second definition should be used for contingency planning

4. See Goldratt and Cox 1992.

purposes. Misusing the definitions can lead to problems. For example, suppose that DOS based on future demand rate is 50 days. The future demand drops by 50 percent so that the DOS now jumps to 100 days. The process performance has not deteriorated; twice as much contingency is in the process. If the new demand rate is likely to continue for a significant period of time, management will try to reduce the inventory level so that DOS drops back to its original value of 50 days. But even if management stops production altogether, it will still take 50 days to return to a DOS value of 50 days. During this time, management may get beaten up because of their "apparent" poor performance, and this may lead them to take inappropriate actions simply to appear to be doing something.

Finally, Little's Law shows us that we cannot focus on inventory, thruput, and cycle time separately. They are not independent. Consequently, it would be a mistake for managers to create three separate improvement teams focused on inventory, thruput, and cycle time individually. In any improvement effort, these measures must be managed as a system.

> Learning Point: Little's Law implies that thruput, work in progress (WIP), and cycle time cannot be managed independently. They must be managed as a system.

Another cautionary note: Because of its apparent simplicity, it is possible to misuse Little's Law. Two potential misuses that readers should be aware of follow:

- Forgetting that Little's Law is an output of a model. It is a result of the dynamics of the process. It does not drive the dynamic behavior of a process. It cannot be used in a dynamic model to predict, for example, the current level of WIP based on the current thruput and cycle time.
- Forgetting that Little's Law predicts *possible* combinations of average WIP, thruput, and cycle time. It cannot by itself predict if any one of these possible combinations will occur. A more detailed representation of the process (such as what a dynamic model provides) is required as well in order to decide which among all the possible combinations may actually occur.

4.3 Queuing systems

The starting point for queuing systems is an extension of the simplistic single-step process to include a queue on the input side of the process. The process in the last section did not need a queue because we assumed that the process had infinite capacity. However, all real processes have finite capacities for processing items. It is possible for an item to arrive at a processing step and not be able to enter the step. Therefore, we must place a queue upstream of the process step. It is important to realize at the outset that such queues are really part of the process. Units entering the process may first spend time in the queue before they have work done on them in the process. We can thus consider a process as having two parts: the part where the units are worked on; and the queue part, where units may accumulate until the work part of the process is ready to receive them. For example, when dining in a restaurant, the work part of the process is sitting down at a table, being served, eating, and leaving the restaurant. However, if people arrive at a rate faster than the restaurant can serve them, a queue builds up in the waiting area of the restaurant. From the perspective of the restaurant patron, both the time spent in the queue and the time spent being served are factors in the overall dining experience. Even if the food is excellent, a patron may stop going to a particular restaurant because too much time is spent in the queue. The same considerations can apply to banking, grocery stores, and so forth. Thus, it is important to recognize that such queues are part of the overall process and must be part of the model.

> Learning Point: Every real single-step process has a queue associated with it on its input side, whether this queue is explicitly identified or not by those managing the process.

(Note that in practice, we also must address the output side of the process. Once items leave the process, where do they go? In general, queuing models tend to ignore what happens to the items once they leave the processing step. However, ignoring the output side may lead to unfair comparisons between alternative methods of process management. We will return to this issue at the end of the next chapter.)

For the remainder of this section, we will refer to the part of the process where work is done as the WORK PROCESS because we want the entire process to include the queue as well.

The role of feedback in queuing models: Continuing the earlier discussion, we know that we must have a queue prior to the work process because, at any instant, the rate at which units enter the process may not equal the rate at which the process can handle the units. In the simple single-step process we have been modeling up to now, we have had no way of representing how many units the process can handle in a given time. That is, we have no way to represent the thruput capability or capacity of the work process. In the simple model for a process, we used a conveyor and set the capacity of the conveyor to infinity. Thus, the conveyor could handle any input rate of units. If we want to reflect a queuing system, we must add a capacity constraint. The simplest such constraint would be a capacity of 1. This means that the conveyor can handle only one unit at a time. If the processing time for the work process step is t hours, say, then the capacity of the work process and hence the capacity of the overall process is (1/t) units/hour. If units arrive at the process at a rate faster than 1/t (the time interval between arrivals is less than t), then the work process will not keep up and units will build up in the queue. In effect, the capacity constraint sets up a feedback between the work process and the queue. If we had included a queue in our previous models, no accumulation of units would have occurred because the work process could handle every unit that arrives. But with a capacity constraint, the queue must look at the status of the work process before it can send a unit to the work process. This is feedback. Thus, we can say that feedback is the first significant basis for queuing models.

> Learning Point: The first significant basis for queuing models is the presence of feedback (because of capacity constraints).

The role of random variation in queuing models: If there is no random variation in the arrival rate of units or the work process cycle time, then the addition of the queue does not add anything of interest to our model. If the arrival rate is less than or equal to our work-step capacity, then the work process can keep up with the arrival rate and the queue will be empty—we will have a steady state, and Little's Law will apply. If the arrival rate is greater than the capacity of the work process, the queue will rise consistently over time and Little's Law will not apply. The presence of random variation in the arrival rate and/or the work process cycle time makes queuing models much more interesting. When the arrival rate is above the work process capacity, then the queue will increase in size. When the arrival rate is below the work process capacity, the queue can fall again. Because a queue is bounded on the low side by zero but can rise on the high side without (any theoretical) bound, the situation is not symmetrical. This means that we can expect that random variation about some average value will not simply cancel out. The net effect of random variation *may be* to impact the average performance of the process. This is different from our previous single-step model, in which random variation did not impact the *average* thruput and/or cycle time. This discussion is based on our intuition. A dynamic model will help us determine the impact of the random variation on the overall performance and check our intuition. Thus, random variation is the second significant basis for queuing models.

> Learning Point: The second significant basis for queuing models is the presence of random variation in both the arrival times of units and the processing times for units.

In elementary queuing theory, the random variation is represented in the model by assuming that the distribution of the time between the arrival of units and the work process cycle time is exponential. The use of these distributions for elementary queuing models is justified in two ways. First, the use of an exponential distribution leads to models for which simple, closed-form solutions can be derived. The reason for this is the memoryless property of the exponential distribution.[5]

5. For example, see Winston 1993.

Consequently, a mathematical model does not have to keep track of all the arrival times because the probability of an arrival in a time interval does not depend on what occurred in previous time intervals. This keeps the math (relatively) simple. The second justification comes from the fact that it can be shown that under certain circumstances, such distributions are to be expected.

Queuing models: In general queuing theory, the work process is often referred to as a server (named because servers are the objects that work on the units and "service" them). The simplest queuing model will have only one server. It is possible to have multiple servers in a queuing model. Typically, servers are assumed to be independent, so the time it takes to process a unit in one server is independent of the time in another server. Thus, the servers behave independently. In ithink, if the processing time distribution of each server is the same, then multiple servers can be modeled as a conveyor with a capacity equal to the number of conveyors. If the processing time distributions are different, then a separate conveyor must be used for each server. Queuing models can be defined by specifying six numbers using the format 1/2/3/4/5/6. This notation is known as the Kendall-Lee notation.[6] The six attributes follow:

1. This specifies the nature of the arrival process. A designation of M denotes that the interarrival times are random variables with an exponential distribution. A designation of G denotes a general distribution, and a designation of D indicates a deterministic situation with no random variation in the arrival times. The mean arrival rate for units is usually denoted as λ so that the mean interarrival time for the units is given by $1/\lambda$. Thus, in steady state, λ represents the mean value of the thruput.
2. This specifies the nature of the service times (the cycle time of the process). The designations are the same as for the interarrival times. The mean service rate for the units is usually denoted by μ units per unit time. Therefore the mean cycle time is given by $1/\mu$.
3. This attribute denotes the number of parallel servers. In our model, this is the capacity of the work process conveyor. In this chapter, we will study models where this attribute is 1.
4. This attribute is the queue discipline. By this we mean the way items are removed from the queue. ithink queues are First Come, First Served (FCFS) or equivalently, First In, First Out (FIFO). Other queue disciplines are possible in queuing systems. For example, arriving units may have a priority associated with them, and they are withdrawn from the queue based on the value of this priority. An order fulfillment process that includes some emergency orders would be an example of when a prioritized queue would be included in the model. When the queue discipline is not important, it is referred to as GD, for general queue discipline. In this case, we do not have to worry about choosing items from the queue in any particular way.

---

6. See Winston 1993.

5. The fifth attribute denotes the maximum number of units allowed in the whole system of queue plus server. These systems have a cap on the number of units that are allowed in the system. This attribute is important because there are systems in which units are not entered into the system if the total number of units in the system is already too large. For example, suppose our system is an order to a supplier. If we see that the total number of orders already being processed and in the backlog is greater than what we think the supplier can handle and still get our order out on time, we may give our order to another supplier.
6. The sixth attribute gives the size of the population from which the units are drawn. In open systems, this number is considered to be infinite. We do not have to track the origin of these items. (Again, an example is customers at a restaurant.) This is an important attribute if the number of available units is of the same order as the number of servers, and if units are recycled. Then as we place units into the system, the number of units available to enter the system is smaller and the probability at which the units arrive at the server will be affected. Suppose, for example, that we want to model the maintenance process for a set of tester units. Such tester units might be used to test the operation of equipment in the field and are sent to maintenance when they show signs of deterioration. Typically, the number of such units will be small. When the maintenance process has been completed on a tester unit, it is returned to service. The more units that are being maintained at any one time, the less likely that a new unit will arrive for maintenance because fewer units are in service.

Note that these six attributes are used to designate queue but are not used explicitly in specifying queuing models in general-purpose modeling packages such as ithink.

The M/M/1/GD/∞/∞ queuing model: The model for this queuing system is shown in figure 4.5.

The elements of this model should be familiar. The capacity of the conveyor WORK PROCESS is set to 1. LAMDA represents the average arrival rate of items, and MU represents the average service rate of items. These parameters are generally given the symbols λ and μ respectively. This means that (1/MU) represents the cycle time for the WORK PROCESS conveyor. Thus, while the model is driven off interarrival times and processing times, we will specify these using rates to be consistent with the large body of literature on queuing theory and applications. Because the conveyor capacity is set to 1, the determining factor in how fast the work process (and hence the entire process) can service items is μ. Thus, the capacity of the process is μ. In queuing models, the ratio λ/μ is known as the traffic intensity, usually denoted with the symbol ρ. It is considered a fundamental parameter of the queuing system.

> Learning Point: The ratio of arrival rate, λ, to the service rate, μ, is called the traffic intensity of a queuing system. It is usually designated as ρ and is a fundamental parameter for the queuing system.

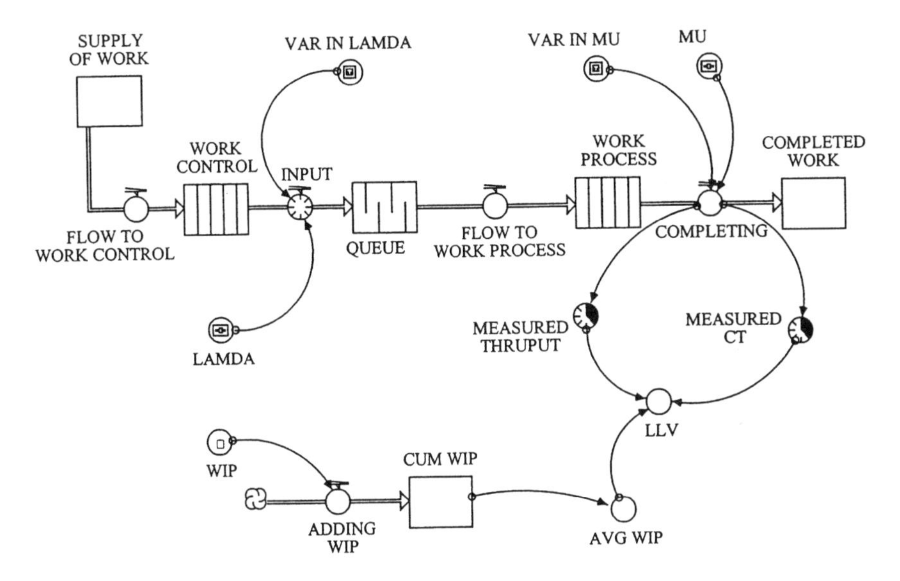

FIGURE 4.5. The M/M/1/GD/∞/∞ queuing model

The converters VAR IN LAMDA and VAR IN MU allow us to turn the random variation in LAMDA and MU on and off. The equations in the flows INPUT and FLOW TO WORK PROCESS are as follows:

INPUT:
(1-VAR_IN_LAMDA)*(1/LAMDA)+
VAR_IN_LAMDA*EXPRND(1/LAMDA)
FLOW TO PROCESS:
(1-VAR_IN_MU)*(1/MU)+VAR_IN_MU*EXPRND(1/MU)

We measure the cycle time and thruput as normal. We have used a variation on the WIP calculation by calculating the average WIP instead of the instantaneous WIP. This is done by integrating WIP over time and dividing by TIME to get the AVG WIP value. Note that WIP now includes both the items in the queue and the work process.

Initially, we set the values of LAMDA and MU to 1 hour and turned off the random variation. Readers can verify the following by running the model at various values of LAMDA and MU.

- With LAMDA = MU ($\rho = 1$), the THRUPUT is 1, the MEASURED CT is 1, and LLV is 1. There is no build up in the queue.
- With LAMDA>MU ($\rho > 1$), there is build up in the queue. No steady state exists and LLV <> 1.
- With LAMDA<MU ($\rho < 1$), the THUPUT equals LAMDA, there is no buildup in the queue and LLV is 1.

These results are intuitive and of no great interest. What is of interest is what happens when we turn on the random variation. Figure 4.6 shows the profiles obtained when the random variation is turned on for both arrival and processing times and LAMDA = MU = 1.

The interesting result here is that even though the average arrival rate equals the average service rate, a buildup of the queue still occurs! This is a nonintuitive result, to say the least. Figure 4.7, which looks at this result more closely, shows the profile for WORK PROCESS for the same simulation run as in figure 4.6.

Early in the simulation, WORK PROCESS is empty or "starved" at times. The conveyor is ready to accept items, but the queue is empty. As the simulation proceeds, the queue builds up and items are always available when WORK PROCESS is ready to accept them. From this point on in the simulation, the rate at which items exit WORK PROCESS must be 1/hour on average. So we have the peculiar situation in which the average input rate is 1/hour, the average output rate is 1/hour, and the queue is still building up. This would appear to defy the law of conservation of material! The profile of QUEUE for the same simulation as the previous two profiles is shown in figure 4.8.

> Learning Point: A queuing system with random variation in the arrival and service times does not exhibit steady state behavior when the traffic intensity is equal to 1.

Note that the inclusion of the words "with random variation" in real processes is unnecessary because there will always be random variation in any system.

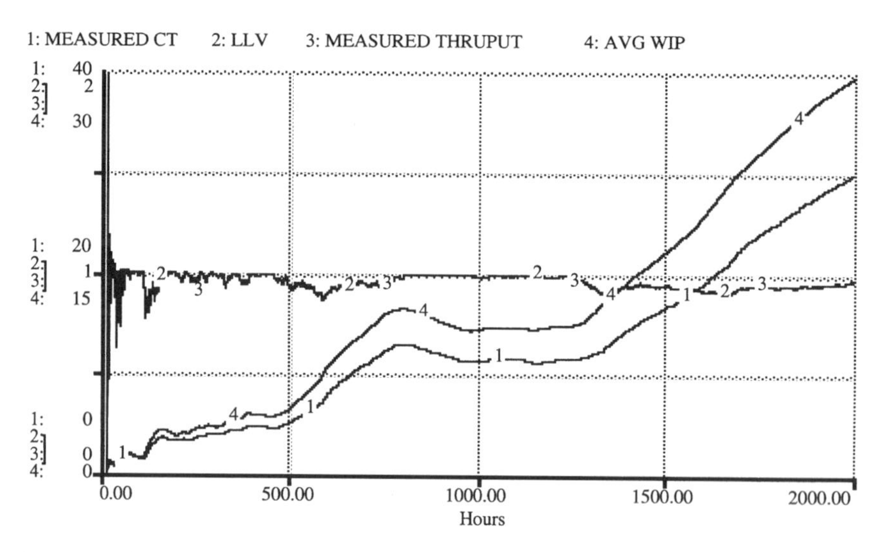

FIGURE 4.6. Profiles with random variation included (traffic intensity, $\rho = 1$)

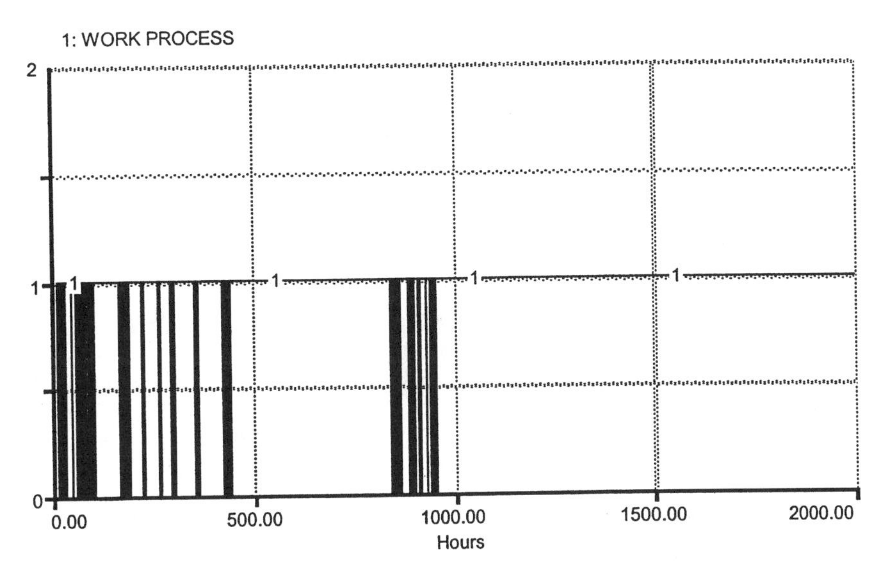

FIGURE 4.7. Profile of WORK PROCESS (traffic intensity, ρ = 1)

So why is the queue building up? We can hypothesize at least four reasons why the queue might be building up:

• Perhaps it is only a startup issue. If we had a large queue already, perhaps the build up would not occur. MODEL 4.5A.STM has been set up to allow the initial queue size to be set anywhere between 0 and 100 items. Even when the initial

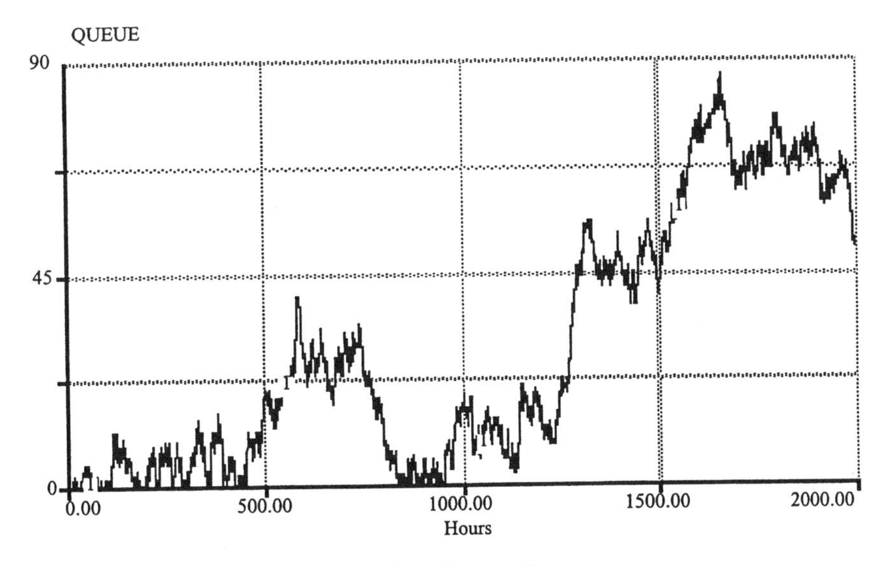

FIGURE 4.8. Profile of QUEUE (traffic intensity, ρ = 1)

value of the queue is set to a large value, running the model will show that the queue still grows.

• Another theory might be that the problem is due to using the exponential distribution. What if we used a different distribution such as the normal distribution? MODEL 4.5B.STM is a modification of MODEL 4.5A.STM, in which the exponential distributions have been replaced with a normal distribution. In this model, we can see the equations for the interarrival time and service time, respectively:

(1-VAR_IN_LAMDA)*(1/LAMDA_MEAN)+
VAR_IN_LAMDA*NORMAL(1/LAMDA_MEAN,LAMDA_STDEV)
(1-VAR_IN_MU)*(1/MU_MEAN)+
VAR_IN_MU*NORMAL(1/MU_MEAN,MU_STDEV)

Readers are invited to run this model and see that the queue still has a tendency to grow (even with the queue initial size being set to a value higher than 0).

• Another more subtle thought is that the problem is because the exponential distribution is an unbounded distribution. That is, in theory the interarrival times and service times can go from 0 to infinity. Therefore, long time intervals may occur and the queue may grow. MODEL 4.5C.STM is a modification of MODEL 4.5A.STM, in which the exponential distributions have been replaced with a uniform distribution. In this model, we can see the equations for the interarrival time and service time, respectively:

(1-VAR_IN_LAMDA)*(1/LAMDA) +
VAR_IN_LAMDA*RANDOM(0.1/LAMDA,1.9/LAMDA)
(1-VAR_IN_MU)*(1/MU) +
VAR_IN_MU*RANDOM(0.1/MU,1.9/MU)

Again, readers are invited to run this model and see that the queue still has a tendency to grow (even with the queue initial size being set to a value higher than 0).

• Finally, perhaps the growth in the queue is because of the combination of random variation in both the arrival times and processing times. If we were to have random variation in arrival times only, then the growth would not occur. All of the previous models can be run with each source of random variation included individually. Again, each source of random variation on its own can lead to unsteady behavior.

When running these models, repeat the simulation a number of times to get a feel for how the model is behaving. We conclude that none of the reasons presented so far can explain the growth in the queue.

So what is the reason for the growth of the queue?[7] It can be shown (see Hopp and Spearman 1996, for example) that if n items are in the queue, the probability

7. The explanation for the growth in queue size for a traffic intensity of 1 is subtle so readers can ignore this section if they wish. The main idea is to understand that the queue can grow without bound.

of the queue growing to $n + 1$ equals the traffic intensity (= 1). Assuming some nonzero finite probability that 0 items are in the queue ($n = 0$), then a finite probability of $n + 1 = 1$ item in the queue exists, hence a finite probability of $n + 2 = 2$ items in the queue exists, and so on, up to infinity. Thus, there is a finite probability that the queue can be infinite in size—so the queue will continue to grow indefinitely. This argument centers on the fact that there is a finite probability of the queue being empty. This is further related to the fact that a queue cannot go negative; our process is asymmetrical.

However, the fact that a queue cannot be negative in practice does not constrain us from asking what would happen if the queue could become negative. Then if the work process conveyor was ready to accept items, it could take an item from the queue and make it negative. MODEL 4.5D.STM is set up as MODEL 4.5A.STM, except that the queue can be negative. To do this, the queue had to be changed to a stock that pushes items into the conveyor if it is empty. (Also, the mode of ithink had to be changed from cycle-time to normal so that the stock could be allowed to become negative. In this model, then, cycle time functions such as THROUGHPUT and CTMEAN do not work). This model is set up to run 100 times so that average behavior can be seen. Over the 100 runs, the value of QUEUE at the end of a simulation run varied from –136 to +178 with an average of 1.5 (which is essentially 0). Comparing that to a similar set of 100 runs for a normal queue that cannot go negative, the value of QUEUE at the end of a simulation run varied from 0 to +198 with an average of 52.5. Thus, it can be seen that the constraint of a nonnegative queue means that the queue will grow unbounded when the traffic intensity is 1. In effect, the fact that the queue cannot be negative creates a lack of symmetry that allows the queue to grow without bound.

> Learning Point: The fact that a real queue cannot be negative means that the queue will grow without bound when the traffic intensity (arrival rate/service rate) = 1.

Of course, a real queue will never become infinite in practice. What will happen is that the queue will begin to rise. Many times you will hear people refer to this in terms of "a rising backlog" when they are discussing orders, shipments, angry customers, or other factors. If management has no understanding of the learning points presented earlier, what may happen is this: To counteract the rising backlog, more resources are put to work in the process. This, in effect, raises the capacity of the work process and the backlog begins to fall. Once it has reached some acceptable level, these resources are removed to get back to a steady state. Then the backlog begins to rise again and the whole cycle is repeated. Hopp and Spearman (1996) refer to this as the "overtime vicious cycle" because the extra resource is often created by making people work overtime. On top of this, each time the backlog rises, people will be assigned to find out the "special" circumstances that caused the increase in the backlog. Their time is being wasted. The rising backlog is simply a result of the "physics" of how queuing systems work. Running the queue so that the arrival and service rates are balanced is an attempt to break the laws of physics. We try this at our peril!

So if we want to run this queuing model under a steady-state condition, then we must have $\rho < 1$. Under these conditions, the queue should build up to a steady state. Figure 4.9 shows the results of running MODEL 4.5.STM with LAMDA = 0.75 and MU = 1 so that $\rho = 0.75/1.0 = 0.75$.

Figure 4.9 shows that the average cycle time and WIP reach a steady state at value of 4.0 and 2.9, respectively. Analytical results are available for this queuing model. Winston (1994) gives the following equations for the average and standard deviation of the cycle time (CT) and the average WIP (W):

$$\mu_{CT} = \frac{1}{(\mu - \lambda)} \qquad \sigma^2_{CT} = \frac{1}{\mu^2} \frac{1}{(1-\rho)^2} \tag{4.4}$$

$$\mu_W = \frac{\rho}{(1-\rho)} \qquad \sigma^2_W = \frac{\rho}{(1-\rho)^2} \tag{4.5}$$

With $\mu = 1.0$, $\lambda = 0.75$, and hence $\rho = 0.75/1.0 = 0.75$, this gives CT = $1/(1–0.75) = 4.0$ and $W = 0.75/(1–0.75) = 3.0$. This is in close agreement with the model, as we should expect. Also note that equation 4.5 is consistent with Little's Law ($\mu_W/\mu_{CT} = \lambda$), as it must be if we are reaching a steady state. It is also consistent with there being no steady state at $\rho = 1$ because as $\rho \rightarrow 1$ both the average and standard deviation values $\rightarrow \infty$. Readers are invited to run the model for different value of μ and λ to check to see if the model agrees with equation 4.5. Remember to run the model a few times at each set of (λ,μ) because the random variation will give somewhat different results each time. Copying the data to a spreadsheet program such as Excel will allow the mean and standard deviation to

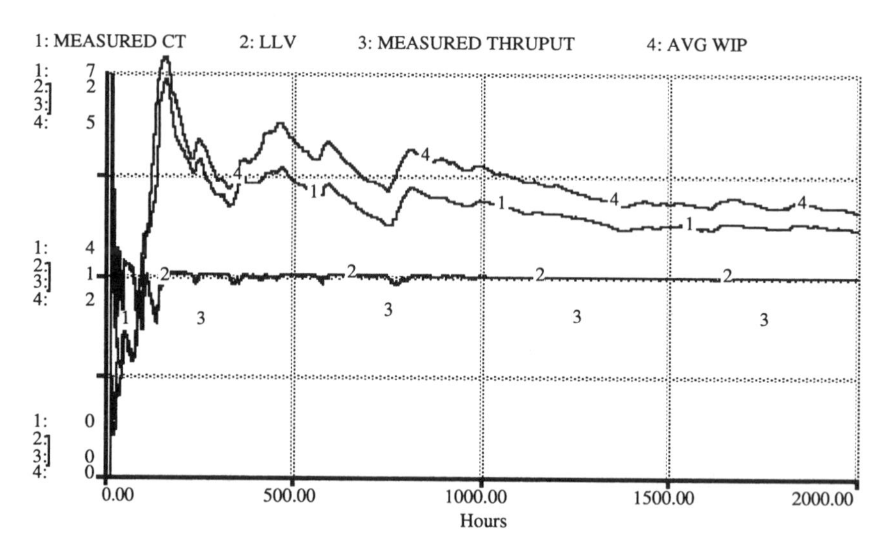

FIGURE 4.9. Queuing model profiles with $\rho = 0.75$

be calculated. Also, note that equation 4.5 represents the long-term behavior of the queuing process. It cannot represent the short-term or transient behavior of the queuing process. Modeling the transient behavior of a queuing process is a complex analytical problem but is a simpler dynamic simulation problem.

We tested equation 4.5 by running the model many times and comparing the predictions of the equation to the average of these multiple runs. To this end, a set of 1,000 runs were made for a particular value of the traffic intensity, and the average and standard deviation of the WIP were measured for each run. Then the average of these variables was computed over the 1,000 runs to get a good estimate of the variables. This was then repeated for values of the traffic intensity between 0.25 and 0.99. The results are shown in figure 4.10.

As expected, the average WIP grows rapidly as the traffic intensity approaches 1. The standard deviation also behaves in the same way, which again is not unexpected. What is curious about figure 4.10 is that for low values of the traffic intensity, the standard deviation is greater than the mean. This situation reverses, however, for higher values of the traffic intensity. Another way of saying this is that the COV[8] of the WIP is >1 for low values of the traffic intensity and <1 for values close to 1. Equation 4.5 predicts that the COV of the WIP should be $1/\sqrt{\rho}$. Because $\rho <= 1$, this ratio should always be > 1. Also, the observant reader may have noticed that the values shown in figure 4.10 are smaller than what is pre-

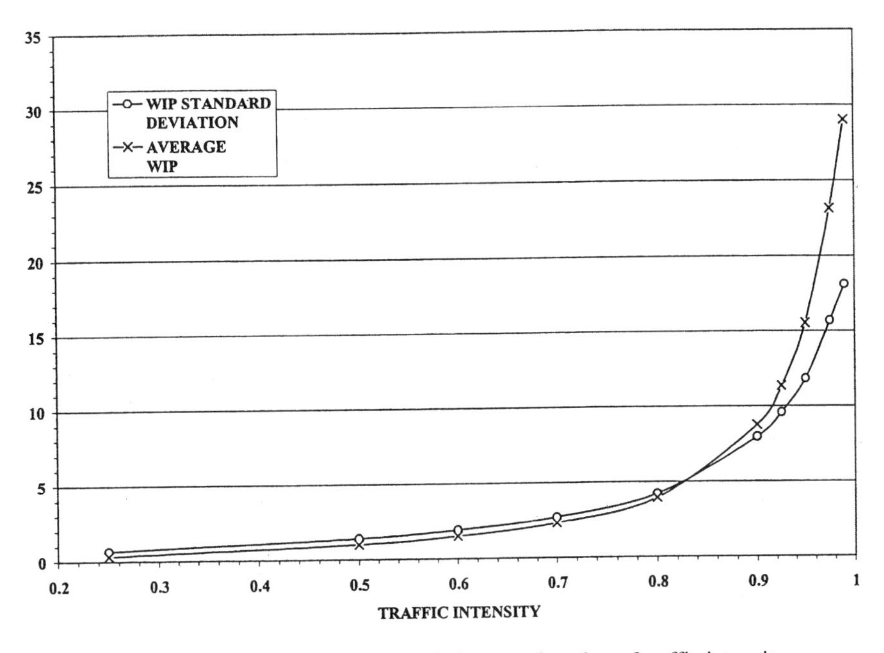

FIGURE 4.10. WIP average and standard deviation as a function of traffic intensity

8. COV, the coefficient of variation = ratio of the standard deviation of a variable to its mean.

dicted by equation 4.5. For example, at a value of $\rho = 0.975$, the average and standard deviation predicted by equation 4.5 are 39.0 and 39.5 respectively. These values are much greater than what is observed from the model. This is a huge discrepancy between theory and simulation, and we must be able to explain it if we are to feel confident about our understanding of how queuing systems behave and the ability of our models to correctly model the queuing system. If we cannot properly model the behavior of the WIP average and standard deviation for a simple queuing system, how can we have any confidence in the model's ability to predict the behavior of more complex queuing systems? So we must understand what is going on here.

Now, because we have used 1,000 runs to get the data points in the chart, it is unlikely that the discrepancy is due to the variation of the WIP average from run to run. In fact, even at a traffic intensity of 0.99, the uncertainty in the average as computed over the 1000 runs is less than 4 percent.[9] So we have a consistent difference between our simulation results and theory. The problem is that we have not given enough time in the simulation to really see the buildup of the WIP: 2,000 hours is simply not enough time to see the full impact of the random variation on the WIP when the traffic intensity is close to 1. But it is enough at lower traffic intensities, hence the better agreement of simulation and theory for these lower values. For example, at a traffic intensity of 0.7, equation 4.5 predicts an average WIP of 2.33 and the simulations predict 2.31. Readers can satisfy themselves by computing the average WIP from equation 4.4 and comparing the result with figure 4.10—the discrepancy between theory and the simulations increases as the traffic intensity increases. So if we want to feel more confident about our model, the simulation runs should be much longer than 2000 hours. Just how long should we run the model? The answer is surprising. Table 4.1 shows data for a series of simulations when the simulation time was 1,000,000 hours[10] instead of 2,000.

TABLE 4.1. Queuing model results for a simulation time of 1,000,000 hours.

Traffic intensity	0.9	0.925	0.95	0.975
Average WIP (average)	9.001	12.304	19.019	38.930
Average WIP (theory)	9.000	12.333	19.000	39.000
Average WIP (minimum)	8.569	11.558	17.867	31.906
Average WIP (maximum)	9.568	13.287	21.009	51.098
WIP standard deviation (average)	9.488	12.728	19.488	38.954
WIP standard deviation (theory)	9.487	12.824	19.494	39.497
WIP standard deviation (minimum)	8.752	11.420	17.097	30.834
WIP standard deviation (maximum)	10.453	14.572	22.820	65.410

9. In statistical terminology, we say that the size of the 95 percent confidence interval is ±4 percent.

10. Yes, we do mean 1,000,000 hours—these runs were done in a simulation package called Extend from ImagineThatInc.

For each value of the traffic intensity, we ran the model 150 times and calculated the average, minimum, and maximum values for the WIP average as well as standard deviation. The agreement between the model and theory is now close, even for the standard deviation at a traffic intensity of 0.99. But look at the range of values over the 150 runs. Even though the average of the WIP standard deviation is 39, close to the theoretical value of 39.5, a single run of 1,000,000 hours could have given a value as low as 30.8 and as large as 65.4. In many ways, this is a frightening result because it says that even when we run a simulation for what appears to be a long time, we could get an erroneous estimate of performance. We can, therefore, find some important learning points in the earlier model.

Learning Points:

- Running a queuing process at a traffic intensity close to 1 is not a good idea—both because of the high WIP and cycle times and also the high variation in these parameters.
- We must be careful when we see an unusual effect in the results of a dynamic model. (The most unusual feature in figure 4.10, namely that the COV of the WIP changes from being >1 to <1 as the traffic intensity changes, turns out to be an issue with the model rather than a real phenomenon.)
- Starting a modeling effort by first analyzing a simple model before going on to a more complex model (such as a multistep process) is highly recommended.

These observations raise an important question: How long should we run a model to obtain useful results? In theory, we may have to run it for a long time to see the average type of behavior in the process. But if it takes 1,000,000 hours to see this behavior, of what real use is this average behavior anyway? In other words, what the model results clearly indicate is that the chances are small that the actual behavior over realistic time periods is close to this ideal averaged behavior. So if we want to use the model to predict average behavior, we may have to use long time periods in simulations. But if we want to look at the potential behavior over real operating time periods, then we should use these operating time periods in our model. We then accept that these shorter time period models may not give results close to the theoretical long-term models—and that this is acceptable. We are simply using the model to answer different questions. Another way to look at this is to say that, while in theory the system has a steady state for traffic intensities less than 1, in practice the system may never look like it is in a steady state when the traffic intensity is close to 1. Thus in the practical application of modeling, the time over which the simulation is run is important. For example, in a run of 1,000 simulations at a traffic intensity of 0.8 with exponential random variation, the maximum average WIP was 11.3 and the minimum was 2.2; the average over the 1,000 simulations was 3.95 (close to the theoretical value of 4.0). Thus a single run of this queuing system over a simulation period of 2,000 hours could give results far removed from expected average behavior. The learning point here is that we should investigate the impact of different simulation periods on the simulation results before we accept them as useful.

> Learning Point: The time period of the simulation is an important parameter of a typical model and should be taken into account when evaluating the results from a model.

The model results and equation 4.5 indicate that as the value of ρ increases toward 1, the impact of the variability is felt more because we need a larger queue to be able to run the queuing process at the steady state arrival rate λ. Thus the closer we want to be able to run the queuing process at its theoretical rate ($\rho = 1$), the more items we will tie up in the queue, and the greater the cycle time of the total process. Looking at this from another viewpoint, if we want to be able to run the queuing process at some rate λ, we must have a work process that can handle the items at a faster rate than λ. The service rate μ must be greater than the arrival rate λ.[11] Basically, we must provide some excess capacity in the system. The more excess capacity we provide, the lower the average queue size we will need to support the required thruput λ. Conversely, for a given amount of random variation, the smaller the excess capacity is, then the greater the queue will be. Thus, we see an interdependency of random variation, average size of the queue, and the amount of excess capacity in the queuing process. Therefore, when managing a queuing process, we cannot manage the size of the queue and the amount of excess capacity independently.

> Learning Point: In a queuing process, there is interdependency between the random variation present, the average size of the queue, and the amount of excess capacity provided in the queuing process.

In effect, the excess capacity acts as a buffer to absorb the impact of the random variation present in the arrival and service rates. More excess capacity allows for more absorption, which means random variation has less impact on the process (as measured by the average size of the queue and the average cycle time). However, we should note at this point that this says nothing about the effect the excess capacity has on the departure of items from the queuing system. This is not an issue here because we are saying nothing about what happens after items leave the queuing system. In subsequent chapters, we will have to address this issue.

Anyone who is familiar with flight delays at airports knows how this works. If an interruption in flights occurs during the day, it can have ripple effects all day.

---

11. The case where we create excess capacity by increasing the number of conveyors will be addressed in chapter 6.

At night, however, it is possible for the airlines to catch up because they have fewer flights. For example, they might normally finish at 12:00 A.M. With delays, they might actually finish at 12:30 A.M. Their ability to make use of a portion of the night hours to catch up represents their excess capacity, and they can use some of this to get back on schedule for the next day. In manufacturing, we see the impact of this phenomenon when we try to push items through a work center at the theoretical rate. Because some random variation always exists, queues will build up. This is avoided by providing more capacity at the work center than is theoretically required. This excess capacity can be provided in many ways—more machines, more people, overtime, and so on. Thus, models such as figure 4.5 can be used to set the amount of queue required at a work center for a given amount of random variation.

4.4 Transient queuing behavior

Because random variation is present, every time we run the model, we will get somewhat different profiles. Equation 4.5 applies only to average behavior; therefore, it cannot represent this phenomenon. This is where the dynamic model is essential. Being able to predict how reproducible the profiles are is extremely important. If the reproducibility is poor from run to run in the model, then we can say that when we run the actual process, there may be widely different possible behaviors consistent with the overall average behavior. Queues may fluctuate widely in size and still be consistent with expected process behavior. Knowing this may prevent people from over reacting to the process performance. Recall from chapter 2 that such overreaction to random variation tends to make process performance worse. In some cases, the reproducibility of the process may be so poor that we decide that it cannot be successfully managed and that some change must be made to the process. Because excess capacity absorbs the random variation better, it would be reasonable to assume that the more excess capacity, the more reproducible the profiles and hence the easier the process will be to manage. To illustrate just how different the reproducibility can be as a function of excess capacity, consider figures 4.11 and 4.12. Figure 4.11 shows five consecutive runs with $\rho = 0.5$ and figure 4.12 shows five consecutive runs with $\rho = 0.8$. These runs are taken from MODEL 4.12.STM (which is basically the same as MODEL 4.5.STM except that the charts have been set up as comparative charts). The scales have been made the same to facilitate comparison.

Remember, these profiles are for the average WIP. Imagine what the variation in actual WIP would be like (readers can run MODEL 4.11.STM to see consecutive runs of the actual WIP). Thus for any given amount of random variation, providing excess capacity will make it easier to manage the process. This is another good reason for providing excess capacity.

We have already indicated that steady state analysis is invalid when the traffic intensity is 1 because equation 4.4 has no solution for this value of ρ. However, we can perform dynamic analysis when $\rho = 1$. Just because the queue will grow

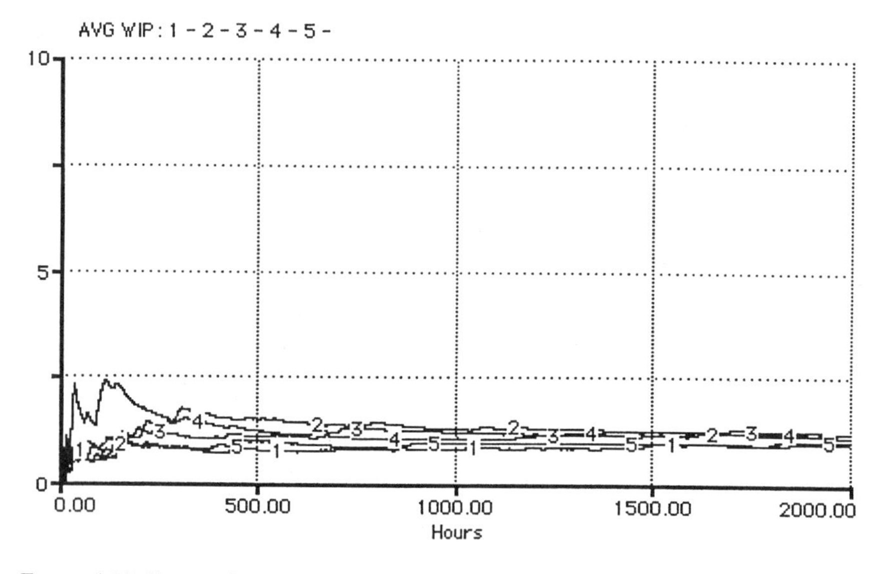

FIGURE 4.11. Consecutive runs for $\rho = 0.5$

without bound under these conditions does not mean that it will grow immediately. (In fact, the queue can drop down to 0 at any time as the simulation progresses.) We know that it will take an infinite time to grow an infinite queue. So it is of interest to know how fast the queue will grow, even when $\rho = 1$. To see what can happen, a set of 1000 runs were performed with $\rho = 1$ and some given simulation time. The final queue size at the end of the simulation and the maximum

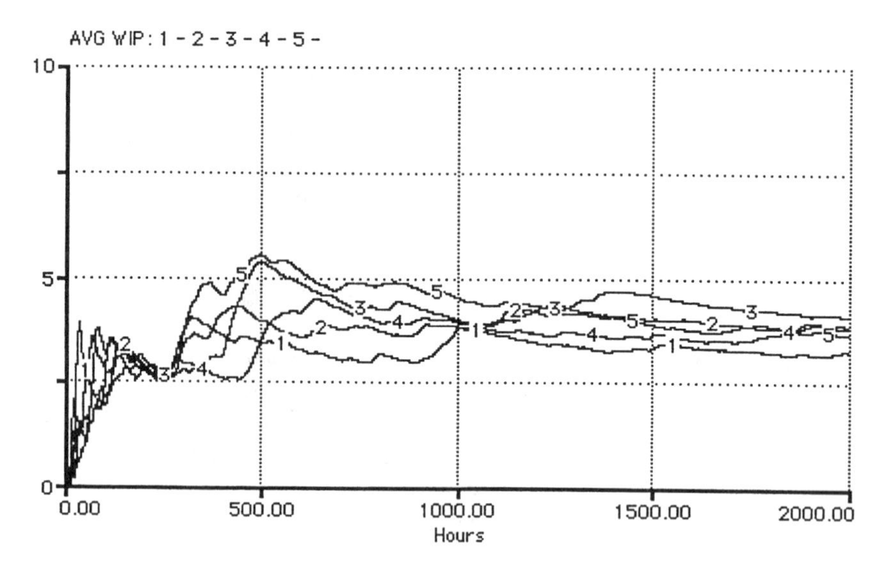

FIGURE 4.12. Consecutive runs for $\rho = 0.8$

queue size during the simulation were recorded. Then the averages of these two values were computed over the 1000 runs. This was repeated for various simulation times between 100 and 1000 hours. The results are shown in figure 4.13.

This graph might be useful if the queuing process is only being run for a short time. In such cases, we could use figure 4.13 to predict how much the queue is likely to grow during the period of operation. If we only intend to run the queuing system for 500 hours, then we can say that on average, the queue will grow to about 25 items and the maximum queue size during this time will be about 40 items. During 500 hours of operation, we had planned to complete 500 units. If up to 25 of those could be in the queue, we should allow more production time (say an extra 25 hours) so that we can empty the queue and reach our production quota. (Of course, this analysis is based on the average behavior of the queue. We know that the actual behavior of the queue can be different from the average behavior. In fact, over the series of 1,000 runs at a time of 500 hours, the largest value for the final queue size was 98 items. This is nearly 20 percent of the production quota. So now you have to explain why you could only complete 80 percent of your production quota even though you were given enough capacity to complete it all in the time period you were given!) In summary, while you could run the queue over short time periods at a traffic intensity of 1, it is not recommended.

> Learning Point: Even running a queue for short time periods at a traffic intensity of 1 is not recommended.

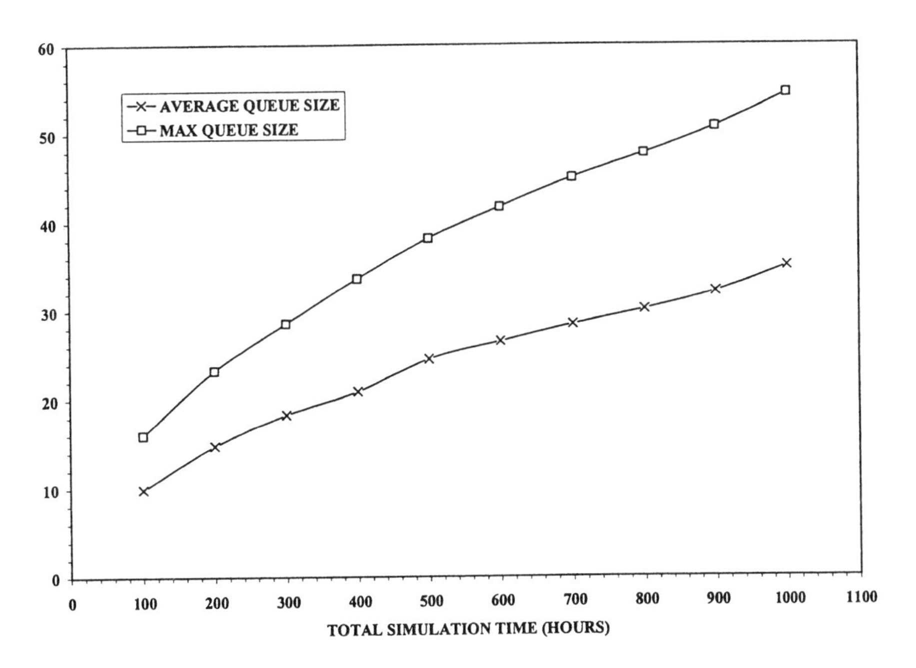

FIGURE 4.13. Growth of the queue over time

These types of analyses—and the issues they can highlight—clearly show the value of having dynamic models of the queuing process.

4.5 Further modeling with queuing systems

Of course, as well as providing excess capacity, we also could reduce the amount of random variation in our process. The model we have been using does not allow us to experiment with the amount of random variation because the exponential distribution has only one parameter and it represents the mean and the standard deviation (so that the coefficient of variation, COV, = 1). If we replace the exponential distribution with a normal distribution, then we can vary the standard deviation independent of the mean. This model is provided in MODEL 5B.STM. MODEL 5C.STM has uniform distributions instead of exponential distributions. Hopp and Spearman (1996) provide an approximate equation for the general purpose queuing system when the random variation is not exponential. Readers can use equation 4.6 to evaluate the results of their modeling.

$$CT = \left(\frac{COV_{AT}^2 + COV_{ST}^2}{2}\right)\left(\frac{\rho}{1-\rho}\right)\frac{1}{\mu} + \frac{1}{\mu} \qquad WIP = \lambda * CT \qquad (4.6)$$

where the thruput is λ by definition. Again, COV is the coefficient of variation (standard deviation divided by the mean). The subscripts AT and ST refer to the arrival time and service time. Because the COV for an exponential distribution is 1, it is left to the reader to show that equation 4.6 is equivalent to equation 4.4 for the exponential distribution. Following are some important learning points from equation 4.6 that readers should verify with MODELS 5B.ITM and 5C.ITM. (Remember that equation 4.6 is only an approximation, so we should not expect calculations based on this equation to agree with the model results.)

- The queue still grows when $\rho = 1$.
- Random variation in arrival rate or service rate alone will cause queues to build up when $\rho = 1$.
- The size of a steady state queue is a function of the square of the COV.
- The rate at which queues grow is a function of the amount of random variation (as measured by a given COV).
- The impact of a given amount of random variation (as measured by a given COV) is the same whether this random variation appears in the arrival rate or the service rate.
- The reproducibility of a profile is dependent on the amount of random variation (as measured by a given COV).

So far, we have measured cycle time and WIP over the entire process, queue plus work process. Sometimes we are only interested in the queue portion by itself; sometimes only in the work process portion by itself. If we only want to model an individual part of the process, we must change where we time stamp items and where we measure the cycle time, thruput, and WIP. MODEL

4.5E.STM looks at just the queue section of the process. For this model (see figure 4.5):

- Items are time stamped in the inflow INPUT.
- Thruput and cycle time are measured at the outflow FLOW TO WORK PROCESS.
- The average WIP is based only on QUEUE.

MODEL 4.5F.STM looks at the work process section of the process. For this model:

- Items are time stamped in the inflow FLOW TO WORK PROCESS.
- Thruput and cycle time are measured at the outflow COMPLETING.
- The average WIP is based only on WORK PROCESS.

Readers are invited to run these models to ensure that they understand the results they get. Also they can compare the results to the following equations:[12]

Queue only:

$$CT = \left(\frac{COV_{AR}^2 + COV_{SR}^2}{2}\right)\left(\frac{\rho}{1-\rho}\right)\frac{1}{\mu} \qquad WIP = \lambda * CT \tag{4.7}$$

Work process only:

$$CT = \frac{1}{\mu} \qquad WIP = \lambda * CT \tag{4.8}$$

with the thruput equal to λ in both cases, as it must be at steady state.

This chapter has provided an introduction to the modeling of simple single-step work processes. Interested readers will find more details in Hopp and Spearman 1996 and Winston 1993. In the next two chapters, we extend our modeling capability to processes with multiple steps.

---

12. The models are set up with exponential random variation so that the COVs are 1.

5

Multistep Serial Workflow Processes

There is no one giant step that does it. It's a lot of little steps.
—Peter A. Cohen, American investment banker

5.1 Introduction

In this chapter, we now extend our modeling to cases in which we want to represent our processes as a series of steps. As we have already pointed out, the motivation for this is to enable us to get a more detailed understanding of how real processes behave. Variables that were inputs to the single-step process models now become predicted outputs of these more complex models. Increased understanding comes at the price of increased detail. As in other areas of life, there are no "free lunches" here.

Following is a taxonomy relating to workflow processes. This list is not meant to be comprehensive but will be a useful starting point:

Type of process: Generally speaking, the two major types of processes are manufacturing and service. Manufacturing processes produce physical products, such as a washing machine, automobile, or cell phone. Service processes produce nonphysical products, such as consultation or information. Service processes may also cover areas that an organization outsources to a more specialized organization. For example, some organizations outsource payroll processes. In this book, we do not make a major distinction between these two types of processes because they fundamentally work in the same way. The basic building blocks of stocks, flows, converters, and connectors will apply in both cases. Of course, the details will be different and the terminology will differ. In manufacturing processes, we will talk about parts, stocks, and so forth. In service processes, we talk about customer service calls, backlogs, and such. While the details may differ, the objectives in each case are basically the same. We want to reduce cycle time, increase thruput, improve customer service, and accomplish other efficiencies.

Discrete, batch, continuous, and flow processes: (we alluded to this in the last chapter) A discrete process is one where the product or service is produced as separated items or units. Examples include complaints handling and tire manufacturing. In a batch process, the items are processed together as a lot. For example, a machine might be able to handle 30 parts at a time, so the parts are grouped in batches of 30. Service processes tend to be modeled as discrete/batch processes. Continuous processes do not produce discrete items. An example would be the continuous production of gasoline in an oil refinery. Flow processes are really discrete processes in which the production rate is so high that it appears somewhat continuous. An example would be the packaging of beer in cans. Modeling flow processes is challenging because such processes are in between the discrete/batch and continuous processes. There are too many units for a model to keep track of, as we can do in discrete/batch processes. But if we model them as a continuous process, we do not represent the discrete nature of the process. In this book, we focus on discrete/batch processes.

Process routing: Routing is a term used to describe how items flow through a process. A serial process is one where all the steps are arranged in a sequence. Each item must go through each step in the prescribed order. A parallel process is one where items can take multiple pathways through the process. A pure serial process is rarely found in practice. Most real processes are combinations of series and parallel subprocesses. One simple reason for this is that all real processes have the potential to fail at each step. When a failure occurs, an alternative manufacturing method is usually required, and parallel processing occurs. Having said this, however, it is possible to consider some real processes as serial at some level, and serial processes are easier to understand than parallel processes. For example, suppose we have a manufacturing process with five steps that are run in a series. After step 1, the items are routed to step 2, and so on. We can model this as a serial multistep process, with each of the 5 process steps being modeled as a step in the model. This approach will work so long as an item cannot bypass a step or cannot be routed back to an earlier step. If these routings can occur, then the serial model will not work. In this chapter, we will consider serial process models. Chapter 6 will consider parallel process models.

Products: Processes may be either single or multiple products. Single-product processes are easier to construct and model given that they have one less variable in the model. In this book, we will emphasize single-product models. It will be left as an exercise for readers to generalize these models to a multiproduct system.

Balance: A process is considered to be a balanced process if all the steps in the process have the same thruput capacity. Put another way, a balanced process is one where there is no excess capacity in the process. It might appear that having a balanced process is a good idea because it says that, as a process manager, we have not wasted money on excess capacity. However, even for a simple single-step process, we saw the error of this thinking in chapter 4. In that case, a balanced process was represented as a process with a service rate equal to the arrival rate (traffic intensity of 1). But we saw that we cannot operate under these conditions. We must provide excess capacity. And as multistep processes are more complex than single-step processes, we can expect that this position will be even more valid for such multistep processes. But in real processes, capacity cannot be added in a continuous fashion. It tends to be added in chunks. We need a 25 percent increase in the capacity of a manufacturing line. We can do this by adding a certain machine at a certain step. However, the only machine on the market will add 40 percent more capacity, so now we have an excess capacity of 15 percent at that step. So we end up creating an unbalanced process. Again, the reason we study balanced processes is that we can learn from them before we look at unbalanced processes. Also, some portions of a process may be so close in capacity that a balanced assumption is valid.

Coupling: We said in chapter 4 that in real single-step processes, a queue is always present before the work process step, whether we acknowledge it or not. In the case of multistep processes, however, the situation is somewhat different. Here we can have situations in which no queue occurs between steps. We might want to avoid having queues for several reasons. For example, the items may require expensive storage. Or the items may degrade if they sit in a storage area before being processed further. A tightly coupled process is defined as one in which there is no queue between the process steps. An uncoupled process is one in which there can be infinite queues between the process steps. A loosely coupled process is an in-between situation in which queues exist at limited sizes. At this stage, the reader should think about the motivation for the use of the term "coupling" in relation to queue size. Later when we analyze some models, the motivation will be clear.

5.2 Modeling multistep processes in ithink

Before we create some models, we will briefly discuss how to create generic models in ithink. This is important because ithink does not contain specially built structures for modeling business processes. It simply provides the basic building blocks of stocks, flows, converters, and conveyors, and it is up to the user to create more complex structures from these basic blocks. This is different than some other modeling packages where some of these more complex structures are already built for the user. The ithink approach offers both advantages and disadvantages. On the one hand, having these structures already built tends to make model building easier and faster. On the other hand, the activity of building the more complex structures from the simpler building blocks is instructive. This latter observation is most relevant in this book because we do not assume any prior modeling knowledge on the part of the reader.

We know that the basic building block for a multistep process will be a set of single steps, each represented by a conveyor. However, we cannot represent a multistep process as simply a connected set of conveyors as shown in figure 5.1.

The reason for this is simple. When an item is ready to exit a conveyor, it must have somewhere to go. If the next step in the process has a limited capacity and is already full, the item will have nowhere to go. ithink recognizes this issue by automatically setting the capacity of the next conveyor as infinite and preventing the user from changing the capacity value. We must add holding points between

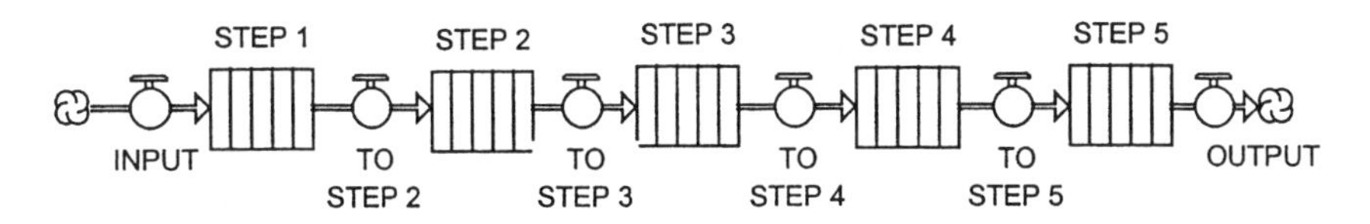

FIGURE 5.1. Serial process represented as a series of conveyors

the conveyors so that items have somewhere to go when they exit the conveyor. Thus the structure of the model must consider the output side of the process, something we did not have to do when considering the single-step queuing processes in chapter 4. The exact form this structure takes will depend on the type of process and will be explained as each is encountered in the remainder of the chapter.

5.3 Specifying models/modeling objectives

We assume that the work processing part of each step can be described as a conveyor with a capacity of 1. If we allow a capacity or more than 1, then in effect we are allowing items to be processed in parallel. Setting the cycle time for each step then specifies each step. The capacity rate of each step is then simply the inverse of the cycle time in items/time. ithink also has an inflow limit number that can be specified. For all the models we will be using, there will be no constraint on this inflow rate. We will leave this number set at its default of INF (∞). We also must specify the holding points. The only specification we will need is the maximum number of items that can be held in these holding points. Finally, we must specify the arrival rate of items to the process.

Once we have specified the process, we can run the model and predict the behavior of the process. We must ensure that our models are set up with structures that will allow us to measure performance data. The main variables we want to predict include:

- The capacity of the process—that is, the maximum thruput we are able to flow through the process.
- The overall cycle time of the process.
- The utilization of each process step. The utilization of a step is the fraction of time that an item is being processed in the step. That is, utilization is the fraction of time a step is occupied with work.
- The thruput bottleneck of the process. The thruput bottleneck is the step that is limiting the overall rate at which items can be processed.
- The critical cycle time step—this step has the greatest impact on the overall process cycle time.
- Which steps should we focus on to create the most improvement for the least effort?

In order to introduce readers to the complexities of multistep processes in as simple a way as possible, we will initially focus on models where the random variation is exponential. We will be able to make use of the memoryless property of this distribution to better see and understand the behavior of the process. We will want to have models where the impact of the random variation is great enough so that we can easily see the impact. The only way we can ensure this is to use a high-traffic intensity (because the coefficient of variation, COV, for the exponential distribution is always 1). However, we must also heed the warning from chapter 4 that we cannot use a traffic intensity that is too high. Therefore, we have chosen a traffic intensity of 0.8 as our base case for analysis. (Note that the traffic intensity for a multistep process is defined as the arrival rate divided by the processing or service rate of the *slowest* step in the process). Once we have developed an understanding of process behavior when the random variation has exponential distributions, we also will want to be able to vary the mean and standard deviation of the arrival and processing times independently. This will force us to use a distribution that is not exponential, a change that significantly impacts the complexity of the behavior of the process.

Note than many scenarios are possible with the models we present, but it is not our intention to cover all of them here. We will look at some general features of the behavior of the process and leave it up to readers to run other scenarios that interest them. Readers are again reminded that when they define a scenario, they should try to predict what behavior they expect; run the model to see what actually occurs; and then reconcile any differences, should they occur. Remember to run the models many times for each scenario to get an overall understanding of the behavior of the model.

5.4 An uncoupled process: An order handling process

We start by looking at an order handling process. The items moving through the process are customer orders, but we will still refer to them as items to reflect the generic nature of these processes. We will assume that we can hold an infinite number of items between steps. An important question is how many steps do we need to include in the process? In a multistep process, it is possible to have entrance and exit effects. That is, the behavior at the first and last steps may be different than in the other steps. If we use too few steps, we may see effects that are purely due to these entrance and exit effects (the model behavior is totally dominated by these effects). This consideration drives us to have many steps in the process. By contrast, we do not want to add unnecessary complexity. This consideration forces us to use as few steps as possible. We have chosen to use a process with five steps. Of course, once we understand this process, extension to modeling larger processes is not difficult.

The model for this process is shown in figure 5.2. The structure shown in figure 5.2 is the important structure from a workflow perspective. The step WIP value is the sum of the STEP conveyor and the QUEUE for that step. Note that to reduce

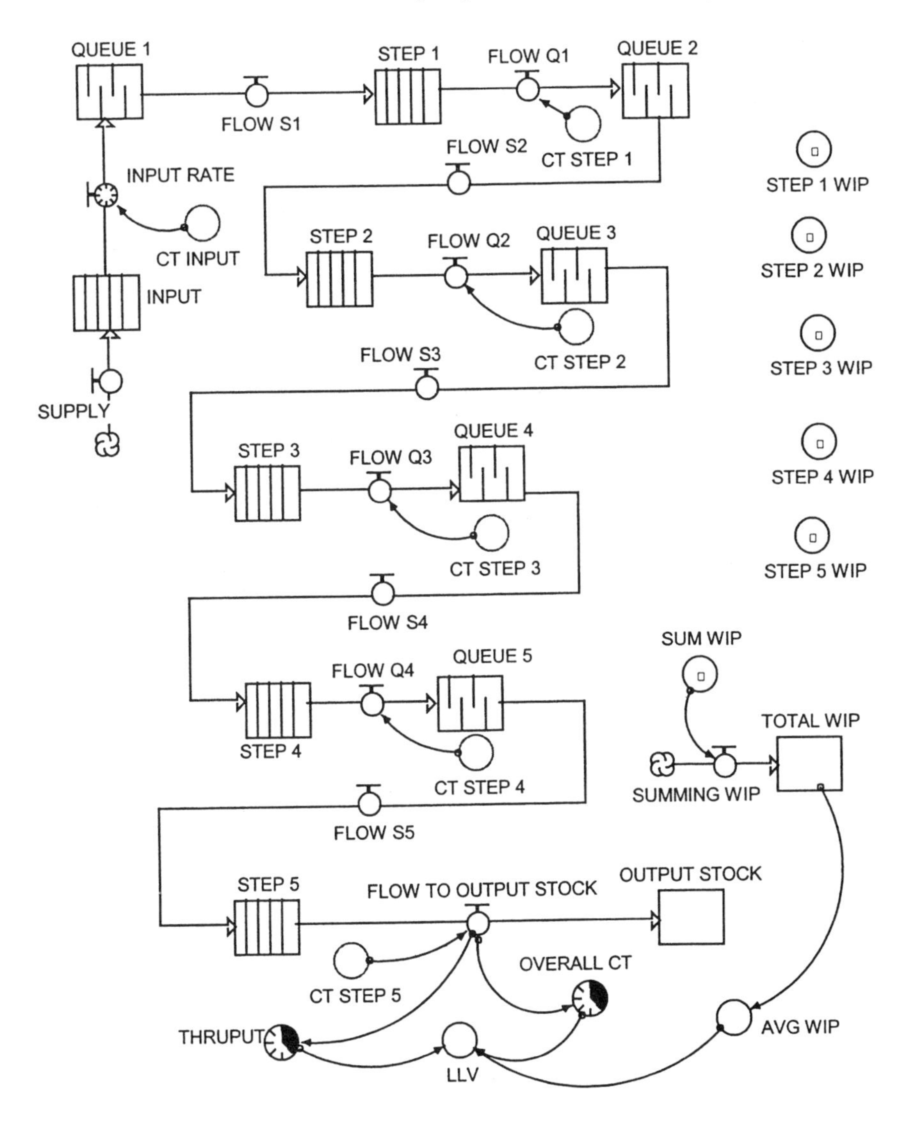

FIGURE 5.2. Five-step product development process

the complexity of the diagram in figure 5.2, we have used the summer functionality in an ithink converter. With this functionality, we can access inputs to the converter without having to draw connections explicitly. For example, here is the equation in the converter STEP 1 WIP:

$$\text{STEP 1} + \text{QUEUE 1}$$

This equation will calculate correctly even though there are no connectors from STEP 1 and QUEUE 1 to STEP 1 WIP. LLV is the Little's Law variable that we described in chapter 4 as a way to check if Little's Law was obeyed in the model. The structure shown does not illustrate all the calculations that must be included to evaluate the performance variables, such as average queue size and average step utilization. However, such averaging structures are easy to implement and can be seen in detail by opening the model in ithink. Note that because the individual steps can process only one order (item) at a time, the average WIP of the step conveyor equals the utilization of the step. Also, the input rate and step cycle time calculations (so that random variation can be modeled) are not shown. Again, inputting an exponential distribution is straightforward (using the EXPRND function in ithink). Switches have been provided in the model so that random variation in the item interarrival and step processing times can be turned on and off individually.

The basic configuration for the model is shown in Table 5.1. All averaging of queue sizes and utilizations are set to 0 at time 200 hours. This is to give the process time to "warm up" and reach a steady state. Therefore any profiles of these variables will be 0 prior to time 200. The basic configuration reflects a balanced process (because all step rates are the same) with no random variation. Under these basic conditions, readers can run the model and verify the following:

- The model runs at a steady state.
- All average queue sizes are 0.
- All utilizations are 1.
- The thuput rate is 1.0 items/hour.
- LLV has a value of 1.
- The process is uncoupled—there is no feedback from any step to a previous (up-stream) step.

These are obvious and expected outcomes and simply help us verify that the model is working. Readers also can run this model in an unbalanced state by changing some of the step cycle times while keeping the input rate cycle time at 1 hour, and can then verify the following:

- If any step cycle time is greater than 1 hour, the model will not run in a steady state.
- Queues will tend to build up before any step whose cycle time is greater than the input rate cycle time unless that step is after the bottleneck step. For example, suppose the step cycle times are 0.5, 1.2, 1.7, 1.5, and 1.3 hours. A queue will build up before steps 2 and 3 (step 3 is the bottleneck). However, even

TABLE 5.1. Basic configuration data for model.

Time range	2,000 hours
DT	0.0625 hours
Mean step/input rate cycle times	1.0 hours
Stdev step/input rate cycle times	0.0 hours
All step and queue initial values	0

though steps 4 and 5 process items more slowly than the input rate, items will only reach these steps at the bottleneck rate of 1 every 1.7 hours. There will be no build up at steps 4 and 5.

- For any given set of step cycle times, we can determine the bottleneck step by setting the input interarrival time to an extremely low-value (high-input rate) and watching where the queues build up. The most downstream step where a queue builds up is the bottleneck step. Using queue sizes to identify a bottleneck step is a useful approach when managing a process. We will see later that we must be careful with this approach when random variation is involved.
- The utilization of a step is simply the bottleneck step cycle time divided by the step cycle time. By definition, the bottleneck step will have the highest utilization. It will be 100 percent if the item input interarrival time is at or below the bottleneck step cycle time.
- There is no critical cycle time step. Reducing the cycle time at any step by the same amount will have the same impact on the overall cycle time.

Again, these results are fairly obvious and help us to verify the model. The more interesting scenario occurs when we add random variation to the model. This will allow us to model the real world, where random variation is always present. Once we add random variation to the model, we truly have a set of connected single step queuing models. We know that the behavior at any step is driven by the arrival time distribution and the processing time distribution (refer back to equation 4.6 in chapter 4, which also shows that the impact of the distributions are represented by the coefficient of variation, COV). Because the processing time distribution is an input to the model, any complexity in the behavior of the process must be due to the arrival time distribution at a step and how this distribution changes as items flow through the process. The arrival time distribution to the first step is an input to the model; arrival times at the remaining steps create the behavior of this process. (Of course, it stands to reason that the arrival time distribution at steps 2, 3, 4, and 5 will depend on the processing times and the arrival times to step 1). So we can break down the overall process performance into the components shown in table 5.2.

We should be able to analyze any results from the perspective of these components. We know that we cannot run the model at a traffic intensity of 1; otherwise no steady state can be reached. Because the queue holds an unlimited number of items, a step cannot cause feedback to any upstream step. Therefore, we expect

TABLE 5.2. Components for overall process performance.

1	The overall performance is the sum of the performance at the individual processing steps.
2	The performance at each step is determined by the arrival time distribution and the processing time distribution at a step.
3	The arrival time variation (as measured by the coefficient of variation, COV) at a step is determined by the arrival time distributions and processing time distributions of steps upstream.
4	The mean arrival time at a step is determined by the mean arrival and processing times of steps upstream.

that the uncoupled structure itself will not cause any drop in thruput across the multistep process. This implies that the mean item interarrival time should be the same at all steps and equal the interarrival time at the first step. This also implies that the main performance measure of interest is the total WIP of the process. Of course, by Little's Law, the overall cycle time is directly related to the overall WIP because the thruput is fixed by the arrival rate of items to the process. Consequently, we do not have to focus on these performance measures separately.

Before we add random variation to all five steps, we will add it to just one step and see what happens. Table 5.3 shows the results from the model when random variation is added to each step individually. Following are the five cases:

Case 1: Mean interarrival time 1.25 hours, mean step processing times 1 hour, exponential random variation in step 1 only.
Case 2: Mean interarrival time 1.25 hours, mean step processing times 1 hour, exponential random variation in step 2 only.
Case 3: Mean interarrival time 1.25 hours, mean step processing times 1 hour, exponential random variation in step 3 only.
Case 4: Mean interarrival time 1.25 hours, mean step processing times 1 hour, exponential random variation in step 4 only.
Case 5: Mean interarrival time 1.25 hours, mean step processing times 1 hour, exponential random variation in step 5 only.

The results in table 5.3 are average results over 1,000 runs of the model (this is true for all results presented in this chapter). We can make a number of important observations from these results:

- Random variation at a step impacts the queue just prior to the step and the queue just after the step. For example, case 3 has a nonzero average queue for steps 3 and 4 but not at steps 2 and 5. This confirms our intuition about how random variation at a step is not propagated backward in the process. But it can be propagated forward downstream in the process.
- The average queue size at step 1 for case 1 is 3.15, close to the theoretical value of 3.2 for a single queue system as given by equation 4.4. The average queue size for queue 2 is only half this value because there is no random variation in

TABLE 5.3. Results with random variation added to individual steps.

	Case 1	Case 2	Case 3	Case 4	Case 5
Overall cycle time (hours)	10.91	11.84	11.77	11.73	9.92
Overall WIP	8.73	9.48	9.42	9.40	7.94
Overall thruput (orders/hour)	0.80	0.79	0.79	0.79	0.80
LLV	0.99	0.99	0.99	0.99	0.99
Average queue 1	3.15	1.60	1.59	1.58	1.59
Average queue 2	1.59	2.44	0.00	0.00	0.00
Average queue 3	0.00	1.14	2.42	0.00	0.00
Average queue 4	0.00	0.00	1.41	2.39	0.00
Average queue 5	0.00	0.00	0.00	1.43	2.35

step 2. The impact on queue 2 comes only from the arrival times from step 1. Thus, queue 2 behaves as if it sees an exponential arrival time distribution from step 1 with a constant processing time for step 2.

- For cases 2, 3, 4, and 5, the average size of the queue at the step where the random variation has been introduced is only about 2.4 items. This is less than the theoretical value of 3.2 if both the arrival and processing times are exponential. Because we know the processing times are exponential, the only explanation is that the steps do not see an exponential arrival time for the items. This also explains why the impact on the queue after the step is only 1.4 items instead of the 1.6 items we would expect if the arrival times were exponential. We would expect the arrival distributions to be nonexponential because the processing times at the upstream steps are constant, but we do not think the value of 2.4 could be arrived at by simple intuition alone.
- We have entrance and exit effects at steps 1 and 5. Step 1 is different because we have set the arrival times to be exponential. Step 5 is different because there is no queue downstream of it in the model.

Next we can look at what happens if we add random variation to all the steps and the interarrival times. In order to make the impact of the random variation clearer, we set the mean interarrival time at 1.25 hours with exponential random variation, and we add the random variation one step at a time. The results are shown in table 5.4. Following are the cases:

Case 6: All step mean processing times 1 hour, random variation at step 1 only.
Case 7: All step mean processing times 1 hour, random variation at steps 1 and 2 only.
Case 8: All step mean processing times 1 hour, random variation at steps 1, 2, and 3 only.
Case 9: All step mean processing times 1 hour, random variation at steps 1, 2, 3, and 4 only.
Case 10: All step mean processing times 1 hour, random variation at all steps.

The results in table 5.4 again show that the uncoupled structure does not impact the thruput of the process.

TABLE 5.4. Addition of random variation to the multistep process.

	Case 6	Case 7	Case 8	Case 9	Case 10
Overall cycle time (hours)	10.91	14.97	18.91	22.87	24.83
Overall WIP	8.73	12.00	15.17	18.33	19.92
Overall thruput (orders/hour)	0.80	0.79	0.79	0.78	0.78
LLV	0.99	0.99	0.98	0.98	0.98
Average queue 1	3.15	3.22	3.18	3.19	3.20
Average queue 2	1.59	3.19	3.17	3.22	3.20
Average queue 3	0.00	1.59	3.23	3.21	3.21
Average queue 4	0.00	0.00	1.60	3.13	3.16
Average queue 5	0.00	0.00	0.00	1.59	3.15

> Learning Point: The uncoupled structure in a serial multistep process does not impact the thruput of the process.

This is the motivation for referring to this process as uncoupled: this structure does not allow feedback to flow upstream in the process. Consequently, the structure does not impact thruput. Note also that the total WIP for case 5 is 20 whereas the actual WIP for the five processing steps is only 5*0.8 = 4 items. This reflects the situation in most real processes, where most of the time items are waiting to be processed rather than being processed. Indeed, one can define the efficiency of a process based on the ratio of processing time (or WIP) to total cycle time (or WIP). In case 5, this ratio is 4/20 = 20%. So on average, the process will only be working on an item 20 percent of the total time the item spends in the process. Actual numbers less than 20 percent are not uncommon in real processes.

It is clear from table 5.4 that the process behaves as a set of independent queues. Adding the random variation to another step adds 3.2 items of WIP to the total. Thus, a linear relationship exists between the number of steps and the average WIP of the process. (Step 5 is the exception, where the additional WIP is only 1.6 items because no downstream queue for step 5 is included in the model). Table 5.4 implies that the exponential random distribution is passed on down the process without change. The arrival times at each step have an exponential distribution with a mean of 1.25 hours (the interarrival time at the first step). This is a result of the memory-less property of the exponential distribution. One might have expected that the queue size would build up from step 1 to step 5 based on the notion that random variation "builds up" as we move down the process. However, this is not the case for exponential distributions.

Figure 5.3 shows the average queue sizes over a typical run for case 10, where random variation occurs in all steps. This is a perfectly balanced process, both in the mean and the variation of the processing times.

One of the interesting observations we can make about this figure is the way the average queue sizes do not always have the same relative position with respect to size. For example, prior to time 500 hours, the queue 2 average size is smaller than the queue 5 average size; later the reverse is true. This means that we must be careful about using queue sizes to determine where the bottleneck step is. Is step 2 or step 5 the bottleneck? We know that there is no bottleneck step (the process is balanced). Therefore, using queue size to determine the location of the bottleneck step can be misleading. In fact, the position of the queue with the maximum queue size jumps around all the time. If we do not realize that this is a result of the random variation inherent in the system, we may interpret these changes as an indication of special situations. Then we ask employees to find the root causes of these changes when, in reality, nothing has changed. We simply make people chase "ghosts" and waste time and resources unnecessarily.

Given the balanced process as a starting point, we would like to know where the best place to improve the process is so that we get the best overall improvement in the process. Should we focus on the initial portion of the process or the

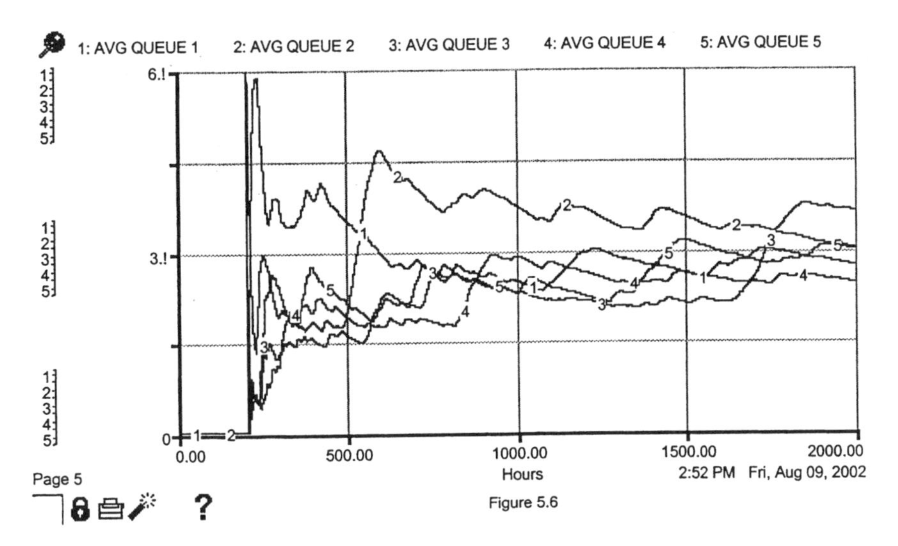

FIGURE 5.3. Average queue sizes for balanced process with random variation in all steps

later stages? In other words, if we were to unbalance our process by reducing the mean processing time at a step, should we lower the mean processing time at a step near the beginning of the process or toward the end of the process? We leave it as an exercise for readers to run the model with various combinations of the mean processing times for the five steps. The result is that it does not make any difference. This result agrees with that of Weber (1979), who showed that when processing times have exponential distributions, the overall cycle time in the process is independent of the order of the steps. Another way of saying this is that the process does not have a cycle time critical step.

> Learning Point: In a serial multistep, uncoupled process with exponential random variation in arrival and processing times, the impact of any step's processing time random variation on the average WIP or cycle time is independent of position of the step in the process.

If we refer back to table 5.2, we can see that for the serial multistep, uncoupled process, the exponential distribution simplifies the analysis for components 1, 2 and 3. (Component 4 is independent of the types of distributions that are used to represent the random variation because the mean item arrival times must be the same as the overall item arrival time for steady state behavior). Component 3 is simple because the arrival time distribution is exponential for all step arrival times. If the arrival times to a step are exponential and the processing time is exponential, then the departure times are also exponential, irrespective of the traffic intensity at the step. Component 2 is simple because the equation for performance, equation 4.5, is exact for exponential distributions. Component 1 is simple because the average behavior at each queue is independent from the other queues: the overall average performance is simply the sum of the individual average queue sizes (plus process step average sizes). This does not mean that dynamic models are of no use for multistep process with exponential distributions. These considerations are just for average behavior and do not help with short-term transient behavior—this is where the dynamic models come into their own.

So what about the situation where at least one source of random variation does not follow an exponential random distribution? As one might expect, the situation here is far more complex. If we refer back to table 5.2, components 1, 2, and 3 are more complicated. Component 1 is more complex because each queue is no longer independent of the other queues: we cannot simply add the average size of each queue to get the overall behavior. Any equation we develop for this aggregation will not be exact. Component 2 is more complex because any performance equations we develop, such as equation 4.6, are not exact. Finally, component 3 is more complex because the interarrival time distribution changes down the process. Any equation we develop for this will not be exact. Thus, we can see that in the general case where at least one of the random distributions is not exponential, three of the four components will involve approximations if we try to describe the process using an approach based on closed form equations.

Of course, the dynamic modeling approach that we are using does not have to use any approximate equations in the model. The only approximation our models will encounter arises from the uncertainty in estimating performance when we can run our model only for a finite amount of time (see chapter 4).

One of the issues with processes for general distributions is that many possibilities can arise. If we fix the mean interarrival time for items to the process as a yardstick with which to measure other input parameters for the model, we still have 11 other variables that must be specified (COV for the item interarrival times, mean processing times for the five steps, and the COVs for the processing times for the five steps). It would take a lot more than this book to cover all possibilities. Instead, we look for generalizations to make about performance. Such generalizations tend to come in the form of heuristics; that is, statements that are true in many or most cases but definitely not true in all cases. Such heuristics are useful when we do not have access to modeling capabilities such as ithink. They are relatively easy to apply and are attractive when a quick answer is required. But in today's business organizations, developing a dynamic model for a process may be a small price to pay to ensure that a heuristic is truly applicable to a real process. One of the simplest heuristics that appears to work well in general is the Variability Propagation Principle, as described by Suresh and Whitt (1990a) and referenced by Hopp and Spearman (1996).

The Variability Propagation Principle (VPP): Increased variability in the arrival process or the service times of a queue tends to propagate to the departure process from that queue and thus to the arrival process of all subsequent queues.

The VPP leads to the simple design heuristic that in a serial queuing system, such as our multistep, un-coupled process, the minimal average WIP is achieved by having the smallest processing time COV at step 1, the next smallest at step 2, and so on, with the largest processing time COV at the last step. Therefore, we should work on reducing variability at the front of a process rather than at the end of the process. Hopp and Spearman (1996) present this as one of their laws of Factory Physics. Suresh and Whitt (1990a) are quick to point out, however, that this heuristic can easily be wrong. Another heuristic that seems appropriate is that we should reduce variability at a bottleneck over reducing variability at a nonbottleneck. This is simply because the impact of variability at a step is greater the higher the traffic intensity at the step. But in the end, we contend that a properly configured dynamic model is the best way to be sure that we are applying these (and any other) heuristics correctly.

Although we cannot cover all possibilities for the general multistep process, we would like to cover one example that illustrates the complexities than can arise. To this end, we have chosen to illustrate the "heavy traffic bottleneck phenomenon" as described by Suresh and Whitt (1990b). We will use the example from this reference to describe and analyze this phenomenon. Suppose we have a process with the following parameters.

- Item interarrival time mean 1 hour, exponential random variation.
- Steps 1 to 4, mean processing time 0.6 hours, exponential random variation.
- Step 5, mean processing time 0.9 hours, exponential random variation.

We know that this process can be modeled as a set of independent queues and that the exponential random variation is passed down from the item arrivals at step 1 to item arrivals at step 5. From equation 4.5, the average WIP at steps 1 to 4 will be 0.6/(1.0–0.6) = 1.5 items, and given that there will be 0.6 items on average in the processing steps, the queues before steps 1 to 4 should have an average size of 1.5–0.6 = 0.9 items. Similarly, step 5 should have an average WIP of 0.9/(1.0–0.9) = 9.0 items. The average size of the queue before step 5 will be 9.0–0.9 = 8.1 items.

Now suppose that the item arrival time distribution to the process is not exponential (it has some COV different than 1). What should we expect to happen in this case? As Suresh and Whitt (1990b) point out, we should expect that the arrival distribution to each successive queue would become more like an exponential distribution. The arrival distribution at step 1 should have less impact on the arrival time at a downstream step (and the processing time distribution should have more effect) the further down we go in the process. Thus we would expect the last step to behave closely to a M/M/1[1] queue. Consequently, we would expect an average queue size of about 8.1 items as occurred earlier. If the arrival time distribution to step 1 has a COV > 1, then the COV at the final step in the process should be > 1 and we would expect the average queue size at step 5 to be a little greater than 8.1. The reverse is true if the COV of the arrival times to step 1 is less than 1. However, what happens is different. The arrival time distribution

1. Refer back to chapter 4 for an explanation of this terminology.

appears to propagate from the input side of step 1 to the input side of step 5. It is as if the steps in the center do not exist. This appears to be a general phenomenon and is referred to as the "heavy-traffic bottleneck phenomenon."

> Learning Point: If we have a serial multistep, uncoupled process where the traffic intensity at one step is close to 1 and the traffic intensities at the other steps are significantly smaller, the queue at the bottleneck step behaves as if the service times at the other steps were all set to 0. This is referred to as the "heavy-traffic bottleneck phenomenon."

We can model this phenomenon using "MODEL 5.2.STM". If we turn off random variation in the arrival times so that they are deterministic (constant), keeping the same mean arrival time of 1 hour, then in effect the arrival times to the process are nonexponential. In fact, the COV is 0.0, the minimum it can be. Table 5.5 shows the results for the two cases:

Case1: Exponential arrival times.
Case 2: Deterministic arrival times.

The "heavy-traffic bottleneck phenomenon" is illustrated by this data. We should expect queue 5 to have an average size of a little less than 8.1. In fact, it is nearly half the expected value. This means that the arrival time distribution cannot be close to an exponential distribution. In fact, it must be closer to a distribution with an effective COV close to 0. Thus the arrival time distribution at step 1 has propagated to step 5 as if the intervening steps were not there. This is the "heavy-traffic bottleneck phenomenon. So, whereas a low-traffic intensity can absorb the variability with respect to its impact on the average queue size, a low-traffic intensity does not prevent the impact of the variability being propagated down the process. This is worth recording as a learning point.

> Learning Point: A low-traffic intensity at a step can absorb the variability at that step with respect to the impact of the variability on the average queue size. However, the low-traffic intensity does not prevent the impact of the variability being propagated down the process.

If we had used an arrival time distribution to step 1 which had a COV greater than 1, the average size of queue 5 would have been much greater than the M/M/1 queue value of 8.1. In fact, Suresh and Whitt (1990b) quote a result where using a distribution with a COV = $\sqrt{8}$ gave an average size for queue 5, which is 3 times the expected value for an M/M/1 queue. This example should make the value of dynamic modeling clear!

Calculation methods are available that try to predict the "heavy-traffic bottleneck phenomenon" as well as other aspects of multistep process performance. These methods try to avoid the need for dynamic models by replacing them with

TABLE 5.5. Heavy-traffic bottleneck phenomenon.

	Case 1	Case 2
Overall cycle time (hours)	14.4	10.1
Overall WIP	14.4	10.1
Overall thruput (orders/hour)	0.99	0.99
LLV	0.99	0.99
Average queue 1	0.9	0.3
Average queue 2	0.9	0.5
Average queue 3	0.9	0.6
Average queue 4	0.9	0.7
Average queue 5	7.6	4.7

calculation methods based on analytical equations. A key aspect of these methods is the equations they use to model the propagation of the item arrival time distributions down the process. These equations are referred to as traffic variability equations and are used to represent component 3 in table 5.2. But as is evident from Suresh and Whitt 1990b and Whitt 1994, traffic variability equations that work in all cases are difficult to develop and the amount of theory required is enormous. And even when they are developed, we must still test them by comparing them to the results of a dynamic model. When the equations lead to different results than those obtained from simulations of the dynamic models, it is the equations that are revised, not the dynamic models. For example, the traffic variability equations used by Hopp and Spearman (1996) can lead to significant errors when applied to processes exhibiting the "heavy-traffic bottleneck phenomenon" (Suresh and Whitt 1990b). It seems that unless we are prepared to spend a lot of energy learning the details of queuing theory, investing our (limited) energy in becoming proficient in developing and analyzing dynamic models of our real processes is advisable.

On a final note, because MODEL 5.2.STM is configured for exponential distributions, we are providing models with other distributions used to model the random variation. Readers can run MODEL 5.2A.STM to see the results for normal distributions and MODEL 5.2B.STM for uniform distributions.

5.5 A tightly coupled process: A fast food restaurant process

In some fast food restaurants, customers enter a single queue and wait to be served in a sequential fashion. The first step is where the order and money are taken, the second is for getting the drinks, and so on. The items in this process are customers. There are no separate queues between each of the process steps. In this situation, the process behaves differently from the uncoupled process we looked at in the last section, because, for the fast food process, customers moving through the process can experience the phenomenon of blocking. Suppose a customer has completed a step and is ready to move to the next step. In the uncoupled process, the customer would simple move into the next queue and wait until the next step is available. However, in the tightly coupled process, there is no queue and the customer must wait until the next step is empty and available. If the next step is occupied and unavailable, the customer must wait in the current step. The customer is blocked from exiting the current step, hence the name blocking. Thus the total time spent in a process step will consist of two parts. The first part is the time that the item was actually being worked on as given by the process step cycle time. The second part is the time the item was waiting until it could move to the next step. In ithink, we still need to have a stock with the conveyor to allow blocking to be modeled, even though this stock is virtual. The basic construct for a step is shown in figure 5.4.

In order for an item to move to the next step, we must check to see if the conveyor and the stock are both empty. We also add in logic that says we must detect an item leaving the step before we add another. Thus, the decision to send on an item also involves the outflow from the step, shown as the outflow TO_STEP_2 in figure 5.4. The equation for the inflow INPUT is this:

IF (STEP_1+STOCK_1) < 1 OR TO_STEP_2 > 0 THEN 1/DT ELSE 0

Note that we have replaced the queues with stocks. The reason for this is that we cannot exert any flow control on the output of a queue. The only place a queue will appear in the model is before the first step. This is to provide a place to hold items (customers) should the arrival rate of items be greater that the capacity of the process, both in the short and long term.

Because we have no step queues in this model, the queue size is no longer something we can focus on. In the case of the tightly coupled process, we instead focus on the utilization of the individual process steps. Because we are only modeling steps that process one item at a time, the step utilization is simply the average of the (step+stock) values. Readers are invited to run MODEL 5.4.STM for various combinations of input rates and step cycle times without any random variation and check to see that the results obtained make sense. (As in MODEL 5.2.STM, switches have been provided so that random variation can be turned off easily.) For example, suppose the model is run with the input interarrival time and

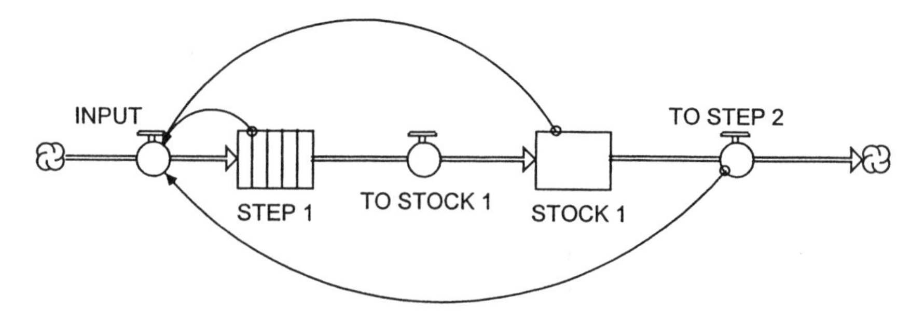

FIGURE 5.4. Model fragment for tightly coupled serial processes

step cycle times all set at 1 hour except for step 2, which has a cycle time of 1.25 hours. Then the overall thruput will be 0.8 items/hours; the utilizations of step 1 and 2 will be 1.0; and steps 3, 4, and 5 will have utilizations of 0.8. (Note that these parameters do not have realistic values for a typical fast food restaurant. However, we use these values to facilitate comparison with the uncoupled process studied in the last section.)

As in the case of the uncoupled process, the process behavior becomes interesting only when we add random variation to the process. The analysis of systems that can exhibit blocking is more difficult than analysis for uncoupled systems (Hoseyni-Nasab and Andradottir 1998; Tsiotras and Badr 1989). For example, even if all the random variation has an exponential distribution, the item arrival time distribution at steps other than step 1 will not be exponential (as it was for the uncoupled case). Thus, even in the case of exponential random variation, we cannot analyze each step independently and then add the results to get an overall impact. However, the simulation of tightly coupled processes is no more difficult than simulation for uncoupled processes. As we did in the previous section, we will first look at the impact of adding random variation (with exponential distributions) in one step only. Table 5.6 shows the results when the item interarrival time average is 1.25 hours and all step average cycle times are 1 hour (traffic intensity 0.8). (The cases are the same as shown for the uncoupled process in table 5.3).

Comparing these results to the equivalent uncoupled case in table 5.3, a number of important differences become apparent.

- The thruput should be 0.8 items/hour but it is only 0.73 items/hour. This means that when the process is tightly coupled and we have random variation, we lose thruput and hence capacity. Because the theoretical capacity of the process should be 1, this process can run at only 73 percent of this theoretical capacity. Considering that the 27 percent loss was caused by random variation in one step only, this is a significant loss. This also implies that the important performance measure for a tightly coupled process is the capacity/thruput of the process. (Note that the WIP in the process is limited by the structure of the process.)

TABLE 5.6. Results for tightly coupled process with random variation added to individual steps.

	Case 1	Case 2	Case 3	Case 4	Case 5
Process CT	5.37	5.73	6.10	6.47	6.27
Total WIP	3.92	4.19	4.46	4.73	4.73
Thruput	0.73	0.73	0.73	0.73	0.73
LLV	0.99	0.99	0.99	0.99	0.96
S1 util	1.00	1.00	1.00	1.00	1.00
S2 util	0.73	1.00	1.00	1.00	1.00
S3 util	0.73	0.73	1.00	1.00	1.00
S4 util	0.73	0.73	0.73	1.00	1.00
S5 util	0.73	0.73	0.73	0.73	0.73
Queue	81.93	81.11	80.41	81.69	81.54

> Learning Point: Random variation in a tightly coupled process reduces the capacity of the process.

- The impact on thruput is the same no matter where the random variation is added. When random variation is added at a step, it impacts all the steps upstream of the step. This is the motivation for describing this process as tightly coupled. In effect, there is a feedback signal from a step to any of the steps upstream.
- The queue average is just over 80 units! This is because the item arrival rate is greater than the process capacity. The model is not being run at steady state. So why is the average LLV so close to 1? The reason is that, anticipating the reduction in capacity, we excluded the queue in the calculation of overall WIP and cycle time; therefore, it also was excluded in the calculation of LLV.
- If we look at any case, we see that the random variation at a step creates two regions. The utilization of the step and all steps upstream of it are 1. All steps downstream have a utilization of 0.73, the thruput rate. The downstream steps are empty and waiting for items about 27 percent of the time. This latter phenomenon is referred to as starving. The upstream steps are blocked (<27 percent of the time). Thus, the random variation at a single step breaks the process into a region of blocked and starved process steps. This means that random variation at a step causes an impact upstream, a phenomenon that we did not see for uncoupled processes in table 5.3.

> Learning Point: Random variation at a step in a tightly coupled process causes blocking upstream and starving downstream.

- Although not shown in table 5.6, random variation in the interarrival times only has no impact on the thruput. Readers can verify that the thruput is about 0.8 and that all the step utilizations are 0.8. This result appears to be related to the queue at the start of the process. This queue absorbs the random variation in the interarrival times. Because all of the steps have the same utilization (0.8) there is no blocking of any of the steps. In effect, the process is behaving like an uncoupled process when the only source of random variation is in the arrival times and when a queue occurs prior to the first step. Of course, if we were to limit the size of this queue, we would see a reduction in capacity.
- The fact that random variation reduces the capacity of a tightly coupled process presents a practical issue when trying to model such processes. We do not know exactly what capacity to expect in any given situation. Consequently, if we get it wrong we will either run the process too quickly and build up the queue, or we will run the process too slowly and not know that we could have run the process faster. The solution is to always run the process fast and let the process establish its thruput, which will then tell us the process capacity. As long as the queue is building up, we know that we are running the process

faster than its capacity. Once we know what the capacity is, we can then modify the item interarrival time to match this capacity. If we ignore the queue in our calculations of cycle time and WIP, then we can ignore it from a modeling perspective. As long as the queue is building up, we can look at the model results as representing what will happen if we run the process at its capacity. Of course, from a manager's perspective, this would not make sense: When running a real process, as we cannot let a real queue build up just to ensure that we are running the process at capacity. Even when we provide enough capacity, the owner of the fast food restaurant must consider the short-term behavior of the process—if care is not taken, the queue might build up to the door! (Of course, probably before this happens, customers will see the long queue and go somewhere else. After all, fast food should mean fast food!) So the system may be self-regulating with the overall impact being a loss of business. Models of the type in figure 5.4 can help managers decide how best to provide capacity to balance excess capacity with loss of customers. This, again, involves some type of economic optimization.

In many situations, we can implement control mechanism to manage the input rate so as to keep the process running at capacity or at some other required thruput. We will return to the issue of material flow control systems in section 5.7.

Table 5.7 shows what happens when we add random variation sequentially to all steps in our uncoupled five-step process. The cases are similar to those in table 5.4.

The results show that by the time we have added random variation to all steps, the thruput has been reduced to just under 0.5 items/hour—a reduction of just over 50 percent! What is interesting about the results is that the impact of adding random variation in a step is not linear. For example, adding random variation to the first step gives a capacity of 0.73 items/hour, a reduction of 27 percent. Adding random variation in a second step gives a capacity of 0.62 items/hour, a further reduction of 11 percent. Continuing, adding random variation in step 3 causes a further reduction of 7 percent. When we add random variation in step 2, the reduction is only 2 percent! This has a significant implication for variability reduction efforts in a tightly coupled process. It says that even if you focus on the

TABLE 5.7. Tightly coupled process with random variation added sequentially to all steps.

	Case 6	Case 7	Case 8	Case 9	Case 10
Process CT	5.37	5.86	6.43	7.11	7.33
Total WIP	3.92	3.63	3.60	3.72	3.78
ThruPut	0.73	0.62	0.55	0.51	0.49
LLV	0.99	1.00	0.98	0.98	0.95
S1 util	1.00	1.00	1.00	1.00	1.00
S2 util	0.73	0.79	0.82	0.85	0.87
S3 util	0.73	0.62	0.68	0.74	0.77
S4 util	0.73	0.62	0.55	0.62	0.66
S5 util	0.73	0.62	0.55	0.51	0.49
Queue	81.93	203.68	276.55	319.63	345.30

biggest area of variability and make a significant reduction in this variability, it may not lead to a significant improvement in performance. We can also show that if we model a tightly coupled balanced process with n steps, the thruput is approximately $1/(T\sqrt{n})$, where T is the processing time of a step. (We leave it as an exercise for readers to modify MODEL 5.4.STM so that these results can be verified.) That is, the thruput scales as approximately $1/(\sqrt{n})$. This has implications when we discuss decoupling a tightly coupled balanced process later in this section.

The other important aspect of the results in table 5.7 is the trend in the step utilizations.[2] Consider case 10, in which there is random variation in all steps. Because the steps are tightly coupled, the impact of the random variation changes as we move down the process from step 1 to step 5. Because the thruput is 0.49 items/hour, step 5 is the last step in the process and there is no downstream step to block step 5, the utilization of step 5 must be the thruput of the process. Step 5 is starved 51 percent of the time and is never blocked. Contrast this to the situation at step 1. Here the utilization is 1. This step is rarely starved due to the queue upstream of the step. However, because the thruput is less than 1 item/hour, we know that step 1 must be blocked 51 percent of the time. This means that the average time units spend in step 1 is greater than 1 hour. In fact, we know that units in step 1 must wait on average 1 hour (approximately) after normal step processing has been completed. Steps between 1 and 5 will be starved for part of the time and blocked for part of the time. These results show that in general we cannot use step utilizations to indicate where the thruput bottleneck is. This process is balanced and has no thruput bottleneck step, but it does have one step (step 1) that is more utilized than the other steps. Clearly, high variation is to be avoided in tightly coupled processes. In many situations inventory is cheaper than capacity, so buffering a tightly coupled process is highly desirable. Of course, when we add a buffer, we are moving to the uncoupled configuration studied in the previous section. In this case, the cycle time in the process will increase. This is sometimes undesirable because the quality of some items can degrade with time.

Learning Point: In a tightly coupled serial process, utilization may not be good indicator of the bottleneck step.

Learning Point: Processes with high degrees of random variation (high COVs) should always be avoided, but this is especially true in tightly coupled processes.

So if we were to decouple the process, should we add the buffer at the beginning or at the end? That is, can we predict the optimum location? We assume that

2. Again, remember that these results represent average behavior.

the buffer has infinite capacity. The important point to realize is that the addition of a buffer breaks the process into two separate tightly coupled subprocesses. Suppose we add the buffer after step 1. In effect, we then have two tightly coupled subprocesses in series, one with a single step and the other with four steps. We know that the thruput will thus be the lower of the thruputs of these two subprocesses. The overall process thruput will be that of a four-step process. We get the same result is we add the buffer before step 5, with the first subprocess having four steps and the second subprocess having one step. Because the thruput scales approximately as $1/(\sqrt{n})$ and the difference in thruput between the five-step and the four-step tightly coupled process will be small, we know that the thruput gain will be small. If we add the buffer after step 2, we now have a subprocess with two steps and one with three steps. The overall thruput will now be that of a three-step process, and we know that adding the buffer after step 2 will be better then adding the buffer after step 1. Continuing this logic to a general n-step process, we would say that the best place to add the buffer is in the middle of the process. It is left as a challenge to readers to modify MODEL 5.4.STM to add buffers after each step and to see the impact. The results are shown in table 5.8.

It is clear that the results in table 5.8 support the analysis presented earlier. The addition of the buffer is best after steps 2 and 3. Note also that the addition of the buffer after step 5 has no impact on thruput, which is what we would expect.

> Learning Point: For a tightly coupled balanced serial process (with exponential random variation), the maximum increase in thruput occurs when a buffer is added into the middle of the process.

It should be obvious from the discussions so far that tightly coupled processes are more complex than uncoupled processes. Indeed, we can say that such tightly coupled processes involve three of the four major sources of system complexity outlined in chapter 2: feedback, random variation, and delays. Uncoupled processes involve only two sources: random variation and delays.

So if we could add excess capacity or reduce variation, which step should we target for the balanced process? In other words, if we want to unbalance a tightly coupled process, where should we focus our efforts? Table 5.9 shows the impact of reducing random variation or adding excess capacity for the tightly couple pro-

TABLE 5.8. Impact of adding a buffer to a tightly coupled process.

Case	Thruput
Addition of buffer after step 1	0.52 items/hour
Addition of buffer after step 2	0.57 items/hour
Addition of buffer after step 3	0.56 items/hour
Addition of buffer after step 4	0.52 items/hour
Addition of buffer after step 5	0.49 items/hour

TABLE 5.9. Tightly coupled process: Results of process improvement.

	Case 10	Case 11	Case 12	Case 13	Case 14
Process CT	7.33	7.43	6.92	7.39	6.40
Total WIP	3.78	3.99	3.77	4.25	3.82
Thruput	0.49	0.51	0.52	0.55	0.56
LLV	0.95	0.94	0.95	0.95	0.95
S1 util	1.00	1.00	1.00	1.00	1.00
S2 util	0.87	0.96	0.85	1.00	0.82
S3 util	0.77	0.83	0.74	0.93	0.69
S4 util	0.66	0.70	0.65	0.77	0.73
S5 util	0.49	0.51	0.52	0.55	0.57
Queue	345.30	320.27	310.23	274.93	255.29

cess (these are equivalent when the random variation has exponential distributions). The cases shown are these:

Case 10: Random variation in all steps.
Case 11: Same as case 10, except no random variation at step 1.
Case 12: Same as case 10, except no random variation at step 4.
Case 13: Same as case 10, except no random variation at steps 1 and 2.
Case 14: Same as case 10, except no random variation at steps 3 and 4.

Comparing cases 11 and 12 indicate that there might be a slight improvement by eliminating the random variation at step 4 rather than step 1. However, the difference is small (thruputs are 0.51 and 0.52 respectively). This should be expected because we said that removing the random variation at one step has a minimal impact on capacity. That is why we tested cases 13 and 14. Here we removed random variation at steps 1 and 2 at the same time (case 13) and steps 3 and 4 at the same time (case 14) in the hope that it would show the contrast even more. Now it is somewhat more obvious that removing the random variation in the downstream steps has a greater impact than removing it at the upstream steps (thruput 0.57 versus 0.55 respectively). This observation makes sense for two related reasons: feedback is an essential element of the behavior of tightly coupled processes, and eliminating the random variation at the downstream steps should reduce the amount of feedback in the process.

> Learning Point: For a tightly coupled balanced serial process (with exponential random variation), equivalent improvements have a higher impact when applied to the end of the process than to the beginning of the process.

What about cases where at least one parameter has a distribution other than exponential? Table 5.2 cannot be applied to tightly coupled processes because the feedback from downstream to upstream steps means that we cannot isolate each step and treat it separately. Our advice is to create the model, run it, and analyze the results. Therefore, no general observations will be made about processes with nonexponential distributions for the arrival and processing times. As in the case of uncoupled processes, we are providing models with other distributions used to model the random variation. Readers can run MODEL 5.4A.STM to see the results for the normal distribution and MODEL 5.4B.STM for the uniform distribution.

If we compare the two types of processes, uncoupled and tightly coupled, we can see an interesting dynamic relationship among the three performance measures of thruput, WIP,[3] and random variation. Once we have random variation in the process, the average thruput and WIP are correlated. We cannot set a desired WIP level and thruput independently. In essence, this is an extension of Little's Law. Unfortunately, the exact relationship among the three performance measures is not simple like Little's Law. We need a model of the process to describe the relationship. But we can see that for a given amount of random variation, the smaller the WIP, the lower the thruput and vice versa. This is an important point. Many managers are asked to reduce inventory levels, and that is an easy thing to do. Simply put a cap on inventory levels. But care is needed, because if we reduce inventory levels, then it is inevitable that the thruput will be reduced. Reducing variation, unbalancing the process by adding excess capacity, or adding buffer capabilities must be considered. There is no silver bullet here. If you want to improve the performance of a process, you must change the process!

5.6 Other configurations

The uncoupled and tightly coupled structures represent two structures for how serial processes can arise in practice. In many real situations, the actual structure does not exactly match these two special cases.

For example, consider the case in which it is possible to hold inventory between process steps but there is a limit on the amount of such inventory. This might arise when the provision of storage capability is expensive and therefore limited. For example, the storage might require special environmental controls such as cold temperatures, humidity, and so forth. Sometimes it is just physically difficult to provide unlimited storage. An example here is a drive-through at a fast food restaurant. Between the window that takes the money and the window that gives the food, there will be limited space for cars, perhaps only one or two. Because this structure is somewhere between the uncoupled and tightly coupled structures, we will refer to this type of process as a loosely coupled process. It should be noted that a loosely coupled process can still exhibit the phenomenon of blocking. The amount of blocking will increase as the maximum allowed queue size decreases until the limit of the tightly coupled process is reached. So how do we model this type of process? In order to move an item from one step to

---

3. Of course, because of Little's Law, we could substitute cycle time for WIP in this relationship.

another, we must look at the quantity of items already in the following storage area. Only if there is space available will we be able to move an item into the storage. If space is not available, then the item must remain in the step, which will block movement of items from the upstream storage area. It is obvious that in our ithink model we must associate two reservoirs with each process step. The first is the stock that allows the item to remain as part of the step. The second is the queue that represents the storage area. It is left as an exercise for the reader to develop the entire model for this type of process using the uncoupled or tightly coupled models already presented as a starting point.

Another situation that can arise in practice involves processes where the items in the process may be stable only at certain stages of the process. For example, suppose items that complete step 1 will tend to degrade in quality until they are further processed through step 2. After completing step 1, it is imperative that the items move quickly through step 2 and get to the storage area after step 2, where they are again stable. This structure creates even more coupling than the tightly coupled structure because now the coupling is across multiple steps. We will refer to this process as having "multistep" coupling. In the ideal case in which step cycle times are deterministic, this situation is not difficult to handle. We simple check to see if step 2 will be free before we send the item into step 1. Because the step cycle times are deterministic, it is easy to project when step 2 will be free and to decide when to charge step 1. However, when random variation occurs in the step cycle times, such projections are impossible. If we base the projection on the mean cycle times, then roughly half of the time, step 2 will not be ready to accept the item. Consequently, degradation will occur. If we know the standard deviation of the cycle times, we could base the projection on the "mean cycle time plus k*standard deviation of the cycle times," where k is some factor that allows us to tune the performance of the process. Obviously, the larger the value of k, the less likely it is that we will degrade items. However, it also means that step utilizations will decrease on average, leading to a decrease in the average thruput. Thus we can use k to define an economic balance between thruput and quality of product. A dynamic model will allow us to define the value of k to achieve the optimum economic return. It is left as an exercise for the reader to create a model appropriate for this process again using the uncoupled or tightly coupled models already presented as a starting point.

Note that in all the cases discussed, except for the uncoupled case, the structure will lead to a loss of capacity. These structures lead to feedback from a particular process step to steps upstream of that step. This feedback reaches its extreme when we have a multistep coupled process where we cannot add an item to the process until the entire process is empty. (In this situation, the process is like a single-step process the mean cycle time of which is the sum of the step mean cycle times.) In effect, this represents the worst-case performance for the multistep process from a thruput perspective. For example, if we consider a five-step process with all mean processing times set to 1 hour, this worst-case performance will be 1/5 items/hour = 0.2 items/hour. So best-case thruput performance is 1 item/hour and worst-case performance is 0.2 items/hour. Note also that this worst-case thruput performance is also the best-case cycle time and WIP performance. In the example, the mean cycle time is 5*0.2 = 1 hour (and hence the mean WIP is 1 by Little's Law), which is the minimum mean process cycle time we can get given the individual step cycle times. This allows us to bracket best- and worst-case thruput performance for a given multistep process and to compare performance for the actual process against this range of best- and worst-case performances. From this point of view, the tightly coupled thruput of 0.49 items/hour from table 5.7 is about halfway between the best- and worst-case capacity performance. We can then refer to the tightly coupled performance as an intermediate range of performance. Using this approach, we can define the variables shown in Table 5.10. Again, remember that the mean cycle time and mean WIP following the opposite direction to the capacity/thruput. The uncoupled process has the best-case thruput performance but the worst-case cycle time performance because as we increase the amount of random variation in the process, the cycle time and WIP grow without bound. Thus, for any given level of random variation, the uncoupled process will have the largest mean cycle time and WIP.

Simply looking at the type of structure can help us to get a quick idea of what type of thruput, cycle time, and WIP performance we can expect to get from a multistep serial process.

5.7 Material control systems

In the models discussed so far, items have been released into the process at some fixed rate. We estimate the process capacity and send in items at a mean rate at or below this capacity. Once the items have entered the process, they move along according to the ability of each step to process the item. As step cycle times fluctuate, the progress of items through the process will be affected and WIP will build up at various points according to the dynamics of the process. Thus the only parameter we really control is the thruput to the process. This type of processing system is referred to as a push system and is the basis for traditional materials requirements planning (MRP). Consider the following when using a push system:

- Capacity is a difficult variable to measure, especially in the short term. Therefore, we may end up sending items into the process too quickly and build up more inventory than required, or too slowly and fail to process as many items as we could have.

TABLE 5.10. Performance rating of multistep serial process structures.

Process Structure	(Capacity) Performance
Uncoupled process	Best case
Loosely coupled	From best case to intermediate case
Tightly coupled	Intermediate case
Multistep coupled	From intermediate case to worst case

- If the bottleneck is in the middle or at the end of a process, we may send in items too fast and build up more WIP than we need (Goldratt and Cox 1992). Then if we have any manufacturing issues, more product is at risk.
- WIP is uncontrolled in a push system. But WIP is the most visible attribute of the process. This means that people are more likely to overreact to fluctuations in WIP than to any other variable in the process.
- When variation occurs in the WIP level, there is no formal mechanism to guide management as to the best way to react to the variation. Therefore, any reaction will tend to be inconsistent and suboptimal.

An alternative way to control a process is to use a pull system. In this case, an item is only allowed to move to the next step in the process when it gets a downstream signal that it is okay to move on. The final step gets its signal from the demand for the item, and then this signal propagates up the entire process until it reaches the first step. At this point, an item is authorized to enter the process. Note that this approach creates a feedback from the downstream part of the process to the upstream part of the process. (However, this is not the same feedback as we saw for a coupled process. We can think of the feedback in a coupled process as intrinsic feedback, in that it is part of the basic process structure. In the pull control systems, we can think of the feedback as being imposed. In this latter case, we can control the feedback.) Hopp and Spearman (1996) give the following particularly concise definition for the difference between push and pull systems:

> A push system schedules the release of items based on demand, while a pull system authorizes the release of work based on system status.

We saw that in our earlier models we fixed on a demand rate, say 0.8 items/hour. Then we scheduled the release of items into the front end of the process based on this rate. A push system would work differently. Suppose we have a process with n steps, with step 1 being the first step and step n being the last step. Orders are used to pull items from the output side of the process. Then authorization to process an item through step n will depend on the status of the process downstream of step n. Similarly, authorization to process an item through step n-1 will depend on the status of the process downstream of step n-1 and so on. Finally, authorization to process at step 1 will depend on the status of the process downstream of step 1. Thus, to make a pull system work, we must decide what attribute of system status will be used to authorize the processing of an item at any step. There are really only three attributes we can choose: WIP, thruput, and cycle time. We have seen that push systems essentially control thruput and let WIP find its natural value consistent with Little's Law and the process dynamics. Pull systems, on the other hand, control WIP levels directly. Consequently, the thruput will find its natural value consistent with Little's Law and process dynamics. So we can see that pull systems have two main components:

- A pull mechanism that allows a feedback signal from the downstream demand to the upstream authorization of production.

- A control mechanism that decides how much production to authorize at any time—this is usually done by imposing a cap on WIP in the process.

For a given thruput, it is contended that pull systems will have smaller WIP and hence (by Little's Law) lower cycle times than equivalent push systems. In order to explore such pull systems, we will create a model of a pull control system. We will look at a particular pull system known as CONWIP, described in Hopp and Spearman 1996 and Marek, Elkins, and Smith 2001.

In a CONWIP system, a cap is set on the total WIP allowed in the process. Once such a system has reached its steady state, an item will not be sent into the process unless another item leaves the process. So even if step 1 is available to process an item, it will not receive an item if the total WIP in the process is already at the maximum WIP allowed under the CONWIP control strategy. Of course, the actual value we use for the WIP cap can be set to any value we want. It is a parameter that we can use to tune the performance of the process. So when an item exits the process to satisfy the demand, a signal will immediately reach the input side of the process to allow an item to enter the process if the first step is available. Of course, if the first step is unavailable, then the item must wait until it is available. Once an item enters the system, it moves through the process as if the process were a push system. Again, it should be emphasized that although the CONWIP model looks like a loosely coupled model (they both have feedback from the downstream to the upstream steps of the process), they are different. The feedback in the loosely coupled model is due to some natural restriction in the process on the size of the WIP. For the CONWIP system, however, the WIP restriction is imposed and can be tuned. In fact, one of the key decisions that we must make in setting up a CONWIP system is the value for the WIP cap. Also, control of when items enter the process is not explicitly exerted in the loosely coupled case. Items will enter the process at step 1 as soon as step 1 is available. In the CONWIP system, the WIP restriction is imposed and an item may not enter the process even if the first step is available to accept the item.

Figures 5.5a and 5.5b show the main structure of the CONWIP model. It has been constructed by starting with the basic uncoupled model, figure 5.2, and making suitable modifications to control the WIP and to monitor demand.

The equation in FLOW S1 is this:

IF SUM_WIP<MAX_WIP THEN 1/DT ELSE 0

This is how we control the inflow to the process based on the WIP cap. The value of the WIP cap is contained in the converter MAX_WIP. (The user can set this value to tune the performance of the process). SUM WIP contains the total WIP in the system. So the control logic for the WIP cap is simple in a CONWIP system. This is one of its attractive features. Notice that while we could leave the queues as queues in ithink, we choose to use stocks and exert flow control on the output. The reason for this is that a queue can only be initialized with a constant value, whereas a stock can be initialized with a variable. So if we want to be able to ini-

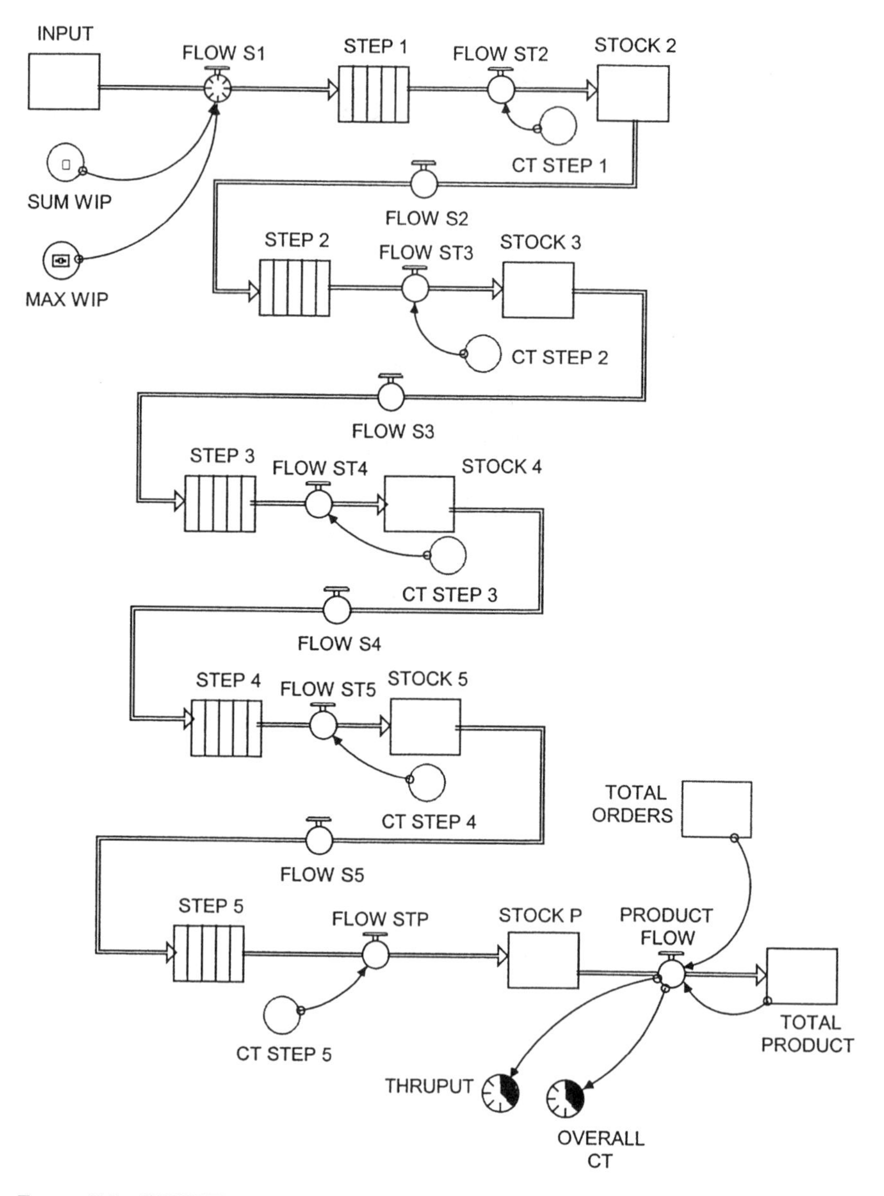

FIGURE 5.5a. CONWIP material control system

tialize the queues in a flexible way, we should use stocks. The ability to initialize the stocks may be useful if we want to reach a steady state faster.

The other major part of the logic in the CONWIP model is the control of the items on the output side of the process. This is done in two parts. First, we must have a section that monitors the orders for items produced by the process. We as-

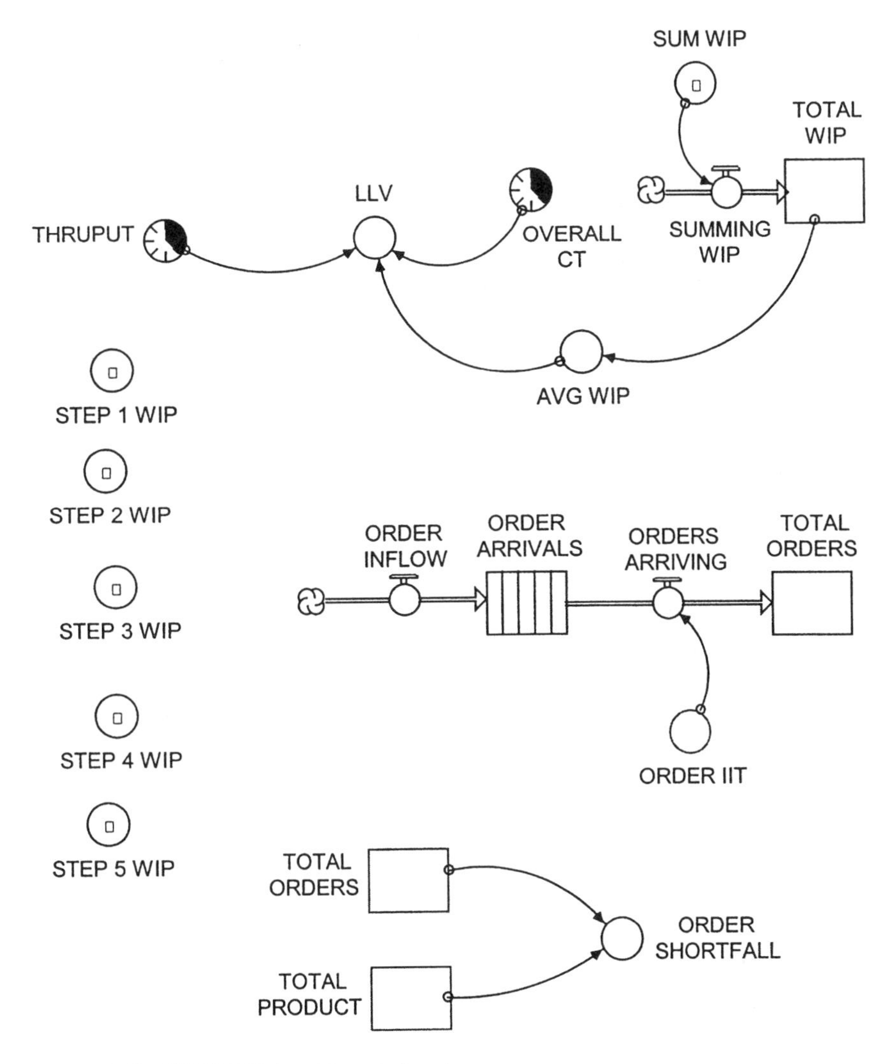

FIGURE 5.5b. CONWIP Material control system

sume that orders arrive as a stream of orders for individual items and that the mean time interval between orders is MEAN ORDER IIT. This is accomplished using the ORDER ARRIVALS conveyor with the conveyor cycle time set to ORDER IIT in the outflow ORDERS ARRIVING. ORDER IIT is calculated as an exponential random variable with mean MEAN ORDER IIT. It would not be difficult to change this part of the model to reflect other more realistic ordering patterns—for example, where orders come in for multiple items on a less frequent basis. These orders are then sent to an accumulator TOTAL ORDERS so that at any time, we know the total items ordered. When items are finished in the last step of the process, step 5, they are put into a stock of product, STOCK P, where they must wait until an order is placed for them. Items in STOCK P are considered part of the process, so the SUM WIP calculation includes STOCK P. We assume that any item can satisfy any order so that we do not have to tie any item to a particular order. This keeps the model much simpler.

The second part is where an item is sent out of the process in the outflow PRODUCT FLOW when the number of items already sent is less than the total orders for items. The equation in PRODUCT FLOW is therefore

IF TOTAL_PRODUCT<TOTAL_ORDERS
THEN (TOTAL_ORDERS-TOTAL_PRODUCT)/DT ELSE 0

This equation ensures that if completed items in STOCK P are available and there are orders to be satisfied, then the items are flowed through to the stock TOTAL PRODUCT. The difference between TOTAL ORDERS and TOTAL PRODUCT is calculated in the conveyor ORDER SHORTFALL. Obviously, ORDER SHORTFALL cannot be less than zero. If items are always available to satisfy orders, then ORDER SHORTFALL must always be zero. Therefore this variable is an easy way to see how well the process is keeping up with orders.

In order to compare the performance of the CONWIP system versus a push system, we need a reference push system. We will choose an uncoupled process with exponential distribution such that the interarrival time has a mean of 1.25, all steps have a mean cycle time of 1 hour, and the traffic intensity, ρ, = 0.8. From section 5.4, we know that, on average, this system behaves like five independent queues. The average WIP for a queuing system with traffic intensity ρ is given by equation 4.4. For $\rho = 0.8$, this gives the average WIP as $0.8/(1.0–0.8) = 4.0$ items. With five steps, this means that on average, the WIP level will be $5*4.0 = 20.0$ items. To compare performance, we ran the CONWIP models at different maximum WIP levels from 10 to 50 in increments of 5. Figure 5.6 shows the results of these runs. Both the average WIP and the thruput are shown, using different axes (as indicted by the arrows) for each.

Figure 5.6 shows that as we reduce the maximum WIP in the system, the thruput eventually drops off as expected. However, the system is fairly robust to WIP reduction for WIP levels below the uncoupled case of 20 items. For example, if we set the maximum WIP at 15, the average WIP is about 13 items, which is 65 percent of the average WIP for the equivalent push system. Yet the thruput is about 0.78 items/hour, close to the expected value of 0.8 items/hour. This robustness has an important practical implication. If we set up a control system by starting out a high WIP level and then decreasing the value of the maximum WIP level, figure 5.6 says that the CONWIP system is somewhat forgiving if we reduce the WIP too far.

In proposing pull systems as alternatives to push systems, Hopp and Spearman (1996) make an important point that, although such pull systems can reduce the average WIP level in a system, this is not the only motivating factor. More important is the fact that, in general, pull systems will have less variability in the WIP than equivalent push systems. This means that they should have less variability in cycle times. Consequently, predictions made to customers about when product or-

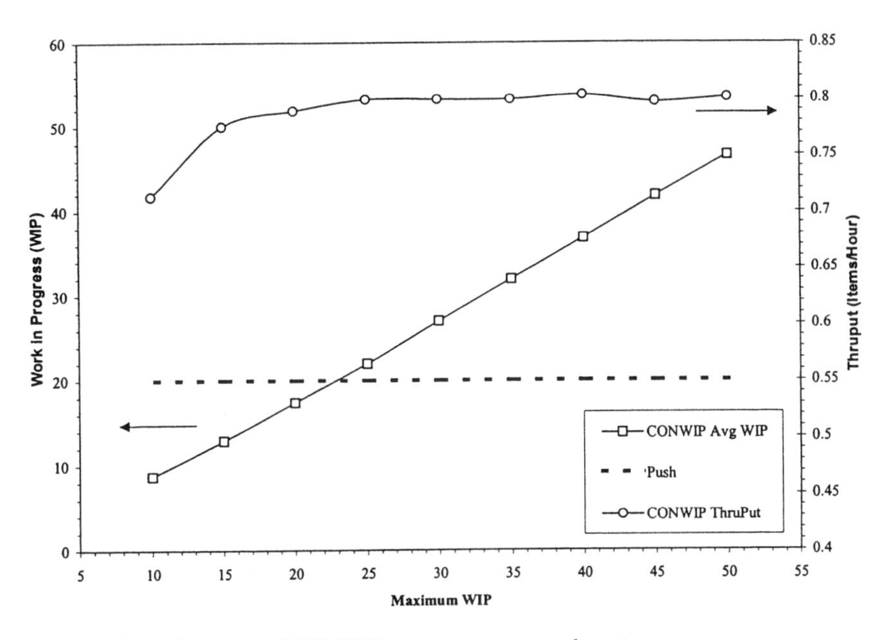

FIGURE 5.6. Performance of CONWIP system versus a push system

ders will be filled should be more accurate in the case of pull versus push systems. So an important performance measure for comparing push and pull systems should be the variation in the total WIP in the system. Because WIP is such a visible attribute of the system, reducing its variation should also reduce the risk of people overreacting to changing WIP levels. In order to have a benchmark for the variation, we go back to equation 4.5, which gives the variance (square of the standard deviation) as a function of the traffic intensity ρ as the following:

$$\sigma_W^2 = \frac{\rho}{(1-\rho)^2} \tag{5.1}$$

Because each step is independent when the random variation is exponential, the variance of the total WIP is simply the sum of the variances of the five steps so that we get this equation:

$$\sigma_{W,Total}^2 = \sigma_{W,1}^2 + \sigma_{W,2}^2 + \sigma_{W,3}^2 + \sigma_{W,4}^2 + \sigma_{W,5}^2 = 5\sigma_W^2 = \frac{5\rho}{(1-\rho)^2} \tag{5.2}$$

Thus for $\rho = 0.8$, the variance of the total WIP is 100; the standard deviation is 10. Figure 5.7 shows the impact of the WIP cap on the standard deviation for the CONWIP system. To calculate the standard deviation of the WIP values, the WIP

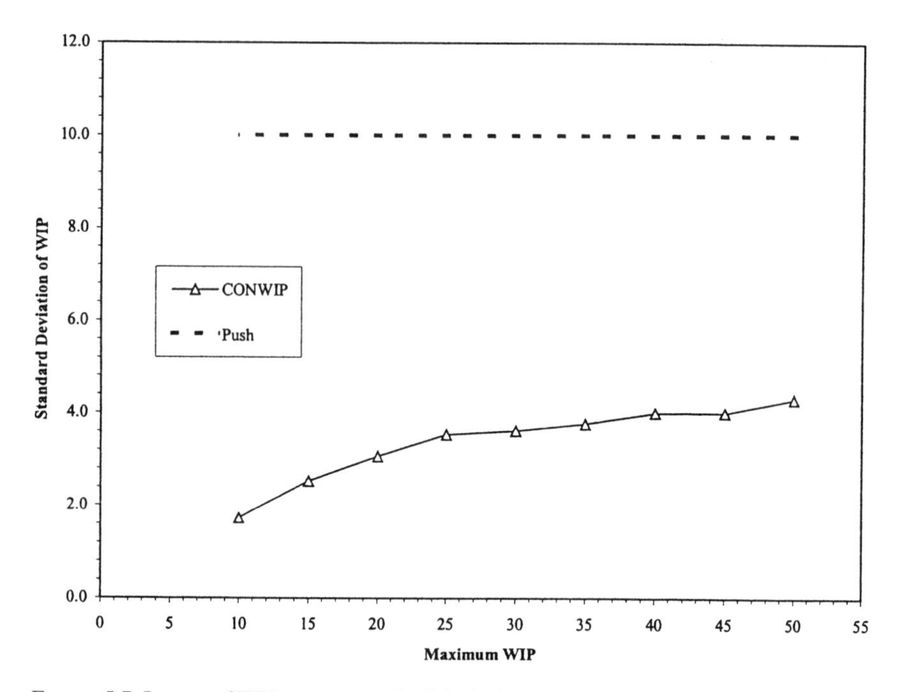

FIGURE 5.7. Impact of WIP cap on standard deviation of the WIP

profile is recorded in a table and then the data are copied to an application such as Excel.

Figure 5.7 shows emphatically that the pull systems have a significant advantage over push systems when measured by variability in the WIP values. In fact, when one looks at the overall impact of the WIP cap from an economic perspective, it might be advantageous in some situations to use a WIP cap that leads to a higher mean WIP if the variation in WIP is so significantly reduced.

> Learning Point: The use of a WIP cap in a pull material control system can significantly reduce the variation in WIP (and cycle time) over the equivalent push system.

In this introduction to material control systems, we do not propose to go into these pull systems in any more detail. One point we would like to make, however, concerns the issue of how to fairly compare push and pull systems. The push models we have used assume that when the process has completed an item, it has somewhere to go, and we do not worry about the details of where it goes. In a pull system, the opposite is true. We assume that when the process requires an item on the input side, it is available. In a real process, we cannot make such assumptions and we have to worry about the following:

- For a pull system, the risk that an item will be required on the supply side and will not be available.
- For a push system, the risk that an item will be required by a customer and will not be available.

It is our contention that a full analysis of any real process may require these parts of the process to be included to prevent reaching an incorrect conclusion because an important feature of the process has not been included in our analysis. Models that focus on the supply side and the customer side will be considered in chapters 7 and 8, respectively.

Another popular pull control mechanism is the KANBAN control system. This is essentially the same as the CONWIP system except that the WIP cap is imposed at each step, not just at the level of the overall process. We will not go into the details here, but readers can consult Hopp and Spearman 1996 and Marek, Elkins, and Smith 2001. We have provided models of a KANBAN system. MODEL 5.8.STM is the KANBAN model with exponential random variation distributions. MODEL 5.8a.STM and MODEL 5.8b.STM are similar except with normal and uniform distributions respectively. Instead of one maximum WIP value, we now have a maximum WIP value for each step. We also have set up the KANBAN models so that they are initialized with the maximum WIP values. Readers are invited to reproduce figures 5.6 and 5.7 for the KANBAN model MODEL 5.8.STM and show the following (note that by equivalent systems we mean the CONWIP maximum WIP is the sum of the KANBAN individual step maximum WIP values):

- The KANBAN maximum WIP values must be set higher than the equivalent CONWIP system to get the same average WIP.
- The WIP/Thruput characteristics are different for the two pull systems. That is, figure 5.6 looks different for KANBAN than for CONWIP.
- For equivalent systems, the variation of WIP for the KANBAN model is better than the push system but not as good as the CONWIP system, especially if the maximum WIP is set high. (Why is this?)
- For equivalent systems, the thruput falls off faster with maximum WIP for a KANBAN system than for a CONWIP system.

Finally, we note that the CONWIP and KANBAN systems used the uncoupled process structure as their starting points. We leave it as an exercise for readers to consider what should be done for the case on tightly coupled processes.

6

Multistep Parallel Workflow Processes

"Cheshire cat," she began. . . . "Would you tell me, please, which way I ought to go from here?"

"That depends a good deal on where you want to get to," said the Cat.

"I don't care much where—" said Alice.

"Then it doesn't matter which way you go," said the Cat.

—Lewis Carroll, *Alice in Wonderland*

6.1 Introduction

In this chapter, we consider models for parallel workflow processes. A serial process was characterized by the following properties:

- All items follow the same path.
- All items are subject to exactly the same operations.
- All items are processed individually.

A parallel process, by contrast, is one in which any one or more of these properties do not hold. Items in the process can take multiple pathways through the process. Unlike Alice, however, we are concerned as to which paths the items take, and our processes contain rules that determine the pathways chosen. In this chapter, we will divide parallel processes into the following three categories:

- Processes in which items can follow alternative routings. Two major subsets of this category follow:
  - The routing may be preordained before the item enters the process. For example, a process may handle three different types of orders: normal, priority, and emergency. When the order arrives, it is immediately classified into one of the three types. The process then handles each type of order differently.
  - The routing may be determined by conditions within the process. For example, there may be a quality check at the end of the process. If the item passes the check, it moves on. If the item fails, it may have to be reworked or discarded.
- An item may imply the creation of a number of items that are processed differently. Many processes consist of a real item and a paper trail. For example,

when an order arrives at a company, it causes the creation of the actual item ordered. Both the order itself and the manufactured item are processed. The process is not considered complete until the order is translated into a shipping instruction and the item is manufactured. Essentially, an incoming item breaks into multiple items that are joined back together at the end of the process. This type of process can be further complicated if many items are created from the initial item. The creation of these items may happen at various points in the process. The joins also may happen at various points in the process.
- Items may be joined in the process and handled as a single item. This is known as batching. Many times working on many items is nearly as easy as working on one item. So items are allowed to collect into a batch and then the entire batch is worked on as an item. (How many times do we hear people say that they let their in-tray build up before they work on it as a unit?)

We will now analyze some situations where parallel processes arise. Note that most (if not all) parallel processes will have sections of serial processing as well. However, in the models presented in this chapter, we will focus on the parallel aspects of the process.

6.2 Parallel queuing models: Designing a checkout system

In chapter 4, we looked at a simple queuing process with a single queue and a single workstation. In this section we expand this model to include parallel processing. This will allow us to address the following situation.

You manage a checkout at a store. It is a small store, and you have only one checkout line. When you started, this appeared to be adequate and customers seem to get through the line quickly. However, all your hard work is beginning to pay off: the number of customers coming into the store is increasing. Because you are customer-focused, you carry out regular customer surveys. These surveys have begun to show that customer satisfaction is dropping, and the main reason cited is that it takes too long to get through the checkout. A study of the current system reveals the following:

- The mean interarrival time for customers to the checkout is 1.25 minutes.
- The mean service time is 1 minute.
- All times are exponentially distributed.

For these conditions, the models in chapter 4 can be used directly to calculate the results for option 1. Using MODEL 4.11.STM, with $\lambda = 1/1.25 = 0.8$ customers/minute, and $\mu = 1.0$ customers/minute, we can show that the mean number of total customers in the queue plus checkout is 4.0 items, and the standard deviation of the total customers in the queue plus checkout is $\sqrt{20} = 4.5$. Under these conditions, we expect that the average number of customers in the checkout line will be 4. With a rate of 0.8 customers/minute, this means that on average, a customer spends $4/0.8 = 5$ minutes waiting for service at the checkout. Of this total of

5 minutes, 4 minutes are spent queuing and 1 minute is spent being served. The study confirmed that this is a good approximation of the actual results. Discussions with customers indicate that the 1-minute service time is acceptable. The 4-minute wait in the queue is the real source of dissatisfaction. Customers would like this reduced to a half-minute wait, on average.

You want to deal with this situation as quickly as possible. You realize that you must add more capacity at the checkout. One way to do this is to add more checkout lines. The question now is how many more you need to add. If one checkout gives an average of 4 minutes in the queue and you want to get to a half-minute wait, your intuition reasons that it could take a few extra checkout lines to achieve this goal. With the cost of cash registers and the cost of labor, this is going to involve both a large initial investment and ongoing expense. The question is this: Is your intuition reliable?

In order to evaluate the impact of adding extra registers, we can use the same model but now set the capacity of the conveyor to values greater than 1. Although this will work, it is not a convenient way to set up the model because you cannot set up the conveyor to have a variable input for the conveyor capacity. The model shown in figure 6.1 is a modified version that allows us to set the conveyor capacity as a variable.

We set the conveyor capacity to infinity (INF) and actually control the capacity via the inflow logic in FLOW TO WORK PROCESS. The equation for this inflow follows:

IF WORK_PROCESS < WORK_CAPACITY THEN 1/DT ELSE 0

This allows us to specify the capacity via the converter WORK CAPACITY. Also, as we are now exerting flow control on the outflow from QUEUE, we must

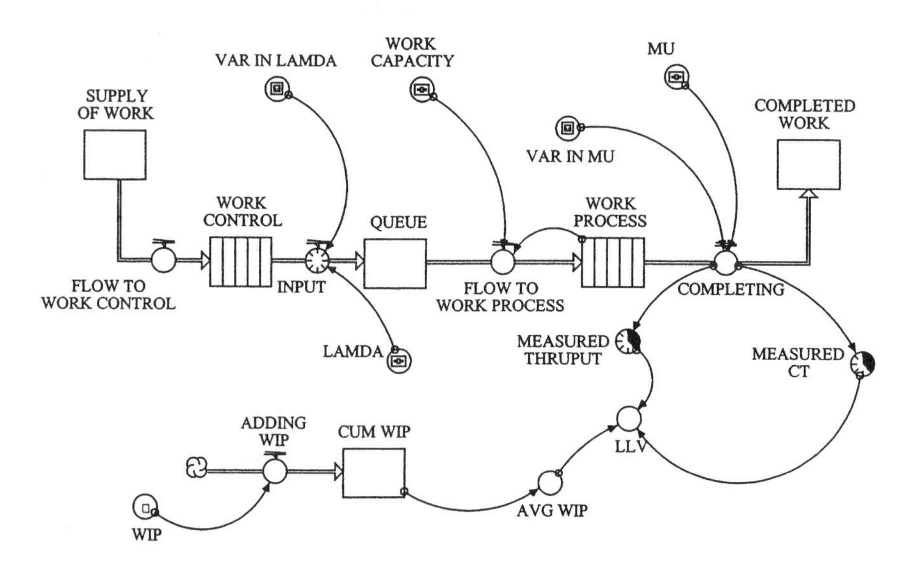

FIGURE 6.1. Parallel queuing system for store checkouts

change it from a queue to a stock. One also should note that with the model in figure 6.1 we are approximating the situation with respect to queuing. We are using just a single queue. (In a real checkout system, each checkout will have its own queue.) If we assume that customers will quickly switch from their queue to another empty queue, then the approximation will be good enough. When the overall checkout system is small and customers can easily see the queues at the other checkouts, this is a good approximation. Table 6.1 shows results for the total WIP and queue sizes as a function of the capacity.

These results show that with just one additional checkout, the average queue size is reduced from 3.2 customers to 0.15. By Little's Law, then, the average time a customer spends in the queue is reduced from 4 minutes to 0.15/0.8 = 0.2 minutes, which is well below the goal of a half-minute. Thus, the original intuition was incorrect.

Table 6.1 also shows another important improvement from adding just one extra checkout line. As well as the average being reduced significantly, the standard deviations are also significantly reduced. If we assume that a worst-case value is the average plus 3 times the standard deviation, then with a capacity of 1, the worst-case value is 3.2+3*4 = 15.2. With a capacity of 2, the worst case is .15+3*.57 = 1.9. From the customer's point of view, the addition of just one extra checkout will significantly improve the situation.

Another situation that can be easily dealt with using this model follows. Suppose we have a two-step process, and we have two alternative design approaches. One design is to have one step done on the first workstation and the second on the next workstation with a buffer in between each station. This is the uncoupled configuration we modeled in chapter 5. The second design is to have two workstations in parallel, where each workstation carries out both steps. Assume that each step takes an average of one hour to process an item, and that items arrive on average every 1.25 hours. Which configuration is better?

In the first configuration, we have two independent queuing systems with a traffic intensity of 0.8. We know the WIP for each system is four items and the average queue size is 3.2. Therefore, assuming exponential distributions, the total WIP and total queue sizes are, on average, 8 and 6.4 items respectively. For the second design, we have two parallel workstations, each with a processing time of two hours. If we run the model with a capacity of two and a processing time of two hours for each step, we find that the average total WIP is 4.4 items and the average queue size is 2.8. Note that the difference represents the average items in

TABLE 6.1. Variation of WIP and queue with capacity.

Capacity	Avg WIP	Stdev WIP	Avg queue	Stdev queue
1	4.0	4.2	3.2	4.0
2	0.95	1.14	0.15	0.57
3	0.82	0.93	0.02	0.17
4	0.8	0.9	0	0.06
5	0.8	0.89	0	0.01

the process steps and is 4.4–2.8 = 1.6 items, 0.8 for each workstation, as it must be by Little's Law. (The cycle time for each step is two hours, and the thruput through each workstation is 0.4 items/hour.) What is happening is that the average number of items in the processing steps is the same in both designs. What is changing is the average number of items in the queues. In the second design, the average number of items in the queue is smaller. The reason for this is that when an item arrives at the queue, it is more likely to have a workstation available when two workstations are present.

> Learning Point: Replacing serial process configurations with parallel configurations can lead to significantly improved performance.

It is left as an exercise for readers to look at equivalent behavior when the probability distributions are not exponential. MODEL 6.1A.STM uses normal distributions and MODEL 6.1B.STM uses uniform distributions.

6.3 Resource implications: The fast food restaurant revisited

In the last section, we looked at a parallel queuing model in which the parallel configuration started from an uncoupled process. Now we look at a parallel configuration for a tightly coupled process. Specifically, we return to the fast food restaurant process that we studied in the last chapter. We saw that the tightly coupled nature of the process caused a drop in thruput. In fast food restaurants, we also see a parallel configuration in use. Instead of the servers performing only one specific task, each server does all the tasks. Then a queue forms at each server. In this configuration, we know that a loss of capacity will not occur. Therefore, a preliminary comparison of the two configurations shows that the parallel configuration should be better. The comparison is somewhat more complicated that this, however. Remember that in serving a customer, the server needs to use equipment resources such as drinks machines. The question now is: Will the conclusion change when we add in the extra consideration of these equipment resources? To proceed, we simplify the modeling by including a system with only two steps. In the parallel configuration, we must then model two lines each with two steps. The important elements of the structure are shown in figure 6.2.

As in the last section, we assume that even though a customer enters one queue, they will switch if the other server becomes free. Therefore, we need to include only one queue in the model. Customers arrive with a mean interarrival time MEAN IIT equal to 1.25 minutes (all distributions are assumed exponential). They are time stamped at the inflow ARRIVING CUSTOMERS. Then they flow through pathways 1 or 2. Each pathway has two steps, A and B. The A steps use equipment Resource A, and the B steps use equipment Resource B. All steps are assumed to have a mean cycle time of one minute. Therefore, each line can serve

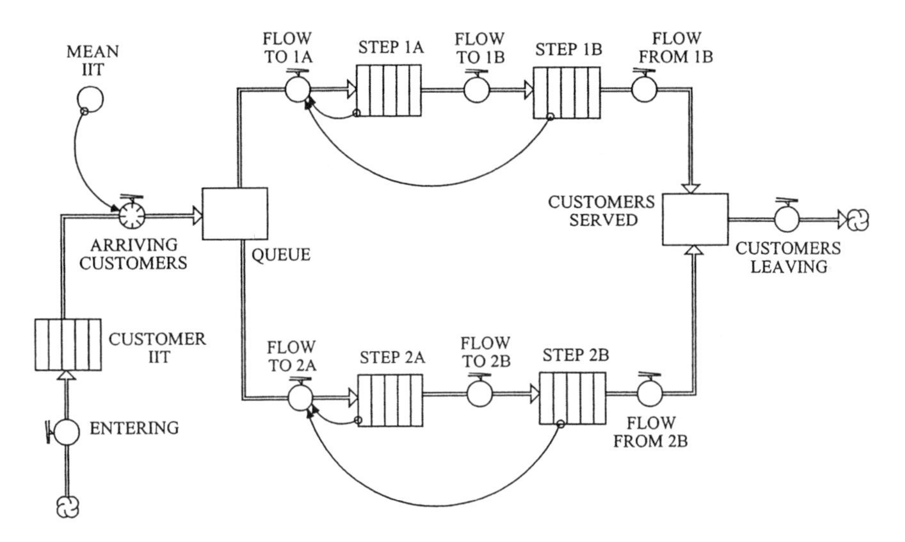

FIGURE 6.2. Basic fast food restaurant parallel configuration

a customer in two minutes, and a customer can be served on average every minute. The equation for FLOW TO 1A is this:

IF (STEP_1A+STEP_1B) = 0 AND STEP_2A = 0 THEN 1/DT ELSE 0

This logic ensures that a customer will not be served until the server is free from serving a previous customer. When serving of the customer is complete, both pathways are joined together in the CUSTOMERS SERVED stock. Then they exit the restaurant via the CUSTOMERS LEAVING outflow. The reason for doing this is to allow the calculation of cycle times and thruput to be done for all customers at one point, the CUSTOMERS LEAVING outflow. We assume that steps 1A and 2A use the same resource, Resource A, and that steps 1B and 2B use the same resource, Resource B. Therefore at any time, we can find the use of Resource A as the sum of STEP 1A and STEP 2A. Similarly, we can find the use of Resource B as the sum of STEP 1B and STEP 2B.

Under these conditions, figure 6.3 shows a typical profile for the average use of resources A and B.

As we would expect, the average values for each resource is about 0.8. There is nothing unusual about this graph. Now look at figure 6.4. This shows the profile for Resource A for the same run as shown in figure 6.3, but over a time period from 500 hours to 600 hours (so that the variation in Resource A is easier to see).

Now it can be seen that while on average we expect 0.8, the amount of resource usage can vary from 0 to 2. Because customers will be in steps 1A and 2A at the same time, the use of Resource A can be greater than 1. Now suppose that there is only one Resource A in the system. Then in practice, when this resource is already in use in STEP 1A, we cannot serve a customer in STEP 2A. Therefore, this latter

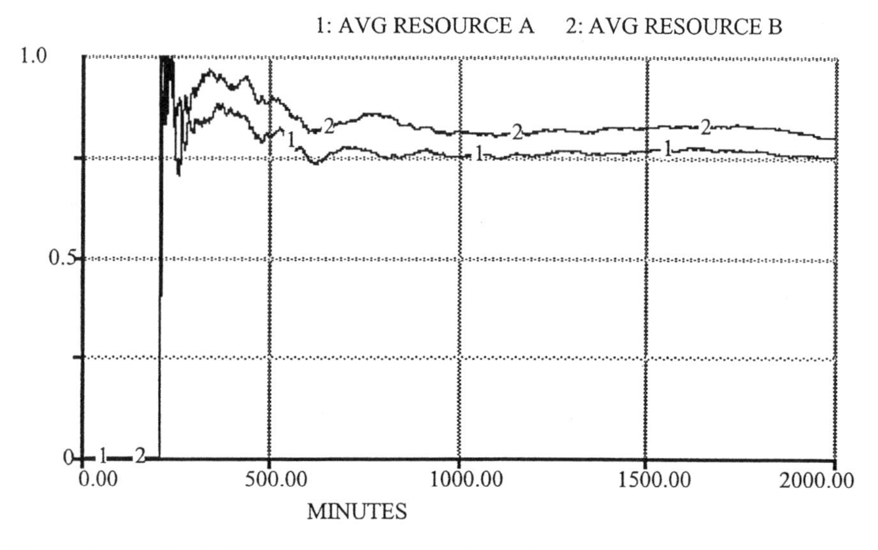

FIGURE 6.3. Profile for the average of resources A and B

customer will have to wait until the resource is available. This means that the thruput will drop. Thus, when we take the equipment resource limitations into account, loss of capacity is possible. In order to take such equipment resources limitations into account in the model, we must add more elements to the model so that items cannot enter step 1A when an item is already in step 2A (Resource A is already being used); similarly, items cannot enter step 1B when an item is already

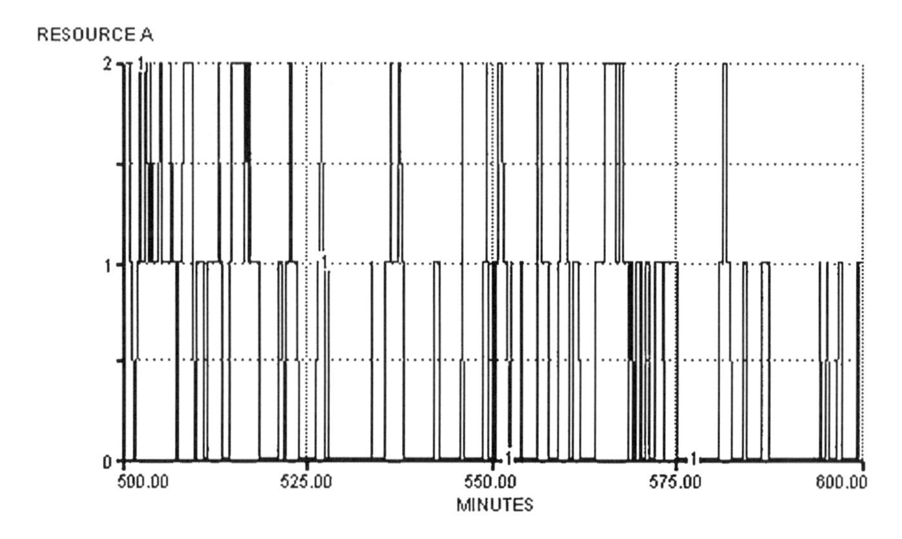

FIGURE 6.4. Profile of Resource A utilization from t = 500 to t = 600 hours

in step 2B (Resource B is already being used). We also must add more queues as shown in figure 6.5. This diagram shows just the portion of MODEL 6.5.STM that is different from MODEL 6.5.STM.

QUEUE 1 and QUEUE 2 are where customers must be "stored" if Resource B is being used by the other serving line. The connections are now more complicated. The equation for FLOW TO 1A is this:

IF (STEP_1A+STEP_1B+QUEUE_1) = 0
AND STEP_2A = 0 THEN 1/DT ELSE 0

The equation for FLOW TO 1B is this:

IF STEP_2B = 0 THEN 1/DT ELSE 0

The first equation ensures that a customer will not be served unless the server is finished with the last customer and Resource A is available. Once STEP 1A starts, STEP 1B has similar logic that only allows FLOW TO 1B to occur when STEP 1A is complete. The logic for FLOW TO 1B is simplified because we do not have to test the status of STEP 1B. We know it has to be empty for a customer to be served as per the first equation. So all we have to do is to test that STEP 2B is empty, indicating that Resource B is available.

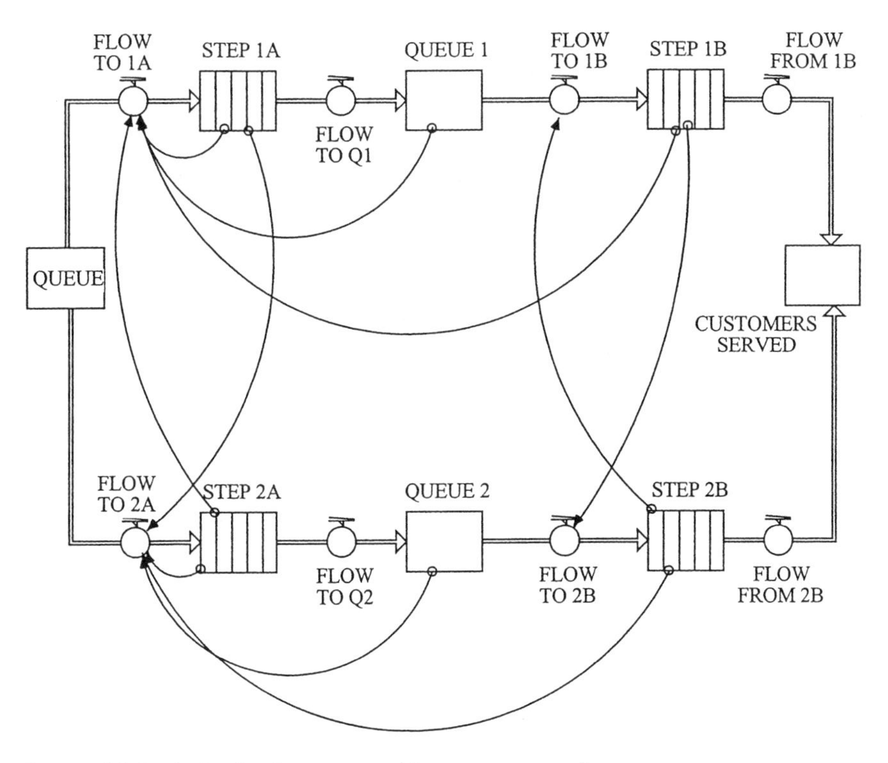

FIGURE 6.5. Basic fast food restaurant with resource constraints

Figure 6.6 shows the profile for cycle time, WIP, and thruput under the same conditions as for the model in figure 6.2.

The average thruput shown in figure 6.6 is about 0.65. Because customers are arriving at a rate of 0.8 per minute, we get buildup of customers in the queue, and the total WIP and cycle time increase without bound. Thus, we have a drop in capacity—in fact we have a significant drop in capacity, from 1 customer per minute to 0.65 customers per minute. We can compare this to the capacity loss in a tightly coupled process with only two steps by reconfiguring the tightly coupled model from chapter 5 to only have two steps. This model is included as MODEL 6.5A.STM. Figure 6.7 shows a typical thruput profile for this two-step tightly coupled model.

The two profiles for THRUPUT are similar. It appears that once we take equipment resource constraints into account, we can get a similar capacity reduction as we did in the case of the tightly coupled process. This shows how important it is to include all the important elements in the model if we are to truly understand process behavior. Ignoring equipment constraints can lead to a different conclusion than we get when they are included. This is always going to be a difficult call for the modeler. Include too few elements of the actual process in the model and you can completely misrepresent the process. Include too many irrelevant elements and the model becomes unnecessarily complicated. Only experience and knowledge can help here.

> Learning Point: The model builder must take great care to include all-important elements in the model. When in doubt, include the element and see the impact.

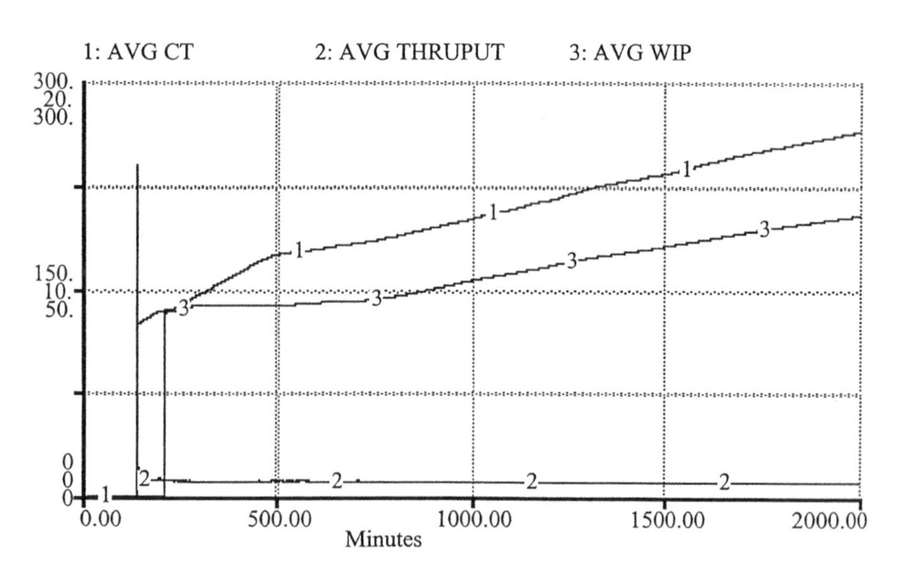

FIGURE 6.6. Results for fast food restaurant model with equipment constraints

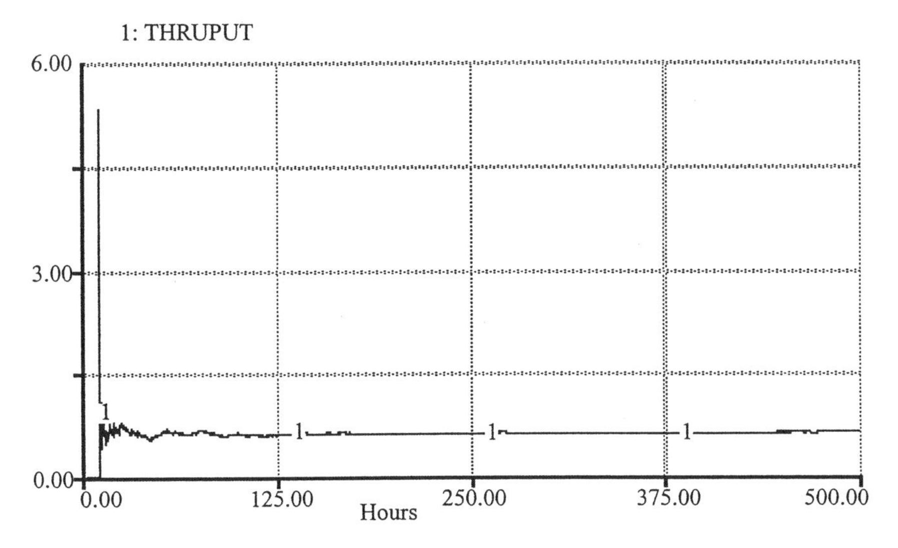

FIGURE 6.7. Profile of THRUPUT for the two-step tightly coupled process

In fact, for real fast food restaurants, the situation with equipment resources is typically much more complicated than we have presented so far. For example, take soft drinks. Typically, drink machines supply many different types of drinks. So even if one server is using the drinks machine to get one flavor, another server may be able to use it if a different flavor is required. Another example is the french fries. Sometimes, servers can work ahead to create a stock of french fry bags. This would allow servers to grab the fries without any resource conflict.

It is also worth noting that in comparing these two fast food scenarios, the length of the customer's wait is only one factor that we must take into account. There may be other performance attributes that are less tangible than this. For example, some people like the idea of having to deal with only one person when they deal with a supplier. This might be motivated by the idea that if something goes wrong with their order, they know exactly who to deal with. So they are less likely to be confronted with one server blaming another and be caught in a finger-pointing "blame game." Therefore, given a choice between equivalent single- and multiple-queue systems, they might prefer the multiple-queue system.

6.4 Telephone call center model: Balking

In a typical call center, a number of people answer calls from customers about obtaining service, reporting complaints, getting updates on orders, and so on. Each caller is routed to an available service person or put on hold. In any real situation, only a finite number of calls can be put on hold. If a call comes in and the number of calls being served and on hold is at the maximum, the call will be lost. The phenomenon of losing a call if a capacity limit is reached in a system is called *balking,* a common phenomenon in business processes. This can occur any time the queue in front of a process has a maximum capacity (which is the case in all real situations).

In designing a call center, the balking rate is an important parameter. This is the fraction of the total calls received at the call center that are lost. Figure 6.8 shows a model for a call center.

This model has been developed using MODEL 6.1.STM as a starting point. The main addition we have to make is the extra outflow from the queue prior to the processing step, LOSING CALLS. The equation for this outflow follows:

IF CALLS_ON_HOLD>MAX_CALLS_ON_HOLD THEN
(CALLS_ON_HOLD-MAX_CALLS_ON_HOLD)/DT ELSE 0

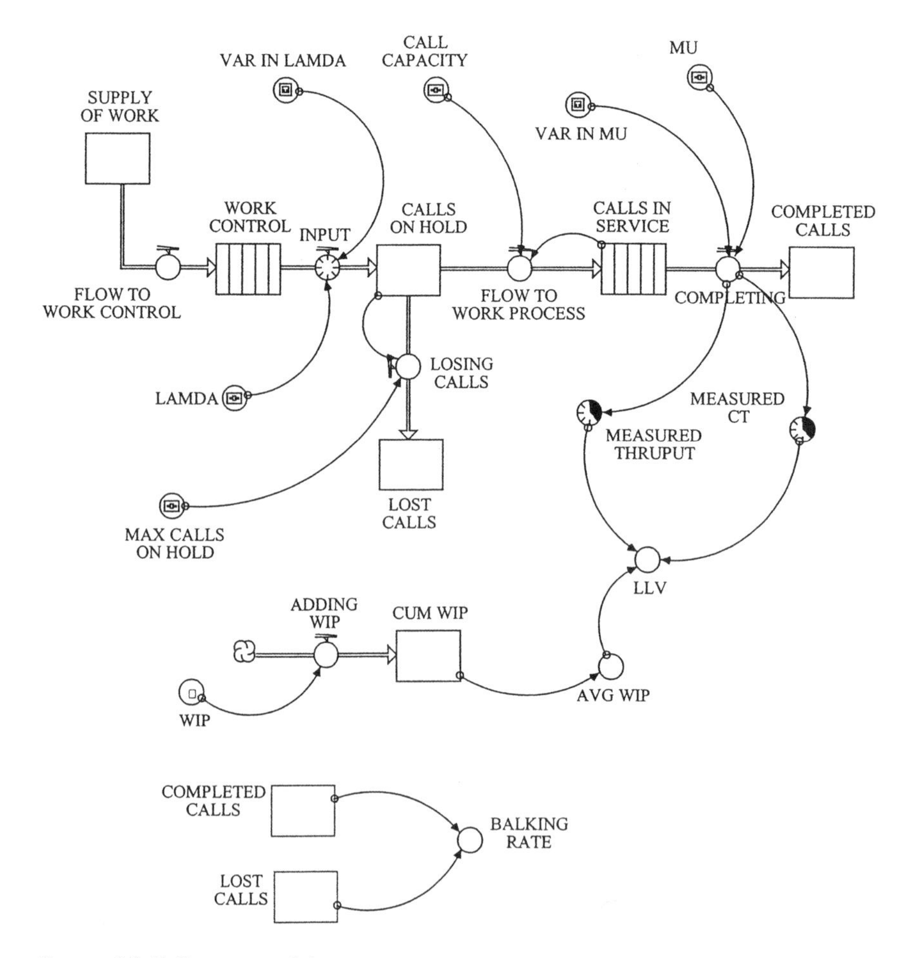

FIGURE 6.8. Call center model

This equation ensures that if the number of calls in the queue is greater than MAX_CALLS_ON_HOLD, the excess calls are flowed in this outflow to the stock LOSING CALLS. This stock thus accumulates the number of calls lost because of balking of calls at the call center. Calls that do receive service eventually make their way to the stock COMPLETED CALLS. The balking rate is calculated in the converter BALKING RATE, which has the following equation:

IF (LOST_CALLS+COMPLETED_CALLS)>0 THEN
LOST_CALLS/(COMPLETED_CALLS+LOST_CALLS) ELSE 0

Before we look at some actual results from the model, we can consider what we expect the balking rate to depend on. The following seems reasonable:

- The value of MAX CALLS ON HOLD. As this value is reduced, we expect the balking rate to go up.
- The traffic intensity (call arrival rate/call service rate). As the traffic intensity tends to 1, we expect the balking rate to increase.
- The value of CALL CAPACITY, the call center capacity. It is not obvious how the CALL CAPACITY should impact the balking rate if we vary it keeping the traffic intensity constant (that is, keeping the product CALL CAPACITY*MU constant as CALL CAPACITY is varied).

Figure 6.9 shows the profile for LOST CALLS and BALKING RATE when the arrival rate, λ, is 1 customer per minute; the service rate, μ, is 1.25 customers per minute (all random variation is exponential); the call center capacity is 1; and MAX CALLS ON HOLD is six customers. Assuming no limitation on the queue size for this system, the average queue length is 3.2 items and the standard devia-

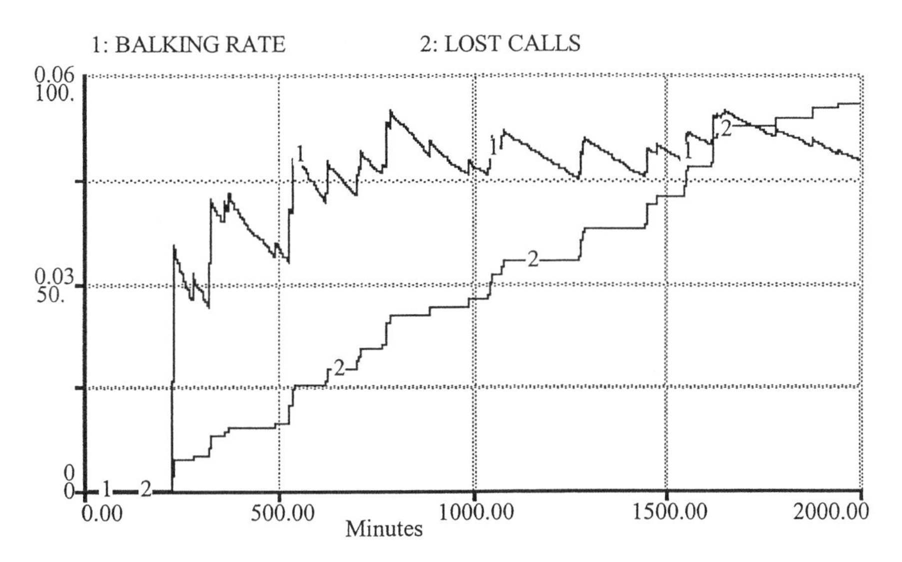

FIGURE 6.9. Call center model: Profile of LOST CALLS and BALKING RATE

tion is $\sqrt{20}$ (from equation 4.5). Consequently, we expect that a maximum queue length of six will impact the performance significantly.

We have lost about 90 calls during the time period of the model, and the balking rate is about 5 percent. So about 1 in 20 calls are lost. An interesting aspect of the profile in figure 6.9 is that the balking appears to occur in clumps. The profile of LOST CALLS stays constant for a significant period of time and then jumps by a large number in a small period of time. By running the model under various conditions, readers can verify that this tends to occur in situations where the overall balking rate is low. This could be important in situations in which the balked customers see each other—for example, those customers who arrive at a restaurant and balk because the waiting area is full. These customers might easily conclude that balking is common and spread the word that the restaurant cannot provide adequate service.

The balking rate represents the probability that a call will be rejected. Therefore, one simple way to think about the balking process is to imagine a pass/fail process whose probability of "passing" is p = balking rate. In appendix A, we give the following equation (equation A.18) for the mean and standard deviation of the number of items passing when N items in total have passed through the system.

$$\mu = Np \qquad \sigma^2 = [Np(1 - p)] \tag{6.1}$$

The maximum value of the standard deviation σ occurs when $p = 0.5$. So the maximum variability occurs in a pass/fail system when the probability of passing = probability of failing = 0.5. This appears to be opposite to what is happening in the call center. Thus the simple analogy to a pass/fail system is inadequate.

Figure 6.10 shows the variation in balking rate with MAX CALLS ON HOLD and traffic intensity for a call center capacity of 1. To generate the data for each combination of MAX CALLS ON HOLD and traffic intensity, the model was run 1,000 times and the average balking rate over the 1,000 runs calculated for figure 6.10.

The variation of BALKING RATE with MAX CALLS ON HOLD and traffic intensity, RHO, is as we expected. The chart also suggests that at least for the range of behaviors we have analyzed, BALKING RATE is more sensitive to MAX CALLS ON HOLD than to traffic intensity. The practical implication of this might be that providing more on-hold capability may have a greater impact on the balking rate than providing more service capacity. Of course, impact on balking rate is only one consideration one would make in determining how to improve a call center. Another might be the impact on the waiting time for those calls that do not get balked. For example, with a traffic intensity of 0.99 and a MAX CALLS ON HOLD value of 1, the overall cycle time for completed calls is about 1.2 minutes. If we improve the service rate so that the traffic intensity is 0.7, this cycle time becomes 0.9 minutes, approximately. If we improve the MAX CALLS ON HOLD value to 10, keeping the traffic intensity at 0.99, the cycle time becomes six minutes. For many customers, six minutes might be too long, and they may close the call themselves. Then the benefits you thought you were

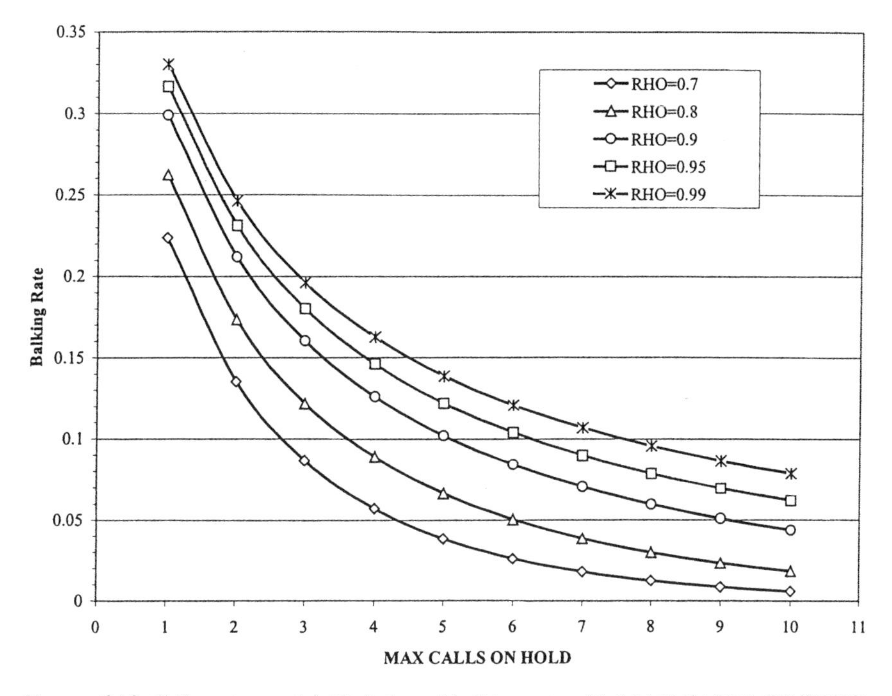

FIGURE 6.10. Call center model: Variation of balking rate with MAX CALLS ON HOLD and traffic intensity

going to get from the reduced balking rate might not be realized at all. So even optimizing this simple system is not quite so simple after all.

Figure 6.11 shows the variation of BALKING RATE with CALL CAPACITY for the case where MAX CALLS ON HOLD is 1. We started with a call arrival rate of 1 call per minute and a service rate of 1.25 calls per minute. This service rate is a combination of a service time for each call, MU = 1.25, and a CALL CAPACITY value of 1. Then we increased the value of CALL CAPACITY to 2 and reduced MU to 1.25/2 = 0.625 calls/minute so that the overall service rate is still 1.25 calls/minute.

Figure 6.11 shows clearly that providing multiple servers will have a significant advantage over the single server system even if the overall service rate (and hence traffic intensity) is the same. Again, this is only from the perspective of balking rate. If the individual servers have a lower service rate, their processing time is longer, and the overall cycle time for completed calls will be longer. We have the same balking rate/cycle time balancing act as we saw earlier.

Earlier we referred to a situation in which a customer ends the call if the call is on hold too long. This phenomenon, called *reneging*, occurs when an event (the reneging event) causes an item entering a queue to leave before service has been received. The most common reneging event is being in a queue for a certain length of time, referred to as the *reneging time.* If an item has been waiting in the

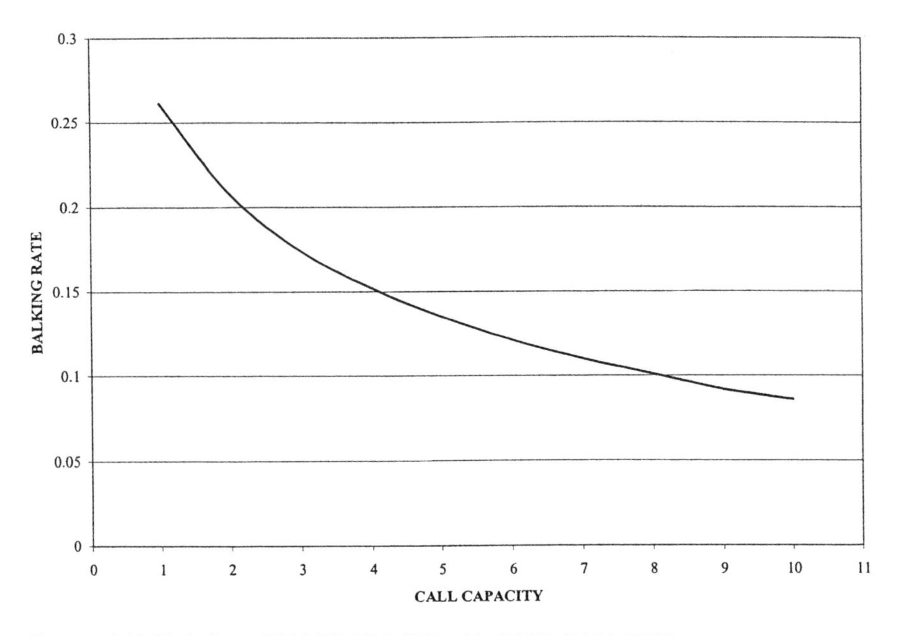

FIGURE 6.11. Variation of BALKING RATE with CALL CAPACITY

queue for a time equal to the reneging time, the item exits the queue without service. Events can also be based on factors other than time. For example, an item might be prepared for an operation before it is put into the queue for that operation. The success of the operations depends on the quality of the preparation. Once in the queue, a check is done to ensure that the quality of the preparation has not degraded because of the time spent in the queue. If the quality has degraded, then the item is taken from the queue. It may be discarded or sent back to be prepared again.

Note that versions of MODEL 6.8.STM have been provided with distributions other than exponential. MODEL 6.8A.STM uses normal distributions, and MODEL 6.8B.STM uses uniform distributions to represent the random variation in arrival and processing times.

6.5 Machine repair model

In any system, equipment will sometimes fail and need to be repaired. When the equipment fails, a repairperson is called to fix the equipment, whereupon it is returned into service. Therefore, any given piece of equipment can be in three states:

1. The equipment is running within its normal operating performance.
2. The equipment has failed and is awaiting repair.
3. The equipment is being repaired.

This process is referred as an equipment maintenance process. More specifically, if a piece of equipment is not repaired until it has failed, we refer to this as a reactive maintenance process. (It is possible to implement maintenance processes that monitor equipment and attempt to detect when a failure is imminent. Then the failure can be avoided by intervening before the failure has occurred. This is referred to as predictive maintenance (see Dodson and Nolan 1995). We will only model a reactive process here.) The two major performance characteristics of a reactive maintenance process are the following:

- The time between successive failures of a piece of equipment. This is usually described by quoting the Mean Time Between Failures, MTBF.
- The time it takes to repair a failed piece of equipment. This is usually described by quoting the Mean Time To Repair, MTTR.

Obviously the mean failure rate is 1/MTBF for a given piece of equipment, and the mean repair rate is 1/MTTR. The failure and repair parts of the process can be represented by conveyors whose cycle times are the Time Between Failures (TBF) and the Time To Repair (TTR). Between these two conveyors we must place a queue that holds failed equipment, if there is no capacity in the repair conveyor to fix them as soon as they fail. The maintenance system is not the same as a serial process with two steps and a queue, however. The processes we have modeled so far in chapters 4, 5, and 6 are open networking models. That is, an item comes in the input side and, once processing is complete, the item leaves the process. The maintenance process, by contrast, is a closed network. It is the same set of equipment that stays in the process. Then when an item leaves the repair conveyor, it must be routed back to the first conveyor. The WIP is this process is fixed by the structure of the process. In a reactive maintenance process, some of the performance characteristics we are interested in include the following:

- The system availability. Assuming that the system cannot be considered available unless all equipment is available, the system availability is the fraction of time all equipment is working and no equipment is in a failed state or being repaired.
- The number of repairmen required to ensure a given level of system availability.
- The utilization of the repairmen.
- The performance characteristics of the queue of equipment waiting to be repaired (average queue size and so on).

The model for this maintenance process is shown in figure 6.12. We assume that we have a fixed number of machines in the process, shown as NUMBER OF MACHINES in the model.

The output of the MACHINES BEING REPAIRED conveyor is routed to the input of the conveyor MACHINES WORKING. Both conveyors have their capacity set to infinity (INF) in the conveyor dialog box. The capacity of the repair conveyor is controlled with the inflow logic in FLOW TO REPAIR PROCESS, which has the following equation:

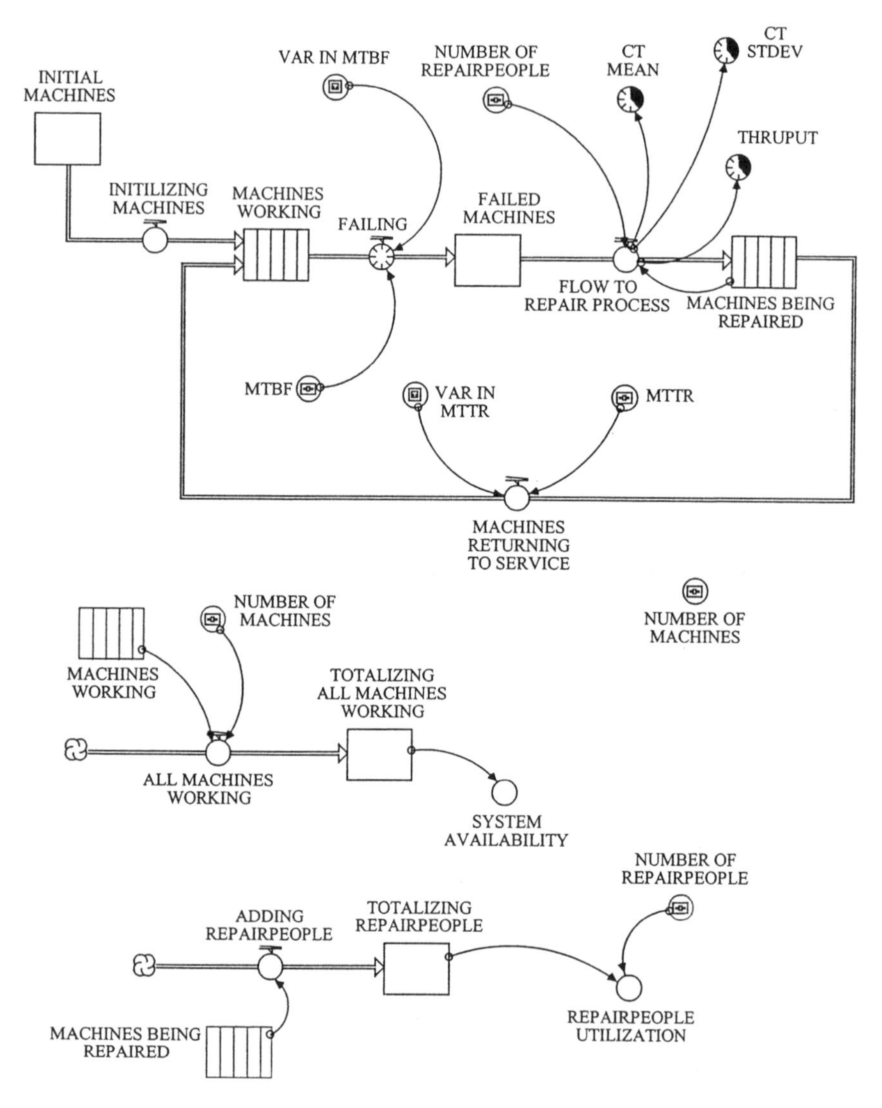

FIGURE 6.12. Machine repair model

IF MACHINES_BEING_REPAIRED < NUMBER_OF_REPAIRPEOPLE
THEN (NUMBER_OF_REPAIRPEOPLE-MACHINES_
BEING_REPAIRED)/DT ELSE 0

A capacity limit on the conveyor MACHINES WORKING would make no sense. Besides, ithink will not allow it because the input is from another conveyor. In order to add variety to our modeling techniques, we are not going to measure the QUEUE characteristics directly. Rather, we are going to measure

THRUPUT and cycle time and infer the queue size from Little's Law. The machines are time-stamped at the inflow FAILING, and then we measure thruput and cycle time characteristics at the outflow FLOW TO REPAIR PROCESS. We also measure the cycle time variation using the function CTSTDDEV in the converter CT STDEV. Although we cannot directly infer the WIP variation from the cycle time variation, we know that a direct correlation exists between the two. So if we make a change that reduces the cycle time variation, we know that it will also reduce the WIP variation.

The lower part of the model is used to measure the system availability by totalizing the number of time intervals where the number of machines in the conveyor is equal to NUMBER OF MACHINES. The equation in the inflow ALL MACHINES WORKING is this:

IF TIME > = 200 AND MACHINES_WORKING =
NUMBER_OF_MACHINES THEN 1/DT ELSE 0

And the equation in SYSTEM AVAILABILITY is

IF TIME>200 THEN TOTALIZING_ALL_
MACHINES_WORKING*DT/(TIME-200) ELSE 0

In these equations, we are allowing the first 200 time units to let the model reach steady state.

The utilization of repairpeople is calculated by first averaging the number of machines in the conveyor MACHINES BEING REPAIRED and then dividing by NUMBER OF REPAIRPEOPLE. This calculation works because we assume that one repairperson can only work on one piece of equipment at one time.

Because this model is a closed network, we have to be concerned with model initialization. We have to get the machines into the process somewhere. The most logical approach is to assume that initially all machines are working correctly. So we want to initialize MACHINES WORKING with the NUMBER OF MACHINES. However, ithink does not allow us to initialize a conveyor with a variable. So we must flow the machines into the conveyor at the start of the simulation. The stock INITIAL MACHINES is initialized with NUMBER OF MACHINES, and the flow INITILIZING MACHINES has the equation 1/DT. Then, every DT, a machine will be added to the conveyor until the stock INITIAL MACHINES is empty. This will give the correct initialization of the model. (Another way to initialize the model would be to set the initial value of FAILED MACHINES to NUMBER OF MACHINES. This would have initialized the machines as all being in the failed state.)

The first observation we can make about the repair process is that it will always run under a steady state (even if there is short-term variation in this steady state). The reason for this is the closed network structure of the process. The machines will be distributed among the three states—working, waiting for repair, and being repaired—such that the rate at which machines exit these steps are the same (on average).

A second observation we can make is that although we can set up processes where the system availability is close to 1, we can never have a system availability of 1. Once *MTTR* > 0, then there is always a finite probability that a machine will be in a failed or repair state, so system availability < 1.

A third observation is that if the number of people is equal to or greater than the number of machines, we have a trivially simple case. Again, this is because of the closed network structure of the process. A machine is either in a working state or a failed state. If it is in a failed state, then there will always be someone available to repair it. Machines never have to queue for repair, so the cycle time mean and standard deviation for the queue are both 0. For this case, we show in Appendix C that the thruput, T_P, the system availability (*SA*) and the repairmen utilization (*RU*) are given by this equation:

$$T_P = \frac{N_M}{MTBF + MTTR} \tag{6.2}$$

$$RU = \frac{MTTR}{MTBF + MTTR} \tag{6.3}$$

$$SA = (1 - RU)^{N_M} \tag{6.4}$$

where N_M is the number of machines in the process. Readers are invited to run the model for the case where the number of repairpeople and the number of machines are both 10 and $MTBF = MTTR = 1$ time period. According to equations 6.2, 6.3, and 6.4, the result should be $T_P = 5$ machines/time period, $RU = 0.5$, and $SA = (1\text{–}0.5)^{10} \approx 0.001$. The model gives the expected results for *TP* and *RU* but gives a value for $SA \approx 0.5$. What is going on here? (Readers should hypothesize a rationale for the discrepancy in *SA* before proceeding.)

In order to understand what is going on, we must remember that a program like ithink works by placing each item in time increments or slots. A quantity Q in a time slot will be represented as a flow of size Q/DT. Suppose all machines start off in different time slots. Because the cycle time through the conveyors is variable, machines that enter the conveyor in different time slots can exit in the same time slot. However, once they move into the same time slot, they will always have the same cycle time associated with them. This means that once two machines exit in the same time slot, they will remain in the same time slot for the remainder of the simulation. Later, a third machine may join the time slot and so on. Eventually, if the model is run for long enough, all the machines will tend to end up in the same time slot. All machines become "joined." Then the machines will all be in working state for the fraction of time MTBF/(MTBF+MTTR). Hence, if MTBF = MTTR, SA will be 0.5, which is exactly what we found. Readers can confirm this by opening the table REPAIR MODEL TABLE and looking at the column FLOW TO REPAIR PROCESS after they have run the model. Scroll to the end to see that the only nonzero value for this flow is 160, which represents 10 machines spread over one time slot of size 0.0625 = 1/16. Also, if our hypothesis is true, then increasing the values of MTFB and MTTR (but keeping them equal) should reduce the value if SA because the probability of items being joined will be smaller the larger the values of MTBF and MTTR (readers can easily confirm this by running the model with values of MTBF = MTTR = 10). Note that the problem of joined items is only an issue for closed network models. Open networks may still have joined items, but these items will eventually exit the process.

Now if this is the proper explanation for the problem, then what we must do is find a way to separate machines once they have become joined. Figure 6.13 shows how this is done. (Another possible approach is to set up the model so that machines actually leave the process and as they do, an equivalent machine is added to the process. Although this will work in principle, it is not pursued here.)

Only the upper half of the model is shown. (The lower half is the same as in figure 6.12.) A stock is placed after the repair conveyor. This stock does not really exist in practice. It is simply a modeling convenience (remember we are building a model so it does not have to be exactly what we see in practice; it just has to represent system behavior accurately enough to be useful). Therefore, we refer to this stock as a virtual stock. The outflow from this stock is 1/DT. This means that if two or more machines enter the stock, they will exit one at a time and be spread out over a number of time slots. Readers are invited to run this model using the previous input parameter values. The value of SA will now be close to the expected value of 0.001. A good takeaway lesson from this discussion is summarized in the following learning point.

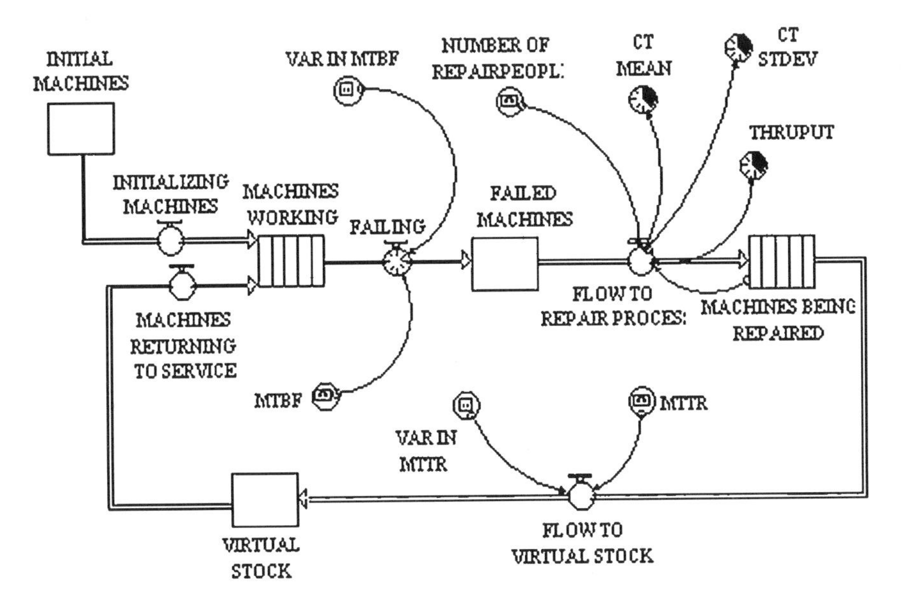

FIGURE 6.13. Improved machine repair model

> Learning Point: It is a good idea to run a model in a simple scenario in which the expected results are known prior to using the model in more complex scenarios so that any issues with the model can be identified easily.

Of course, in many real machine maintenance and repair situations, the value of *MTTR* is much less than *MTBF*. Consequently, we do not have to provide the same number of repairmen as machines. We can imagine a balanced repair process as being one where the following equation applies:

$$\frac{Number\ of\ Machines}{MTBF} = \frac{Number\ of\ Repairmen}{MTTR} \tag{6.5}$$

Table 6.2 shows the results when the number of machines is 10, the *MTF* is 10, and the number of repairpeople and *MTTR* are varied such that equation 6.5 holds.

As we reduce the number of people (and the *MTTR*) from 10 to 1, the system availability begins to rise so that the fraction of time that all machines are working is increasing. Also, as the number of repairpeople increases, the chance that a machine will fail and that a repairperson is not available also decreases. Therefore the average size of the failed machine queue (and its variation as measured by *CT* sigma) increases as well. The only good trend in the results is the increase in repairperson utilization indicating that both the *SA* and *RU* values are getting better. We have higher *SA* and more efficient use of the repairpeople. It should be obvious from these results, however, that in general a balanced process will not give us the values of *SA* we need to run a real maintenance process. Values such as 0.22, the best in table 6.2, are simply too low. In order to increase the value of *SA*, we must unbalance the process. We can do this in two ways. One alternative is to provide more repairpeople than is required by the balanced process as given by equation 6.5, but still have each repairperson work on a single machine at a time. The other is to allow the extra repairpeople to work together on a single machine.

TABLE 6.2. Repair process: Results for a balanced process.

Number of repair people	MTTR	Thruput	RU	SA	CT Mean	CT Sigma
1	1	0.79	0.78	0.22	1.73	2.05
2	2	0.76	0.76	0.12	1.16	1.73
3	3	0.73	0.73	0.06	0.74	1.40
4	4	0.69	0.69	0.03	0.44	1.07
5	5	0.66	0.65	0.02	0.23	0.76
6	6	0.62	0.62	0.01	0.10	0.49
7	7	0.59	0.59	0.01	0.04	0.28
8	8	0.55	0.56	0.00	0.01	0.13
9	9	0.53	0.52	0.00	0.00	0.03
10	10	0.51	0.50	0.00	0.00	0.00

TABLE 6.3. Results for repair process: Adding extra repair people.

Number of repair people	MTTR	Thruput	RU	SA	CT Mean	CT Sigma
1	1	0.79	0.78	0.22	1.73	2.05
2	1	0.90	0.45	0.37	0.16	0.45
3	1	0.91	0.30	0.38	0.02	0.12
4	1	0.91	0.23	0.38	0.00	0.03
5	1	0.91	0.18	0.38	0.00	0.00

This should improve the MTTR of the repair process. So we can model this alternative by keeping the number of repairpeople equal to 1 and reducing the value of MTTR. Table 6.3 shows the results for the first alternative. (The number of machines and MTBF are both 10.)

Adding the first repairperson gives a significant benefit, increasing SA from .22 to .37. However, additional repairpeople have minimal impact on SA. Also, RU gets smaller and smaller as more repairpeople are added. In effect, the extra repairpeople will have nothing to work on. These results indicate that it is the ability to fix the machines fast (as indicated by the value of MTTR) that drives the value of SA. So we should expect better results for SA with the second alternative. The results for this alternative are shown in table 6.4.

The results in table 6.4 clearly support the idea that reducing the MTTR gives better results. With MTTR = 0.2, which is equivalent to the case of five repairpeople in table 6.3, the value of SA is 0.8 (in contrast to 0.38 for the first alternative). Note that in both cases, the values of RU in equivalent cases are the same, as they must be. So the effectiveness of the repairpeople in the second alternative is greater than in the first.

> Learning Point: In the repair process described in figure 6.14, the availability of the system, SA, is driven more by the ability to return machines to service faster (as indicated by the value of MTTR) than by the number of repairpeople provided.

TABLE 6.4. Results for repair process: Reducing the value of MTTR.

Number of repair people	MTTR	Thruput	RU	SA	CT Mean	CT Sigma
1	1	0.79	0.78	0.22	1.73	2.05
1	0.9	0.82	0.74	0.26	1.33	1.71
1	0.8	0.85	0.68	0.32	1.00	1.40
1	0.7	0.88	0.61	0.39	0.72	1.10
1	0.6	0.90	0.54	0.46	0.50	0.84
1	0.5	0.93	0.46	0.54	0.32	0.61
1	0.4	0.95	0.38	0.62	0.19	0.42
1	0.3	0.96	0.29	0.71	0.10	0.26
1	0.2	0.98	0.20	0.80	0.04	0.13

Of course, this analysis assumes that increasing the number of people working on a repair job will reduce the value of MTTR by a factor equal to the number of people required. In practice, the relationship between MTTR and the number of people assigned to each job will not be so simple. However, even if the relationship is not quite as good as a simple inverse relationship, the results in table 6.4 indicate that using the extra repairpeople to reduce the value of MTTR may still be the better strategy.

Of course, having to run a process with a people utilization of 20 percent to get a system availability of 80 percent is unappealing. In many situations, we want repairpeople to have some spare capacity for training and other activities, but they should not need 80 percent spare capacity! One way to drive down the need for people resources is to reduce the variability in the time between failures and the repair times. We leave it as an exercise for readers to run the models with other distributions. MODEL 6.13A.STM uses normal distributions for the failure and repair times, and MODEL 6.13B.STM uses uniform distributions. Be aware that many enhancements could be made to this model. For example, in order to increase system availability, we could add redundant equipment. Then, if a machine fails, the redundant machine takes over, and the impact to system availability is minimized. Indeed, if the redundant machine is a live backup, the impact to system availability will be zero. This is obviously the kind of backup you want in a nuclear facility. It is clear that for this strategy to be effective, the repair time for the machine should be less than the time to failure. Otherwise the redundant machine may fail before the first machine is repaired. If the system is critical enough that high system availability is required, a second backup machine may be installed, and so on. Obviously, a balance between the need for system availability and the economics of the capital investment in machines must be considered to find the optimal approach.

6.6 Batching: A laboratory analysis model

In a typical analytical laboratory, samples arrive from a manufacturing operation for analysis. The analyst takes the sample, prepares it, and then loads it onto the analysis equipment. Once the analysis is complete, the results are obtained, verified, and sent to the manufacturing operation that requested the analysis. If the laboratory analyzes in-process samples, then it is important that the sample results be analyzed as quickly as possible because the manufacturing process is probably waiting for the results before proceeding to the next step. Or it may be waiting to adjust the process based on the analysis results. If the laboratory analyzes final product samples, the urgency with the sample turnaround may not be as great. The reason for this is that the analysis results are typically required to determine the quality of the product so that it can be released to a customer (either internal or external). Other activities must be done as well to get the product released, however. Turning the sample analysis results around quickly will not have a major impact on overall cycle time if these other activities are cycle time limiting. In this situation, laboratory management may be interested in balancing cycle time requirements with analyst/equipment productivity.

One way to do this is to use a batching strategy in the laboratory. Some analytical instruments are capable of handling more than one sample at a time, so samples can be batched onto the instrument. The net result is that batching the samples can impact the capacity of the laboratory and the cycle time of the samples through the laboratory. The actual impact in any given case is going to depend on the amount of the measurement analysis time that is batch dependent versus batch independent. Suppose we have to weigh out the samples as part of the sample preparation. Then the weighing time is batch dependent. In other words, the amount of time spent weighing samples will scale as the batching number. Consequently, batching will have no advantage. We also may have to prepare the instrument before running the sample, but the preparation time may be the same no matter how many samples are batched. This preparation time will be batch independent. It is the amount of batch-independent time that provides the opportunity associated with batching. This opportunity must also be compared with the disadvantage of waiting for all the samples to accumulate before creating a batch. This will add cycle time to the analysis process.

An important part of any batching strategy is the batching decision. That is, when do we decide to take the samples we have and make a batch from these samples? This decision is based on the occurrence of some event. The simplest event would be that we specify up front the batching number: when the number of samples equals the batching number, we create a batch and process them. Other events are possible, as can be seen in the following two examples:

- The event is based on reaching a certain time. We might decide that on a certain fixed time interval, we batch with whatever number of samples we have at the time. For example, when the laboratory analyst arrives at work each day, he or she might look at what samples are in the queue and create a batch with whatever samples are there.
- The event might be based on the status of the manufacturing and/or analysis process. For example, suppose that manufacturing normally sends in a sample every 6 hours. A sample has just arrived but is one short of the batching number. A call arrives from manufacturing to say that there is an issue, and production is being halted. The laboratory now realizes that there will be a significant delay before the next sample arrives. Therefore, they proceed to batch the samples already in the laboratory queue.

Figure 6.14 shows an example of a laboratory analysis process with batching. This model is set up using exponential distributions for the arrival and service times. The service time (the analysis time) is assumed to be all batch-independent time. Thus the service time (and so the service rate) will be independent of the batching number.

This model is basically a modification of the parallel queuing model in figure 6.1. The major difference is in the flow control to the conveyor SAMPLE ANALYSIS. The equation for the inflow FLOW TO ANALYSIS PROCESS is this:

IF SAMPLES_WAITING > = BATCHING_NUMBER AND
SAMPLE_ANALYSIS = 0 THEN BATCHING_NUMBER/DT ELSE 0

The batching process joins the batched samples so that they enter and exit the queuing process at the same time. This approach of looking at the batching system as a modification of figure 6.1 is useful because it shows the batching system as a more complicated form of a basic parallel queuing system. We assume that the instrument can only process one batch at a time; hence, a new batch can enter only when the conveyor SAMPLE ANALYSIS is empty. The WIP in the process is still calculated in the converter WIP as the sum of the stocks SAMPLES WAITING and SAMPLE ANALYSIS. A point of interest about this model is that unlike other models we have discussed, this model takes physical items (samples) and produces logical items (results). Thus we do not have to explicitly worry about any queuing issues downstream of the process. We are simply producing results that are sent back to the manufacturing area once they have been created.

Looking at figure 6.14, it is clear that the WIP and cycle time performance is governed by the WIP and cycle time performance in the queue for batching, SAMPLES WAITING, and the analysis process SAMPLE ANALYSIS. Although not shown here, the queue for batching, SAMPLES WAITING, is really composed of two parts:

Part 1: The time samples are waiting to create a batch.
Part 2: The time a batch may be waiting because the analysis process is busy with the previous batch and is not ready to accept the batch.

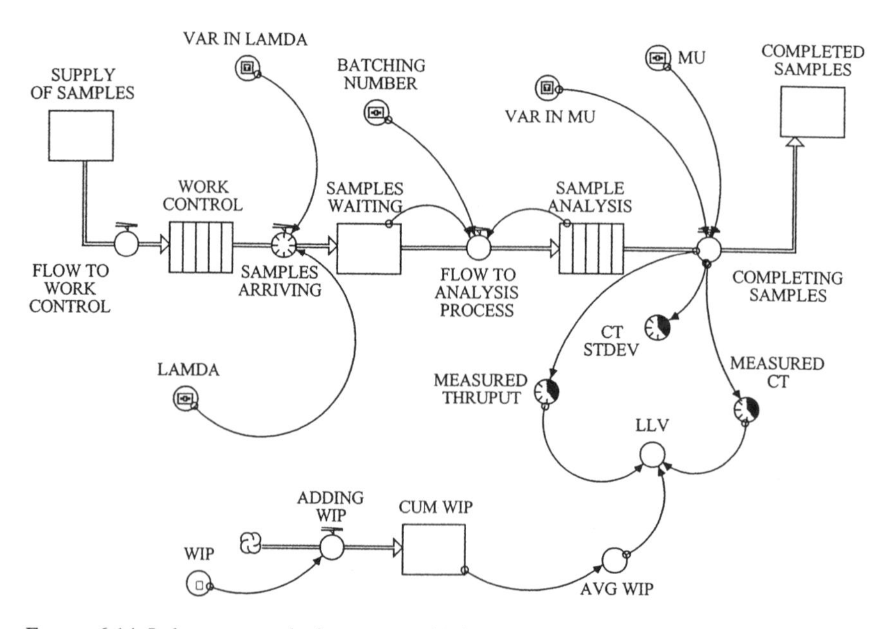

FIGURE 6.14. Laboratory analysis process with batching

When the batching number is 1, we have the simple queuing system discussed in chapter 4 and these distinctions are unnecessary. When the batching number is greater than 1, this distinction is useful in understanding the behavior of the batching process in terms of the queuing process. Suppose that the batching number is denoted B_N, the sample arrival rate as λ samples/unit time, and the analysis rate as μ batches/unit time. Then the arrival rate of batches is λ / B_N batches/unit time. The effective traffic intensity of the analysis process, ρ_A, is given by the following equation:

$$\rho_A = \frac{(\lambda / B_N)}{\mu} = \frac{\lambda}{B_N \mu} \tag{6.6}$$

Therefore, we can calculate the Part 2 process performance as a queuing process with a traffic intensity given by equation 6.6. Note that for the analysis process to have a steady state performance, $\rho_A < 1$, which means that in situations where $\mu <= \lambda$, we must use batching ($B_N > 1$).

A little thought will show that the average WIP for part 1 is given by $(B_N - 1)/2$. Thus the total WIP in the system is given by

$$WIP_{TOTAL} = WIP_{Part1} + WIP_{Part2} = \frac{(B_N - 1)}{2} + B_N \frac{\rho_A}{1 - \rho_A}$$

$$= \frac{(B_N - 1)}{2} + \frac{B_N \left(\frac{\lambda}{B_N \mu} \right)}{1 - \left(\frac{\lambda}{B_N \mu} \right)} = \frac{(B_N - 1)}{2} + \frac{B_N \lambda}{B_N \mu - \lambda} \tag{6.7}$$

For a given arrival rate λ, the average WIP performance (and hence the average cycle time, because Little's Law must still hold), is determined by the batching number and the dependency of the batch service rate, μ, on B_N. The simplest situation is where μ is independent of the batching number. (In general, the analysis process is likely to have some batch-dependent components, so this case is not found often in practice.) Figure 6.15 shows the results for the average WIP from the model and equation 6.7 when $\lambda = 0.8$ samples/unit time and $\mu = 1.0$ samples/unit time.

The agreement between the model results and equation 6.7 are good. In particular, both predict an optimal value for the batching number at a value 2. The reason for this optimal value of the batching number can be seen from equation 6.7. For a low value of the batching number, the effective traffic intensity of the queuing process dominates the WIP calculation (the second term in equation 6.7). The WIP values are being driven by the queuing processes described in chapter 4. Indeed, for $B_N = 1$, the first term is 0, as we would expect. As the batching number increases, the effective traffic intensity of the queuing process drops, and the impact of this term drops. Now the first term begins to dominate. As the batching number increases, we have to wait longer for enough samples to arrive so that a batch can be completed. (It is easy to show by differentiation that if we treat B_N as a continuous variable, the optimal batching number, B_{NOPT}, and the corresponding minimal average WIP, WIP_{MIN}, are given by this equation:

$$B_{NOPT} = \rho\left(1+\sqrt{2}\right) \qquad WIP_{MIN} = \rho\left(\frac{3}{2}\sqrt{2}\right) - \frac{1}{2} \tag{6.8}$$

Setting $\rho = \lambda/\mu = 0.8/1.0 = 0.8$ gives $B_{NOPT} = 1.9$ and $WIP_{MIN} = 1.8$, so that the actual optimal batching number is 2, in agreement with figure 6.19.)

The results in equation 6.7 are an approximation even for the case of exponential random variation because the arrival time distribution of batches to the queuing process will not be exponential, even if the arrival time distribution of samples is exponential. If the batching number is B_N and the sample arrival rate is λ, then the arrival time distribution of batches will be nonexponential with mean B_N/λ and standard deviation $\sqrt{B_N}/\lambda$.[1] The coefficient of variation, COV,[2] for the arrival time distribution of the batches is thus $1/\sqrt{B_N}$. In order to calculate the WIP, we should then use equation 4.6, which is an approximation that can be applied to nonexponential distributions. Using the equation for exponential distributions, equation 4.5, will tend to overestimate the average WIP values, which is exactly what we see in figure 6.15. We leave it as an exercise for readers to apply equation 4.6 to the batching process.

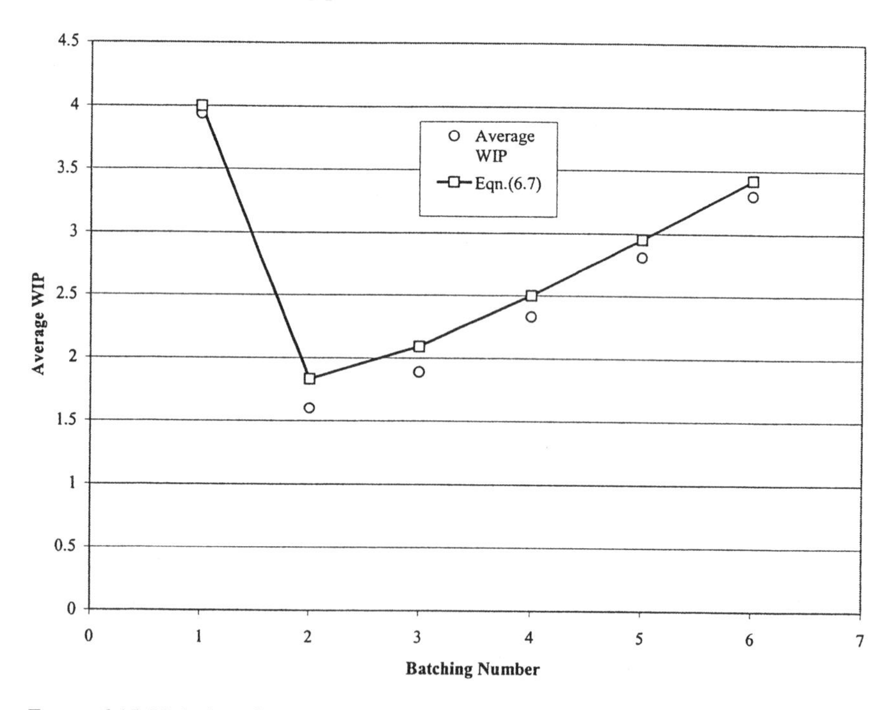

FIGURE 6.15. Variation of average WIP with batching number

1. Because each sample arrival time is independent.
2. COV = standard deviation/mean.

Because the impact of increasing the batching number is to reduce the effective traffic intensity at the queue as well as the COV of the batch arrivals, we should anticipate that increasing the batching number will reduce the variation in the process. Figure 6.16 illustrates this by showing the impact of the batching number on the average WIP profile. Figure 6.16 shows five runs when the batching number is 1 (the upper five profiles) and five runs when the batching number is 5 (the lower five profiles).

In effect, we are using the capacity of the analytical instrument to absorb the random variation in the arriving samples and the analysis times. One can anticipate, then, that the impact of the batching number will depend on the amount of random variation in the process. MODEL 6.14A.STM and MODEL 6.14B.STM are provided so that readers can look at the performance of the process with normal distributions and uniform distributions, respectively.

The observant reader might ask why we do not use the cycle time standard deviation as measured in the converter CT STDEV to show the impact of batching number on the variation in the process. The reason for this is that the way ithink works. When a set of joined items exits a flow, the average of the cycle times for these items is used in the calculation of the standard deviation function CTSTDDEV. Because the standard deviation of an average of a set of values is lower than the standard deviation of the individual values, this effect alone will reduce the value of the cycle time standard deviation as measured by ithink as the batching number increases. Thus, it will be difficult to make a proper comparison of the cycle time standard deviations using the built-in ithink function. Another way to proceed would be to break the samples apart (as we did in the repair model of the previous section) by adding another stock and measuring the cycle time standard deviation after that stock. However, this does not correctly address the issue either.

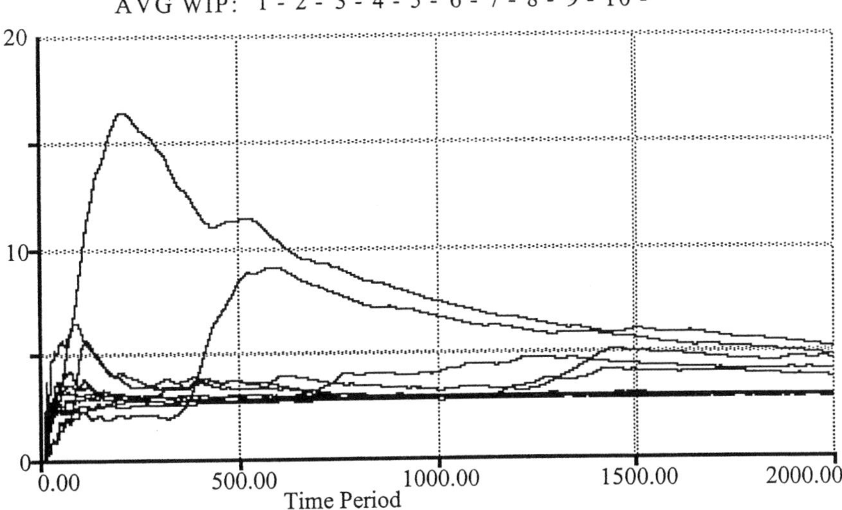

FIGURE 6.16. Impact of batching number on variation in average WIP

We said earlier that the analysis time may consist of two components, a batch-independent component and a batch-dependent component. MODEL 6.14C.STM has both batch-independent and batch-dependent components included in the calculation of the service rate, μ. The converter MU in figure 6.14 is no longer a given value but is calculated from the batching times according to the following equation:

$$\mu = \frac{1}{(T_{BI} + B_N T_{BD})} \tag{6.9}$$

where T_{BI} is the batch independent time and T_{BD} is the batch-dependent time. If $T_{BD} = 0$, then we have the model in figure 6.14. Readers can verify that if T_{BD} is increased while keeping the sum $(T_{BI} + T_{BD})$ the same, then the average WIP and its variation increase. The reason is that an increasing amount of batch-dependent time means that the impact of the batching number on the effective traffic inten-

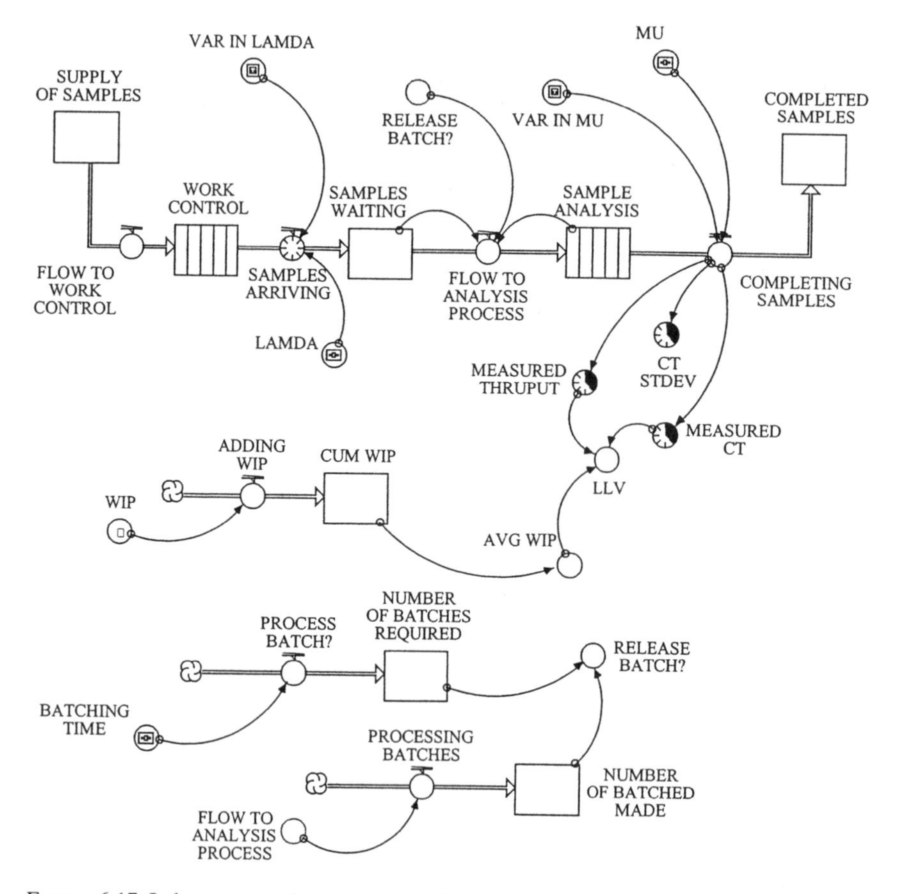

FIGURE 6.17. Laboratory analysis process with batching based on time

sity (equation 6.6) is lower. Consequently, the average WIP and its variation are higher.

As a final model, we leave the reader with a batching model based on batching time, not on batching number. The model for this process is shown in figure 6.17.

Every batching time, an item is flowed into the stock NUMBER OF BATCHES REQUIRED. The equation for the inflow PROCESS BATCH follows:

PULSE(1,BATCHING_TIME,BATCHING_TIME)

which shows that the process requires a new batch every BATCHING TIME. The stock NUMBER OF BATCHES MADE accumulates the number of batches actually made by using the FLOW TO ANALYSIS PROCESS to indicate when a batch has been produced. The equation in this flow is the following:

IF FLOW_TO_ANALYSIS_PROCESS > 0 THEN 1/DT ELSE 0

Then the converter RELEASE BATCH? has this equation:

IF NUMBER_OF_BATCHES_REQUIRED >
NUMBER_OF_BATCHED_MADE THEN 1 ELSE 0

Now the flow logic in FLOW TO ANALYSIS PROCESS equation is this:

IF RELEASE_BATCH? = 1 AND SAMPLE_ANALYSIS = 0 THEN
SAMPLES_WAITING/DT ELSE 0

This logic ensures that once the requirement for a batch has been created, it will remain until the actual batch has been sent to SAMPLE ANALYSIS. But if the batch cannot flow until after the batching time because the conveyor SAMPLE ANALYSIS is not empty, it will batch all available samples when the conveyor becomes free. Also, if the stock SAMPLES WAITING is empty when it is time to create a batch, a batch will not be created until a sample arrives at the stock. This means that in situations where the probability of this stock being empty is significant (the sample arrival rate is much less that the service rate), the time between batch arrivals will not be constant equal to the batch time. Therefore, we cannot model this type of batching as if it is a queuing process with a deterministic arrival rate (COV of the arrivals = 0). We leave it as an exercise for readers to run this model and compare its performance with the batching by number model in figure 6.14.

7

The Supplier Interface: Managing Risk

He that runs the risque deserves the fair.

—Susanna Centlivre, *A Bold Stroke for a Wife*

7.1 Introduction

The perception and calculation of risk are extremely important issues in business today. We frequently fly, drive, or cross a busy street with subtle calculations of risk versus benefit. We use our own history and that of others to inform us of risk.

For example, imagine that every day at 11:00 A.M., you arrive at your favorite store and wish to park in a city-metered lot for exactly two hours. The cost of parking is $1/hour. The parking police visit the lot at random every two hours, but they are four times as likely to come during the first hour as during the noon hour. The fine for an expired meter is $5. How much meter time should you buy? The answer is one hour, as this is the point at which meter savings equals the probability-weighted fine.

There are countless examples where such calculations could be made if we had enough information. In this chapter, we present a novel way of viewing the risks involved in choosing between two apparently similar ways of solving the same problem. It is based on an analogy with the theory of statistics. Risk considerations can be mathematically placed in a hierarchy of statistical moments: first moment, second moment, third moment, and so on. Our analogy to these moments is that managers can use a similar hierarchy in structuring their decision-making process. Although most managers are those who dwell almost exclusively on the demands of the first moment—the mean value of the variables that affect their decisions—managers who rise within the organization consider the moments further into the hierarchy. Let us elaborate.

Too often we have noticed that when faced with a difficult decision for an operating system, managers focus on the average behavior of that system. To most of this class of managers, getting the inputs arranged to meet the average output is the best they feel they can and need to do. The issues of output or demand variation (second moment), the skewness of that variation (third moment), or the problem of rare disruptions (fourth moment) are deemed beyond their control and therefore, beyond their concern. Yet it is these very issues that bring many systems to unexpected difficulty when they are not anticipated in the operating plan.

We use a simple dynamic model to explain the dilemmas encountered as these issues appear sequentially. Such models give invaluable insights into the way the dynamics of systems produce such complexity. Each dilemma may be identified with one of the moments of statistical analysis. We see this process as a useful construct to explain the range of complexities that arise in system management. Such an explanation is essential input to good management decision making. The process is also useful for elaborating and checking the operational view of the systems for which one is responsible.[1]

Our model displays the ideas behind the concept of managing by the moment. The manager is faced with determining the appropriate balance of supply and demand for electricity. He has responsibility for two communities whose demand is variable. He tries to do the best possible job of matching supply and demand with either of two options: a single large plant, serving 1,000 people;[2] or two plants, each half the size of the single large plant and each serving 500 people.

7.2 First-moment managers

First-moment managers focus on the mean: Does the supply (expressed as a single value) equal the demand (again a single value)? In the case of our two communities and the choice of one or two plants, the decision for this manager is obvious indifference: either one large one or two small ones of half-size each. In either case, the mean supply is matched to the mean demand: end of story. The manager would choose the option with the lower cost. In our case, this is not so clear. One 1,000-mw plant is probably cheaper than two 500-mw plants, but the larger plant requires high voltage transmission lines and will be "less local" out of necessity. The small plants are situated close to their demand. This type of manager recognizes that demand will outstrip supply about half the time but does not recognize the distinction in the variation between the two options.

> Learning Point: Most managers manage by observing only the mean value of the critical variable in their process.

7.3 Second-moment managers

Second-moment managers compare the mean demand with the two supply options and realize that the mean supplies meet mean demands, but they go on to

1. Bingham, A. and B. Hannon, "Managing by the Moment," Draft, November 2000.
2. Consider these numbers as representing thousands of people.

ask if the variance will suggest a difference in continuity of supply. That is, will one of the two options be more dependable than the other, given that the demand is going to fluctuate and may sometimes exceed supply? This manager is aware of the second-moment issue. He asks whether or not the two-plant system fails to meet demand more often over the years than the single-plant system.

We construct a model of both systems to explore these options. Let the demand for each of the two cities be identical but partially random: 450 + random(0,100), giving an average demand of 500 units per city (CITY 1 DEMAND, CITY 2 DEMAND). The supply of each small plant is 500 (SUPPLY), and that of the single large plant is 1000 (2x SUPPLY).

In the two-plant system variables CITY 1 COUNT and CITY 2 COUNT, we ask for a 1 whenever the demand at either city exceeds its own supply. In the single-plant system we ask for a 1 in ONE–PLANT COUNT whenever the combined demand is greater than SINGLE SUPPLY, which is twice the single-plant supply. The shortfalls are counted in TWO-PLANT BLOUTS for the two-plant arrangement, and in ONE-PLANT BLOUTS for the single-plant arrangement. The blackout frequency ratio is simply the ratio of the number of one-plant to two-plant blackouts. The impact ratio is the ratio of the number of blackouts in the single-

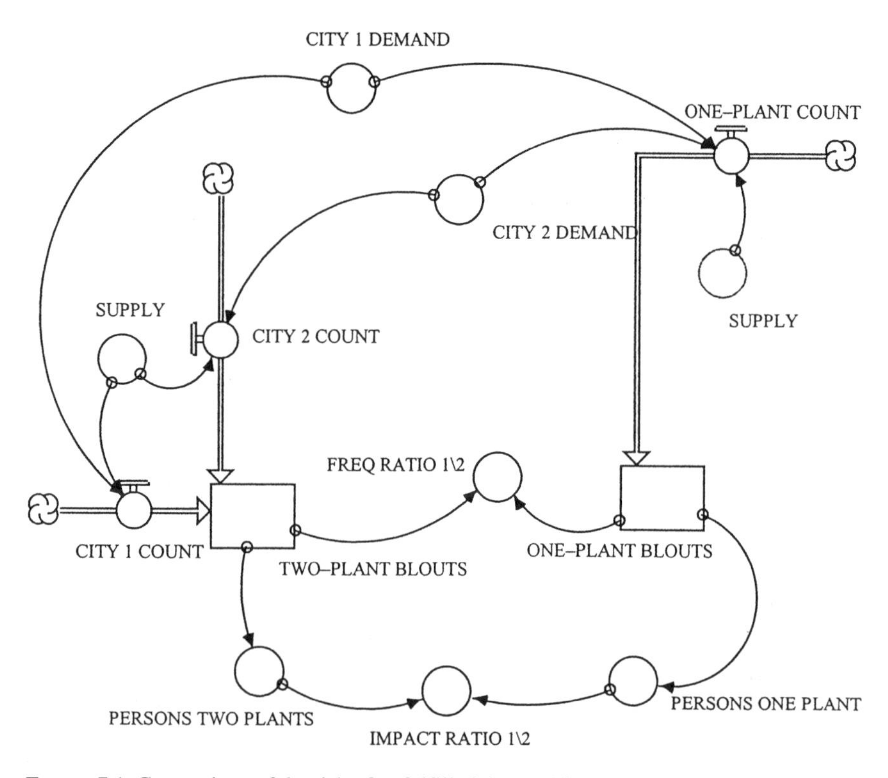

FIGURE 7.1. Comparison of the risk of unfulfilled demand for a single large generator compared to such a risk for two half-sized generators, one devoted to each city

plant system times 1,000 (thousand) people, to the number of two-plant blackouts times 500 (thousand) people. The results are shown in figure 7.2, curve 1.

The blackout frequency ratio is 1/2 and the impact ratio is 1. It is clearly much riskier in terms of shortfall *frequency* (that is the number of times demand exceeds capacity) to generate from the two smaller plants than from the large single plant. The number of people impacted by blackout is the same in either system. Given this data, the manager would opt for the single-plant system.

Now suppose that our manager asks what blackout frequencies and impacts would occur if the two small plants were interconnected such that when the demand in one city exceeded its supply and the other plant had available capacity, they could transfer electricity between them, avoiding a blackout in that period. As we shall see, this limited share protocol reduces the two-plant blackout impact to less than that of the single plant. The blackout frequency in the single-plant system is still less than that of the two-plant system. However, this protocol reduces the two-plant frequency to less than that of their noninterconnected condition.

The logic in CROSSOVERS =

if (CITY_1_DEMAND – SUPPLY > 0 and SUPPLY – CITY_2_DEMAND > 0 or CITY_2_DEMAND – SUPPLY > 0 and SUPPLY – CITY_1_DEMAND > 0) and CITY_2_DEMAND - SUPPLY < = SUPPLY - CITY_1_DEMAND then 1 else 0

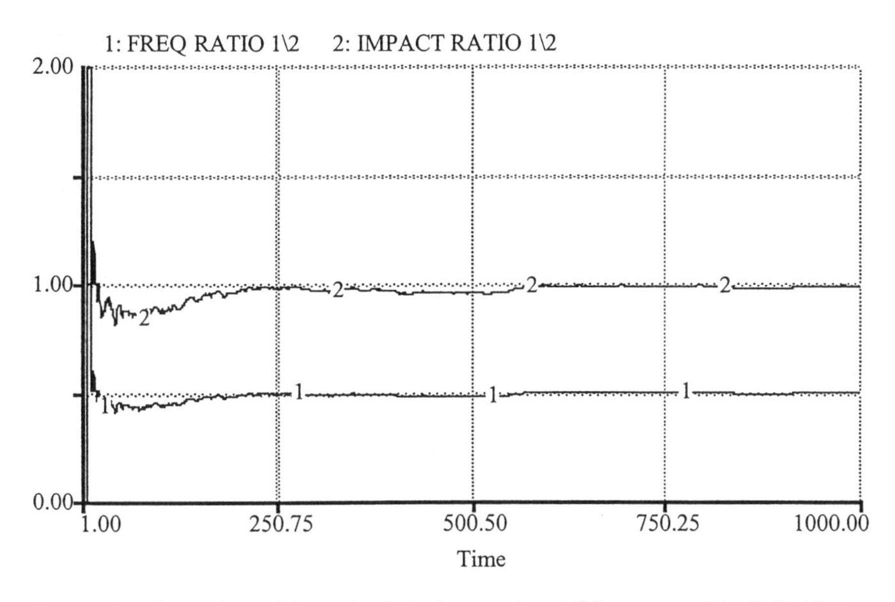

FIGURE 7.2. Comparison of the ratio of blackout or shortfall frequency of ONE-PLANT to TWO-PLANT BLOUTS (curve 1) and the ratio of the impact (people affected) by these blackouts (curve 2): Typical values are 1/2 and 1, respectively

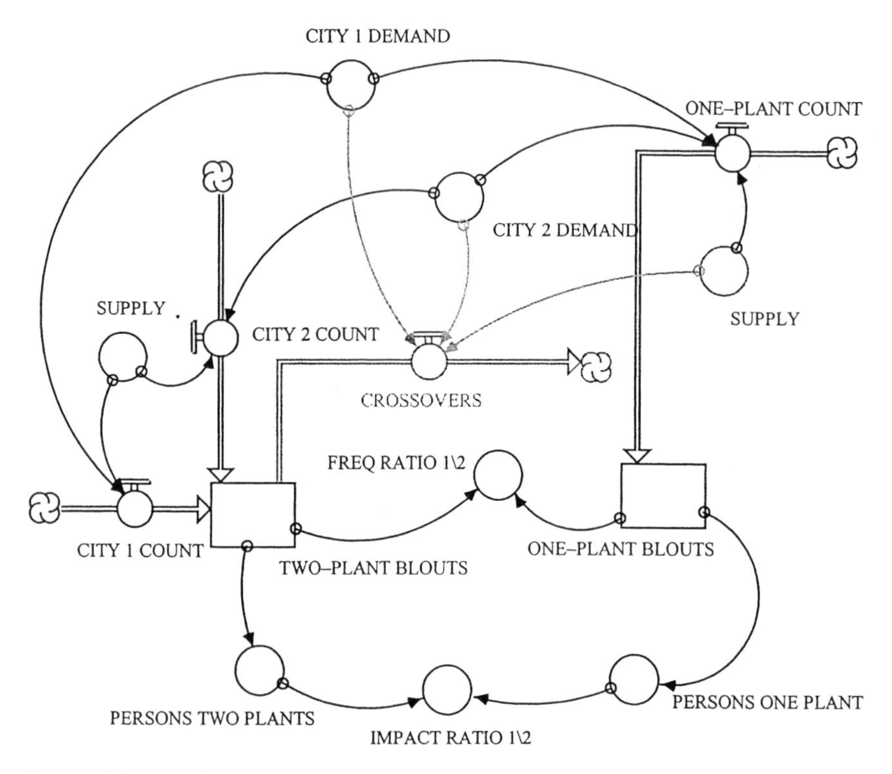

FIGURE 7.3. A model to elaborate on the importance of variability: The lighter elements were added to the model in figure 7.1

This statement controls a correction to TWO-PLANT BLOUTS. If either city's demand exceeds its supply, a 1 is added to TWO-PLANT BLOUTS through the two city count variables. CROSSOVERS take a 1 away from TWO-PLANT BLOUTS if the excess demand in either city could be met by the excess supply in the other city during that time period. Interestingly, the addition of a transmission line between the two cities changes the IMPACT RATIO 1/2 to favor the two-plant layout, as can be seen in figure 7.4. The second-moment manager has a dilemma in this case. If the two small plants are interconnected, the single-plant system suffers blackouts less frequently but the two-plant system blackouts affect fewer people.

Our second-moment manager must pay attention not only to the shortfall frequency but also to the impact of those shortfalls. He needs a decision criterion to resolve the choice. He could look to the third moment of the two systems for an answer.[3]

3. Of course, we believe that a good manager looks at all four independently, even if the examination of previous moments does not give ambiguous results.

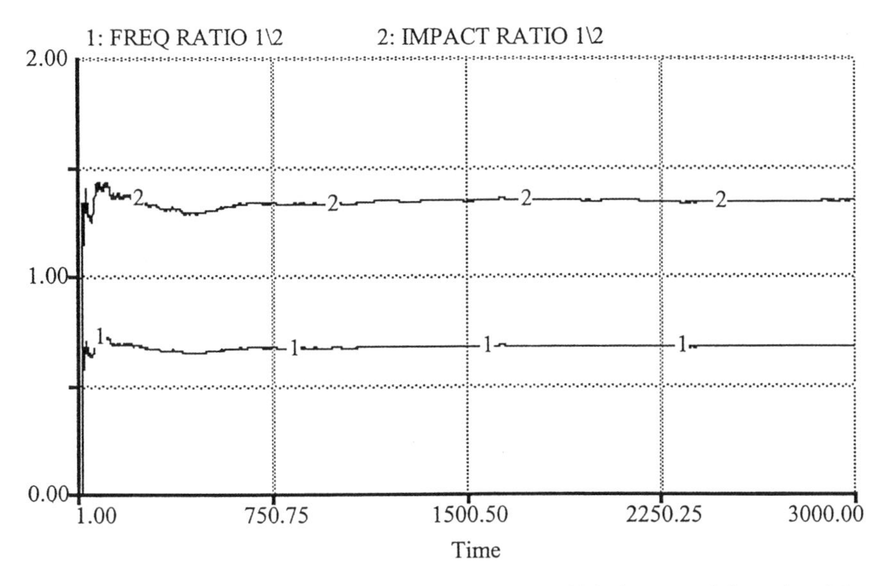

FIGURE 7.4. A display of the ratio of the frequency (1) of blackouts, and the ratio of the number of people impacted (2) by blackouts, those of single plant blackouts/two-plant (cross-connected) blackouts: The IMPACT RATIO is 4/3 while the FREQ RATIO is 2/3

> Learning Point: Some managers give attention to the variance of their critical variables. Complications can arise that are difficult to resolve without a model of the process.

7.4 Third-moment managers

Third-moment managers pay attention to the mean, the variance, and the skewness of the data that could affect their decisions. Suppose that the CITY 1 DEMAND and CITY 2 DEMAND were skewed such that excess demand was more likely to occur than excess supply. That is, suppose that CITY 1 DEMAND = CITY 2 DEMAND = 450 + RANDOM(0,200). The single-plant supply is 500, so excess demand is three times as likely as undercapacity demand. The FREQ RATIO1/2 drops from 2/3 (figure 7.2) to 0.59 and the IMPACT RATIO 1/2 drops from 4/3 to 1.18. These ratios are for the situation in which the two plants are cross-connected, but this feature makes little difference with such a skewed demand. So the upshot of such skewness toward excess demand is that the two-plant arrangement is less desirable; the conclusion is the same as that drawn from an examination of figure 7.1. If the situation is reversed and the demand is three times as likely to be under capacity [CITY 1 DEMAND = CITY 2 DEMAND = 350 + RANDOM(0,200)], then these ratios become 0.73 and 1.46, respectively.

Thus, when the chances are greater for excess capacity, the two-plant system is favored even more from an impact perspective (still disfavored from a frequency perspective, but slightly less so).

When the demand exceeds the supply, the customers are disadvantaged. When the supply exceeds the demand, the company stockholders are disadvantaged. A good company manager, a third-moment manager, is aware of the mean demand and its variation, and whether or not it is more costly to allow demand to exceed supply or vice versa. With the introduction of a cost factor, the variation of excess demand/excess supply becomes skewed.

> Learning Point: Few managers are concerned about the skewness of the variability of their critical variables. Such attention can produce subtle insight.

7.5 Fourth-moment managers

Generally, good managers work to avoid surprises, those infrequent events that might lead to especially significant costs. In our example, a fourth-moment manager might fear that a large-scale electricity shortfall could cause sudden and significant social reaction. These risk-averse managers prefer randomness without a tail on the distribution of demand. For example, they would prefer the uniform distribution of demand to a Normal or Poisson distribution: in doing so, this manager avoids the infrequent but calamitous events. However, when the demand variation actually has a long positive tail, fourth-moment managers will need an emergency plan for unusual events. Plans might include using an emergency generator or having the ability to buy power quickly from a neighboring company. For example, it could be that blackouts occur whenever the combined demand exceeds the capacity by a fixed amount, say 50. In this case the single plant fails when the demand is 1,050. For either of the two plants to fail, however, the demand in either city must exceed 550, a less likely event. The fourth-moment manager would prefer the two-plant system in this case because the latter condition is obviously less likely to occur.

A manager who emerges from "managing by the mean" or from the status of the first-moment manager, will view the variation of demand and supply, the possible skewness of supply/demand variation, and the consequence of large but unlikely events. When electric utilities were regulated monopolies, the customer alternatives did not exist. Consequently, avoiding the impact of power interruption was the responsibility of the regulatory commissions. With deregulation of this industry, the role of the consumer is more important but still far from fully clear. Under current deregulation plans, the old monopoly power will be reduced and customers will be able to change generators. Then the cost of customer dissatisfaction will be calculable, and even the economics may favor the two-plant system. It is surprising how this simple model could produce so many complexities.

> Learning Point: Very few managers have contingency plans for those rare, extreme events involving their process.

This little demonstration should add to your understanding of the importance of greater management sophistication and its crucial dependence on dynamic modeling. Moment management is really a formal method of managing risk in an enterprise. The intriguing part of this concept is that it tends to assure a manager that a hierarchical framework covers the most likely possible nuances of complex system management. Those of you in the world of management, whether in the private sector, the government, or academe, should be able to benefit from such a powerful perspective. The next time you are part of a discussion during which a group is trying to reach a difficult decision, notice who focuses their thoughts on which *moment* of the problem.

Another aspect of risk management deals with the *responsibility* for risk. For example, pollution from a plant may be exposing the nearby public to a substantial health risk. The emission rates are within the current Environmental Protection Agency (EPA) regulations, but the company holds special information about these emissions that indicate the risks are significantly higher than those that were the basis for the EPA standard. The company is legally protected by the regulation but faces a moral dilemma. Perhaps that dilemma disappears when the moment analysis is considered. The mean level of the pollution is within the standard. But what are the variations in such emissions? Are the company's internal costs with respect to the operation of the plant skewed? That is, is it possible for the internal costs of an event, accompanied by temporary emissions well over the standard, to be extremely high? Is it possible that an extreme emission event could so concern the surrounding public that they mount a well-organized publicity campaign that unearths the company's internal prior knowledge of the danger, tarnishes the company's national reputation, and leads to a tightening of the standard? By examination of the deeper moments, the decision to shift to a nonpolluting alternative process might be made, quietly eliminating the moral dilemma.

Inherent in such considerations is the nature of the risk. Is the risk voluntary or involuntary? It is common for risk managers to question why a public should be so concerned over such a risk from the pollution and apparently not so concerned about the greater chance of getting more seriously injured crossing the street in busy traffic, or getting hit by lightning, and so forth. Such a management view misses the concept of involuntary risk. The public is not agreeing prior to the management decision to emit this substance: it is a risk they bear involuntarily. Furthermore, the risks are additive. That is, the risk of impact on health from the pollution is *added* to the risk of crossing busy streets, getting hit by lightning, and so forth. Adding the pollution impact does not lower the risks the public is already taking. Finally, the average person in the public takes steps to counter these risks—they insist on streetlights and driver's license exams and they stay indoors during thunderstorms, for example. So the management of a polluting industry can expect an unusual reaction from the surrounding public when they have the opportunity to express themselves. A first-moment manager, especially, would be surprised by such a reaction.

On the other hand, the burden of such a pollution risk might be borne voluntarily by some of those in the surrounding public. Employees of the company use cognitive dissonance to allow themselves to live under such circumstances: they are thus able to keep such contradictory thoughts out of their mind. They inherently are inclined to speak against further pollution control, especially if the company indicates that such control would be so costly that they would close the plant rather than reduce the pollution. Good managers are aware of the power of cognitive dissonance and its effect on their employees.

Still, some companies have found that a focus on pollution reduction, with the realization that all pollution standards will be tightened eventually, leads to a general increase in creativity within the company, resulting in a host of new and profitable products.

This brief discussion of risk could not begin to cover the full range of such considerations for managers at all levels. The literature is ripe with advice on risk management. We want to bring some fresh perspectives to the concept of risk management and to stress the importance of dynamic modeling to reveal the nature and depth of risk.

8

The Customer Interface

> The authors and leaders of opinion on mercantile questions have always hitherto been of the selling class.
>
> —J.S. Mill, *Political Economy*

8.1 Introduction

One of our most difficult issues involves understanding the complex way in which the world of the customer connects with the world of retail business. The vagaries of pleasing customers on the one hand while dealing with suppliers on the other can sometimes seem overwhelming. Charge too much or not have the article in stock, and you may never see the customer again. Order too much or too often, and you may run up the inventory cost, causing profits to vanish. This is a balancing act if ever there was one. How can you walk that tightrope and survive? Reliance on mental models of a process fraught with randomness, feedbacks, and delays makes the most astute of mental management models unreliable. We will demonstrate how the computer modeling process unfolds and show you how to develop a more and more accurate picture of the dynamics involved. In a real sense, the mental and computer model of the process become one.

Consider a business that orders widgets from vendors and then sells the widgets to customers. First, realize that we must represent these customers by some sort of mathematical description for the dynamic appearances of potential sales. We must do the same to represent the vendors' responses to our call for more widgets. We connect these two entities with an inventory, and we realize that items in inventory are costing us more the longer they sit there. We continue the development of the inventory problem until we are satisfied that we have captured most of the physical realities of the problem. The next step is to consider the overarching reasons we are running this business: to make a profit, to meet desired profit goals, or to expand market share. We will then need the tools to size our business to optimize the desired goal. These tools are of two general types. We need the methods to cast our otherwise physical problem into a financial one. Then we must understand the technical procedures to optimize the financial goal.

In the first part of this chapter, we develop a model based on inventory control, the Make-to-Stock model. Then, we develop the model for the Make-to-Order process, which simulates companies that have no inventory.

The first part is aimed at describing the physical aspects of the inventory system. Next, we discuss the objectives and decide which to use for the exposition of method. Finally, in appendix D, we examine the technical aspects of optimizing the parameter choices to achieve a specified goal.

8.2 Controlling the inventory level: Make-to-Stock model

Inventory or Make-to-Stock models are useful for modeling the flow of relatively low-priced, high-volume items.

Anyone in any business finds the pace of sales does not match the rate at which orders arrive. The only way to keep from turning away unsatisfied customers is to have an inventory of unsold items in the store. This inventory acts as a buffer between incoming orders and outgoing sales. A simple inventory model would look like figure 8.1.

We will consider the procedure of controlling the stock level by proper ordering procedures, and will develop such models and show how to optimize the system to meet physical objectives. Then we will model the process of efficient response to sales orders. Ultimately, we want to control the inventory to reduce the number of times customers find the store unable to sell them the desired product.

Consider an initial stock (STORE) of seven widgets. The widgets are sold at a rate of one per day, with a weekly shipment of seven. What do we expect the steady state to look like? Receiving is represented by the built-in function PULSE (7,7,7), which mean seven units will arrive on day 7 and every seventh day thereafter. The STORE variation is a bit more difficult to predict. First, we can assume that we will not run out of widgets because the long-term average of RECEIVING is equal to the steady rate of SELLING. So we would expect the STORE to rise from zero on the day of the incoming shipment to seven less the one unit sold that day (that is, to a peak of six). Then STORE should decline steadily to 0. This is just what the model does, as shown in figure 8.2.

Is this model complete? It behaves as we expected, but does it reflect reality? Try running the model with a different starting stock in the store (for example, try 10). What happens? The peak values of STORE are now different. They are elevated by three (the difference between 10 and the original seven starting STORE

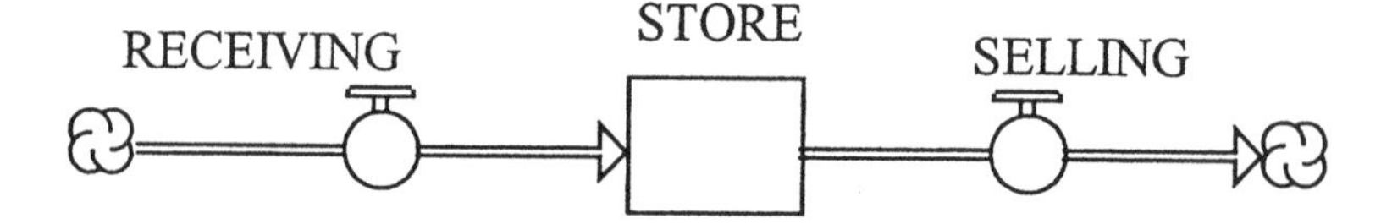

FIGURE 8.1. The beginnings of an inventory model

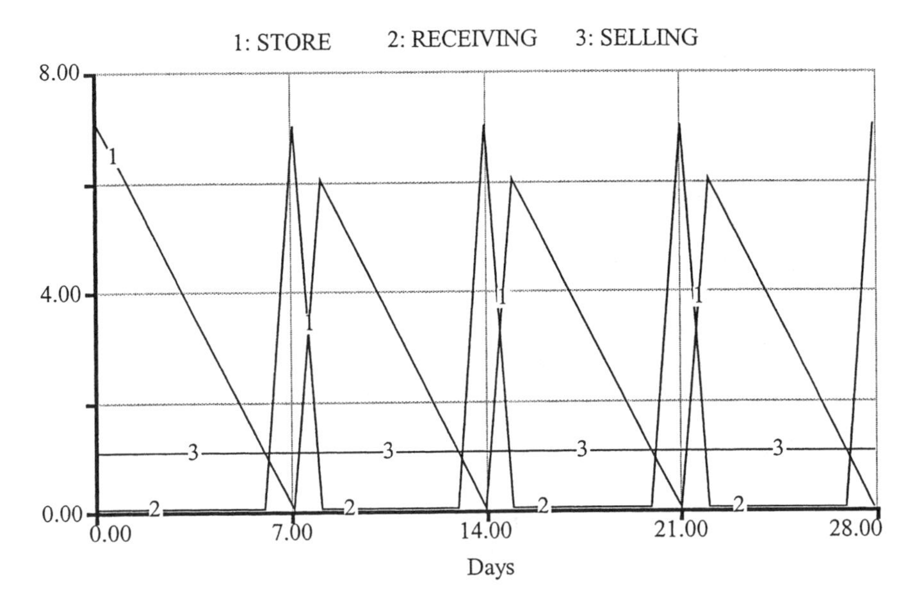

FIGURE 8.2. The plot of RECEIVING, STORE, and SELLING for the simple inventory model: The DT is 1

level). This will not do: we have an inventory that is dependent on the initial value of that inventory. Later in this chapter, we find that this dependency disappears when the model is developed with controls.

Reset the initial STORE to seven, and then change the day of the first received shipment to three. So RECEIVING is now Pulse (7,3,7). What happens? The STORE variation is still six as before, but the peak is 10, while the minimum is four. We have found another unrealistic anomaly of our model. By the third day, when the first shipment arrived, the STORE had dropped to four. On that day, a shipment of seven arrived and one item was sold. So the new STORE became 6 + 4, or 10 in the fourth period. This cycling continued unchanged, leaving us with an unnecessarily large inventory.

A further reality check shows us that the model lacks actual selling variation. Make SELLING a random variable with the same mean, say, SELLING = POISSON(2). This built-in function will ensure a positive integer set of random choices with a mean of 2. Rerun the model and note what happens. The STORE level drops to 0 routinely. (Note that the default setting in the STORE dialogue box prevents negative values). This means that customers are being turned away due to lack of inventory—generally not a good practice. Change the first entry in the RECEIVING Pulse function until the STORE level is 0 instantaneously. You should find that you must nearly double the RECEIVING rate from 7 to 14 to eliminate the empty store problem. At the same time, the peak STORE level increases significantly, from 6 to about 15.

> Learning Point: A random variable exaggerates the levels of important variables in the model. Randomness in a process makes mental modeling difficult.

We are getting closer, but the model is still unrealistic. Consider now the fact that SELLING is not only a variable but also is not always equal to the DEMAND. We might define the daily DEMAND rate to be, for example, either a normal distribution or a Poisson distribution. We choose the latter for our example, as it has been shown that this distribution describes many consumer patterns. Add DEMAND to the model and define it as POISSON(2). In addition, let there be a minimum stock level maintained at the store so that every customer can always see the product and perhaps place an order for later delivery. Set this minimum STORE level at 1. SELLING now becomes a conditional statement: IF STORE – 1 – DEMAND > 0 THEN DEMAND ELSE STORE–1.

Again the store level drops to the minimum possible level (1, the "presentation" level). Run the model several times to get a good feel for the range of possible variations in the curve. Raise the pulse level in RECEIVING until the presentation level is reached only infrequently.

> Learning Point: The inclusion of conditional statements that constrain the model dynamics realistically can produce unexpected results. Such constraints are difficult to incorporate in our mental models.

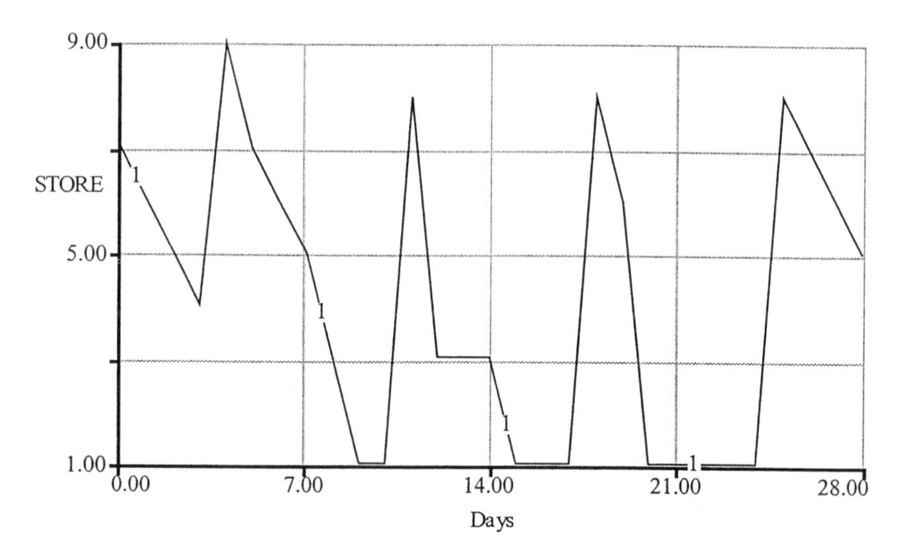

FIGURE 8.3. The model in figure 8.1 with DEMAND (= POISSION(2)) driving SELLING with a minimum presentation stock of 1, an initial STORE level of 7, and RECEIVING set at PULSE (7,3,7)

Our model is now more realistic on the SELLING side and much less so on the RECEIVING side. To add realism here, we add a stock called SHIPPED and turn it into a conveyor (rather than the default choice of reservoir). We must also specify the transit time (the delay) for the widgets between the order placement and their arrival in our store. This is the time between ORDERING and RECEIVING at the STORE. We make the transit time seven days and enter this number into the dialog box in RECEIVING. Think of the conveyor as a continuous moving sidewalk with seven slats or platforms chained together in a line. We must specify the initial distribution of widgets on each slat. To simulate the ORDERING process, we will use a simple idea. We will order seven whenever the STORE level drops to two. Once that order is placed, it will arrive in the store seven days later. In the meantime, SELLING will take place independently as long as more than one widget remains in the STORE. Figure 8.4 shows how the model should look.

A typical result can be seen in figure 8.5.

The model now shows a high STORE level with rather frequent out-of-stock conditions, indicating that the rule for ordering is a rather poor one. These rules (or controls, as they are some times called) will be elaborated later in this chapter, once we have decided on the proper objective or goal for the model.

The model is exhibiting a definite cycle of about 36 days. This cycle is a result of the delay or transit time and its interplay with the ORDERING process. The ORDERING process is a function of the STORE level, which in turn, is a function of SELLING, which in turn, is a function of the DEMAND.

> Learning Point: The delays or lag times in models can produce counterintuitive results. Delays make mental modeling exceedingly difficult.

Tinker with this model. Change the transit time and the amount ordered, noting the effects each has independently and jointly on the STORE level.

Change the level of ORDERING from a constant 7 to a RANDOM (4,10), a random function with the same mean. To ensure that you get whole widgets, write this function as INT(RANDOM(4,10). Note how much increased variation in the STORE level such randomness has wrought.

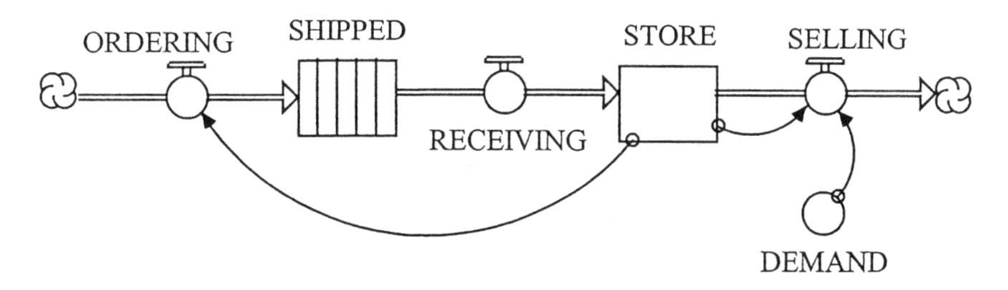

FIGURE 8.4. The model expanded to include a delay after ordering

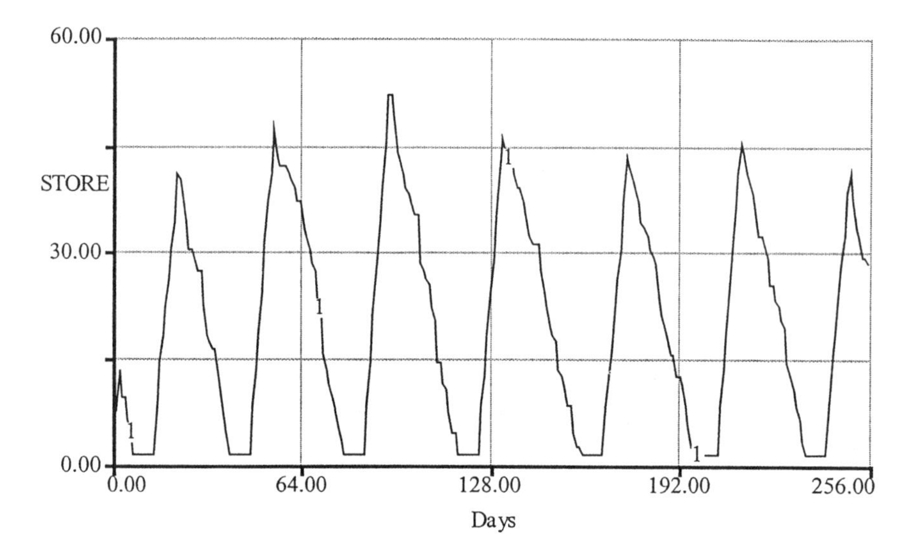

FIGURE 8.5. The variation in the STORE level for the model shown in figure 8.4

It should be clear by now that the ordering process must be more rational. To do this, we must define two ideas. First, we must establish a proper set of controls. These controls represent all of the downstream data on sales, inventory, and shipping practices. We will do this in the next section of this chapter. With a fully representative set of ordering controls, we have done all that can be done by way of the controls. Second, we must define our objective. Do the managers of this store want to meet physical objectives, such as a target inventory or no more than a specified number of disappointed customers? In the following sections, we show how the model can be optimized for physical and financial objectives.

Choosing the controls: The critical skill in the development of any model is to distinguish between those variables that are to be assumed as given and unchangeable, those which the manager must treat as objectives, and those which can be used to control the system to optimize one's objectives. This should be done continuously as the model is developing. To fully explain the development of the appropriate controls, we will start with the model shown in figure 8.4 and add the controls to that model.

Actual store operators develop a target inventory (STORE TARGET) and place orders when the STORE level drops below this target. This is a formal statement of what we were doing earlier. Let this target and the ordering rate become the set of controls for the inventory model. Both are measurable and can be implemented with great practicality (see figure 8.6).

Now the model should look like this: Initially, an order placed seven days ago is just ready to exit the conveyor. Otherwise, nothing is in the conveyor at the start of the model. As before, RECEIVING is where the delay is specified. The condi-

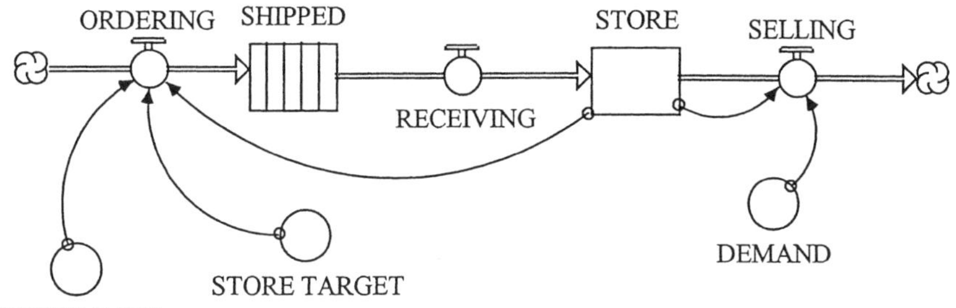

FIGURE 8.6. The model with a delay in the arrival of orders: The timing of the orders is based on the difference between the STORE level and a STORE TARGET level of inventory; DEMAND = POISSON(2); STORE TARGET = 2; ORDER RATE = 7

tional statement on ORDERING (= IF STORE 0 –1 – STORE TARGET < 0 THEN ORDER RATE ELSE 0) replaces the former statement there. The remainder of the model is unchanged. What do we expect to happen to the STORE inventory levels? It is difficult to even guess. The model now contains randomness, a delay, and explicit feedback controls. The results are quite unpredictable.

As before, these results show a cycle of about 30 days with STORE levels reaching to nearly 50 items, and with periodic times when the STORE is empty. This is not a desirable situation: a store that is selling on average two items per day must provide shelf space for 50 items. There must be ways to improve upon this situation.

Note that the key trouble-making parameters in this model are the mean selling rate and the arrival delay time after ordering. These parameters might be thought of as given quantities although, of course, they can be improved on by offering sales and by working closely with the supplier. The other parameters and the conditonal statement make up the controls for the model. We can change the ORDERING RATE, the STORE TARGET and the nature of the condition statement in ORDERING. Change the controls to try to reduce the peak STORE level without causing an increase in the empty store time.

Note in figure 8.7 that ORDERING is high when the STORE level is low and low when STORE is high. Make a table of data on STORE and note also how the STORE never drops below the specified minimum level of 1. During this time of minimum STORE inventory, SELLING is necessarily 0. This condition represents the possibility of customers denied. SELLING might be 0 because DEMAND is 0, but this is not the only condition for 0 sales. When customers are turned away, the firm is losing future business because of customer disappointment. This is a cost to the business, although it is a difficult cost to pin down. A denied customer may be a customer who will never return. In this case, the DEMAND should be reduced accordingly.

One issue should become clear: the number of widgets in transit (i.e., the number in SHIPPED) should help control how much we order. We must modify the

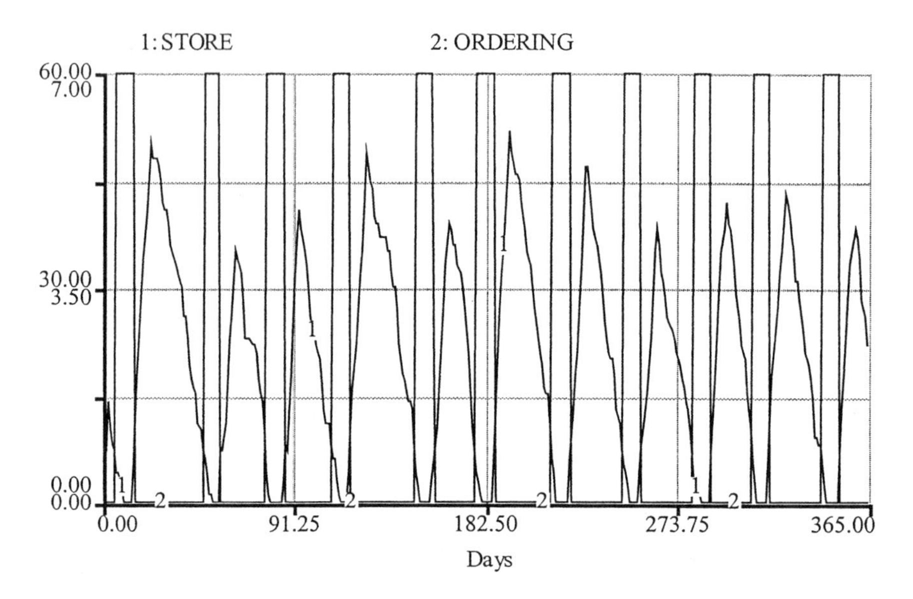

FIGURE 8.7. Typical results for the model in figure 8.6, a model with seven-day delays in ordering

ORDERING procedure to acknowledge the in-transit effect. The ordering effect must also contain information on actual sales, the two target stocks, and the reaction sensitivities to these targets. The actual SELLING is modified by the reactions of the manager to the two gaps, those between the targets and the actual values, to produce the control equation for ORDERING. Thus, the equation is this:

ORDERING = SELLING – round(REACTION_STORE*(STORE – STORE_TARGET) + REACTION_SHIPPED*(SHIPPED – SHIPPED_TARGET))

The four controls can be determined statistically for any manager, or as we will show in the discussion related to figure 8.8, they can be determined to optimally satisfy an objective of the firm. At this point, also add the DEMAND (POISSON(2)) and the conditional statement in SELLING that limits the minimum STORE level to a presentation level of 1. (See the discussion for figure 8.4.)

The flow SELLING should be delayed in this model because the actual data on selling takes some time to compile and relay back into the ORDERING process. If we assign a weekly cost for each unit of STORE inventory and a cost for each turned-away customer, a delay of one week in the SELLING information in the ORDERING calculation raises the total average cost by 12 percent. This is a significant change and underscores the important role that delays play in the business world. Adding similar delays for the information on STORE and SHIPPED made little difference in the average total cost.

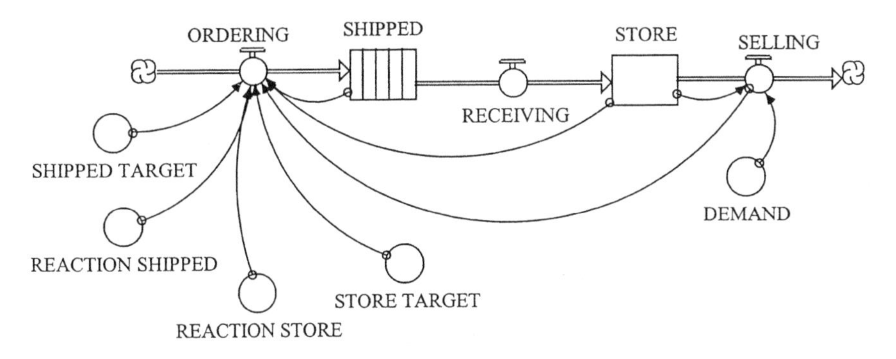

FIGURE 8.8. The model with the controls on ORDERING and SELLING

> Learning Point: To properly control the process, we must know the current sales rate and the current number of items in transit and in the inventory. These are measurable data in this process.

We realize that each manager is likely to have different reaction coefficients and likely to choose different store and shipped targets to control the ordering process. For example, the owner might be insistent on never running out and might have an avid fear of turning away customers or directing them to a substitute. In this case, the fearful manager will maintain high targets for both store and shipped inventories and have high reaction coefficients.

Realizing this, a perceptive owner will want to set the targets and the reaction coefficients and then direct the manager to behave in accord with an ordering rule. But what criterion is the owner to use for setting these numbers? The owner could set a goal of not turning down more that two customer per month and find the targets and coefficients to accomplish this goal—a form of satisficing behavior. The owner might set a sales goal and raise the goal periodically, as a kind of market-share-increasing strategy. The most perceptive owner might calculate all of the potential costs and revenue associated with the business and determine the targets and coefficients that maximize the economic value added (EVA), a quantity closely related to profit. (See appendixes B and D). In this case, the model would be called an optimizing model. We do this in appendix D, in wihch we first optimize the physical model and then add the cost functions to optimize the expected value added, a close relation to profit.

8.3 The Make-to-Order process: Customer interface

In a modeling sense, this version is much easier to represent. Make-to-Order processes are for the larger, more expensive consumer items. Cars are one example. When you buy a car, the one you really want is unlikely to be on the dealer's lot. Ultimately, you specify the one you want and the dealer promises you a delivery time.

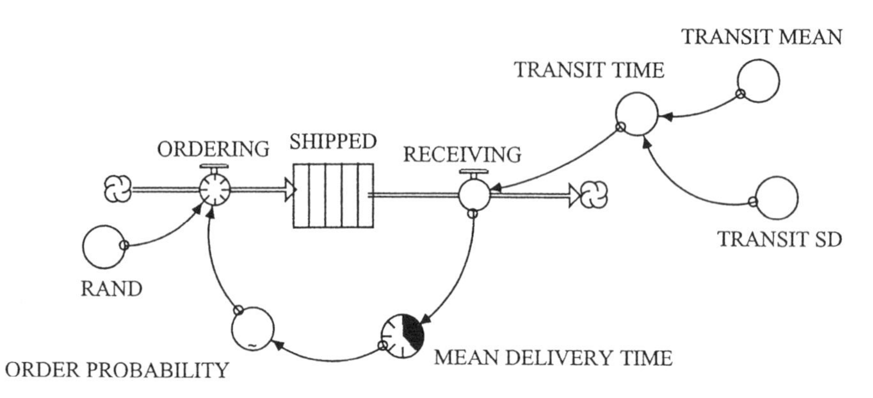

FIGURE 8.9. The model for the Make-to-Order process

If it is more than a week, you might well go to another dealer. Other people in your circumstances might wait longer, or they may be more impatient than you are. So the customers must be represented by a probability of ordering that is a function of the expected delivery time. These probabilities are important experimental data. Once the order is placed, the delivery time is not certain. We could represent the transit time as normal distribution around a given mean time and standard deviation.

This process can be represented with a highly truncated version of the Make-to-Stock model, shown in figure 8.9.

Once an order is placed, it is time stamped and sent into a conveyor called SHIPPED. Immediately on entering the conveyor, the order is assigned an integer TRANSIT TIME (the delivery truck only comes once a day), chosen from a normal distribution with a TRANSIT MEAN and TRANSIT SD. We use the INT (integer) function to ensure that a whole integer TRANSIT TIME is chosen. When the order gets off the conveyor (arrives at the store), the MEAN DELIVERY TIME (CT-MEAN in ITHINK) is calculated. This time has an impact on the customers through a graphical function called ORDER PROBABILITY, shown in figure 8.10.

Once the ORDER PROBABILITY is determined, the process of ORDERING can begin. We compare a randomly chosen number from 0 to 1 (RAND) with the ORDER PROBABILITY. If RAND is less than the ORDER PROBABILITY, then we order 1 item. Otherwise, nothing is ordered. The expression for doing this follows:

IF RAND < ORDER_PROBABILITY THEN 1 ELSE 0

The loop is now complete. We can vary the TRANSIT MEAN to find the corresponding likelihood of ORDERING, as shown in Table 8.1.

TABLE 8.1. Variation of order probability for different transit mean times.

Transit Mean	Order Probability
4	0.98
5.5	0.96
7	0.90
8.5	0.84
10	0.77

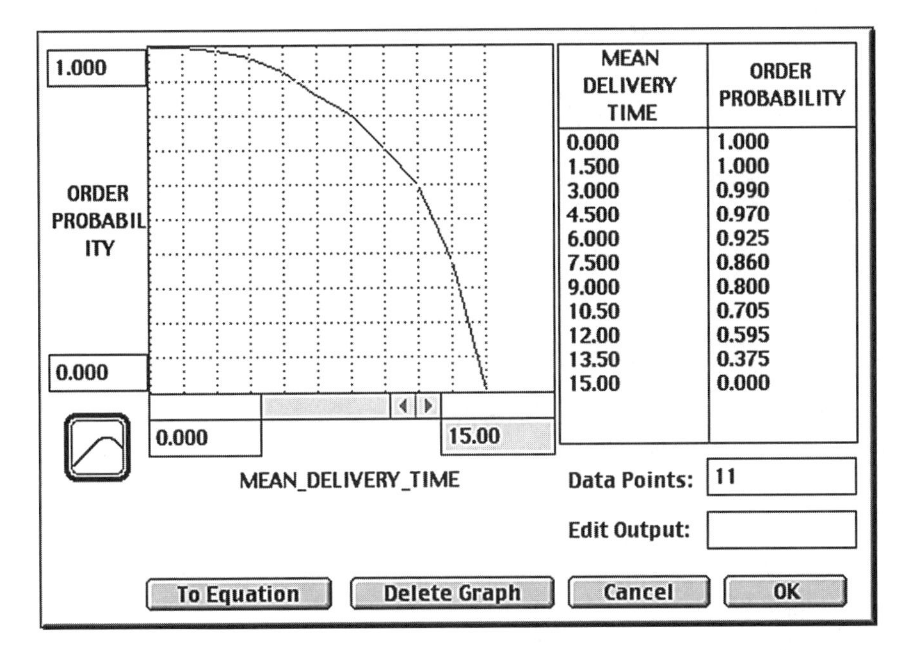

FIGURE 8.10. The dialog box for the graphical function that determines the ORDER PROBABILITY given a MEAN DELIVERY TIME

The ORDER PROBABILITY is declines rapidly as the TRANSIT MEAN time lengthens. When the TRANSIT MEAN is set at 16, no ORDERING takes place at all.

The final step in this process is to calculate the cumulative DISCOUNTED PROFIT for the Make-to-Order process. Note that we present a different basis for the optimization here than we used in appendix D (EVA, Expected Value Added). The EVA is common in modern business practice. The method here is similar but more common in economic studies.

Learning Point: Customers leave as the wait-time grows longer.

The model for such a PROFIT calculation is shown in figure 8.11. The HANDLING COST includes the vendor's cost, plus the transit costs, plus any costs associated with short-term storage and retailer handling. It is simply 20*RECEIVING. The LOST CUSTOMER COST is calculated at 10*(1–ORDER PROBABILITY)*TIME. The most that a customer here can order is one item per time period, and we assume that the lost-order cost is 10 per customer. The DISCOUNTING PROFIT is simply the SELLING PRICE*RECEIVING – HANDLING COST – LOST CUSTOMER COST*EXP(– INTEREST RATE* time). The INTEREST RATE is 0.003 per week, which is about 15 percent per year. For a TRANSIT MEAN of seven days, this gives us an average cumulative DISCOUNTED PROFIT of about 4,230 with a standard deviation of about 265,

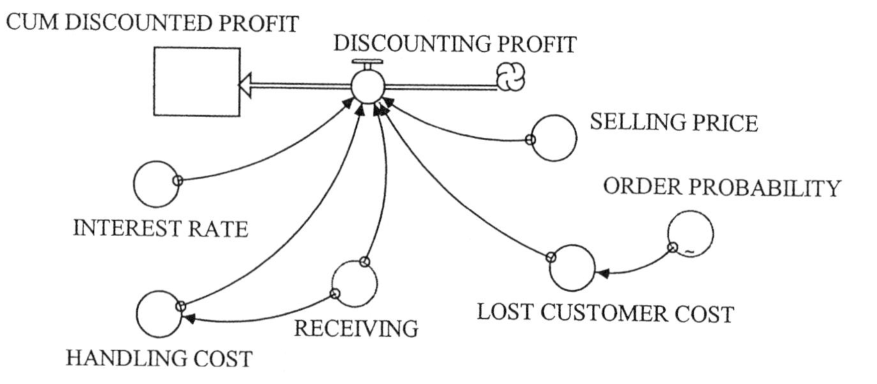

FIGURE 8.11. The additions to the model in figure 8.10, needed to calculate the costs, revenues, profit, and cumulative discounted profit (see appendix D for a definition of PROFIT and a different basis for the optimization process)

the averages of many repeated runs. Note that here we are taking weekly time steps rather than the daily steps of section 8.1.

It is not the process of ordering itself that determines if one should model the system as a Make-to-Stock or a Make-to-Order process. One can order books from, say Amazon.com and find their delay times will not vary greatly from the delays experienced by another customer. Make-to-Order modeling is done when the delivery time is quite variable between producers, giving the consumer a choice of where to go when the items are expensive and when the order rate is small. A car dealer, for example, will not keep a full variety of cars of all colors and models in the inventory. It is simply too expensive. Dealers choose a different strategy by listing all their inventory on a regional information network and trade cars among several dealers. Such a trade can be effected faster than a factory order and implemented if the dealer senses that the customer is in a hurry. So these dealers are operating a hybrid system that is part Make-to-Order and part Make-to-Stock. We leave it to the student to build an example of a hybrid model in which parameters such as price and rate of demand can dictate which process is to be used.

A natural question arises for those in the Make-to-Order businesses. How might I vary the parameters involved to increase my cumulative DISCOUNTED PROFIT? I could vary the selling price and determine the effect on the PROFIT measure. This would require a knowledge of the demand curve for my product and is a rather obvious step. Suppose we take a slightly more subtle approach. I will try to decrease the transit mean time (of course, at a cost), and see what the effects are on the cumulative DISCOUNTED PROFIT. Somewhat to our surprise, cutting the TRANSIT MEAN to four days and increasing the unit handling cost by five changes the cumulative DISCOUNTED PROFIT by an extremely small amount. Clearly, our company is indifferent to this kind of improvement. Much of this indifference comes from the shape of the ordering curve. For values of the TRANSIT MEAN less than about 7, the ORDER PROBABILITY does not change much. Redraw this curve as a convex and redo these tests.

Let us now suppose that it cost five more per unit sold to halve the variability of the transit time (to halve the standard deviation, but leave the average transit time at seven days). In this case we find a significant drop in the cumulative DISCOUNTED PROFIT (3,950 average), with a slightly reduced standard deviation on the PROFIT (240). Clearly, it is not economically worthwhile to make these changes in the delivery time statistics. Experiment with efforts to reduce handling costs and find the importance of such changes. Do this with a convex curve for ORDER PROBABILITY.

Optimization in the Make-to-Order format is possible by asking what the cumulative DISCOUNTED PROFIT maximizing SELLING PRICE would be. The reader is urged to follow the procedures used in appendix D to this chapter and find such an optimal price for this model.

9

The Tradeoffs Among Quality, Speed, and Cost

I have bred her at my dearest cost
In Qualities of the best.

—William Shakespeare, *Timon of Athens*

9.1 Introduction

Every experienced manager knows that the world is filled with tradeoffs. The quality of a product may be changed if the speed of production—the thruput—is changed or if the production effort or cost is changed. Lower the cost, speed up production, and quality often suffers. But a slowing of production may not raise product quality and instead, may raise costs. Managers are constantly looking for ways to cut costs without sacrificing quality.

In this chapter, we lay out a simple model of a production process without feedback. The model is of a production process that makes widgets of acceptable and unacceptable quality. The rejected widgets are passed through a reclaiming process during which some can be brought to acceptable quality.

The speed or thruput of the production process depends on the time of production, inspection, warehousing, and shipping of the quality parts, and also on the extra time taken in reprocessing and reclaiming the rejected product. The cost of the product is clearly a sum of production and reprocessing costs.

Lowering the cost is a kind of subobjective for the manager. Net profit is a better indicator of the ideal process configuration. It is conceivable that cost reduction can be carried so far that profits fall. It is also possible that a high cost configuration is the most profitable. Therefore, we calculate the profits for such an operation so that we can optimize the process.

One of the costs left out of this model is the consumer cost of using low-quality products. Although consumers may be slow to learn to choose between price and quality, some do eventually. This cost is difficult to determine, yet the sensitivity of profit to various levels of this quality-cost relationship could be determined. This activity is suggested for the reader.

Once the model is developed, we fashion a set of experiments designed to expose the tradeoffs between speed of thruput, quality, and cost. By doing this, we hope that we will reveal the complex interlinking between these three goals of the production process.

9.2 Model development

Our goal is to develop the simplest model that contains the most general aspects of a typical production system. With such a model, we can convey the familiar modes of the system and still demonstrate the subtleties of the tradeoffs.

We begin by setting up a stock in the form of a conveyor, called PRODUCTION, into which LOADING is delivering a constant maximum LOADING RATE of 400 assemblies per hour. Our process takes 10 hours (PRODUCING) to produce widgets from these assemblies. Once produced, the widgets pass through an INSPECTION station where some are REJECTING and the rest are PASSING. Successful widgets are sent to the WAREHOUSE. Rejected widgets are sent through a REPROCESS station (a five-hour conveyor) where we are RECLAIMING some and LOSING the remainder.

When the assemblies entered PRODUCTION, they were time-stamped (in LOADING). This allows us to keep track of the thruput rate (or SPEED through the entire system) as both forms of widgets (PASSING and RECLAIMING) pass through the WAREHOUSE and exit our system at SHIPPING. The QUALITY of the product is defined to be a ratio of the SHIPPING rate to the combined SHIPPING plus LOSING rate.

Figure 9.1 shows what the model looks like. The fraction rejected by the inspection process is called the REJECT FRACTION. As the LOADING RATE increases, we find from experience that the REJECT FRACTION increases according to the graph in figure 9.2.

The LOADING RATE also drives the graphical variable LOSING, somewhat proportionately, as shown in figure 9.3.

To calculate the TOTAL COST with a (constant) PRICE, the REVENUES, and ultimately the PROFITS, we need an addition to the model, as shown in figure 9.4.

The amount being produced (PRODUCING) is combined with the unit production cost (PROD UNIT COST) and the reprocessing cost (REPRO COST) to

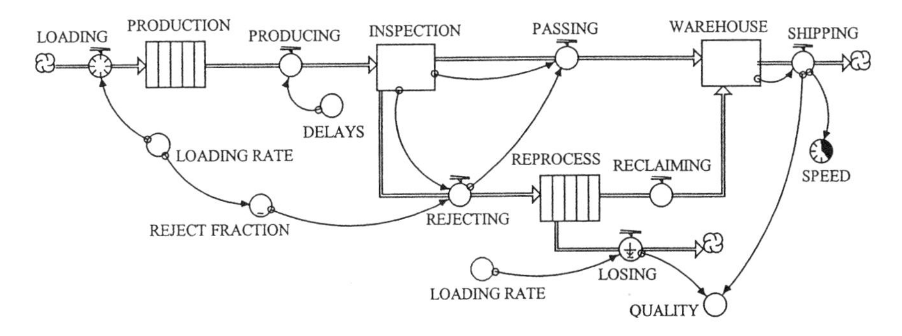

FIGURE 9.1. The production-inspection-reclaiming-shipping process

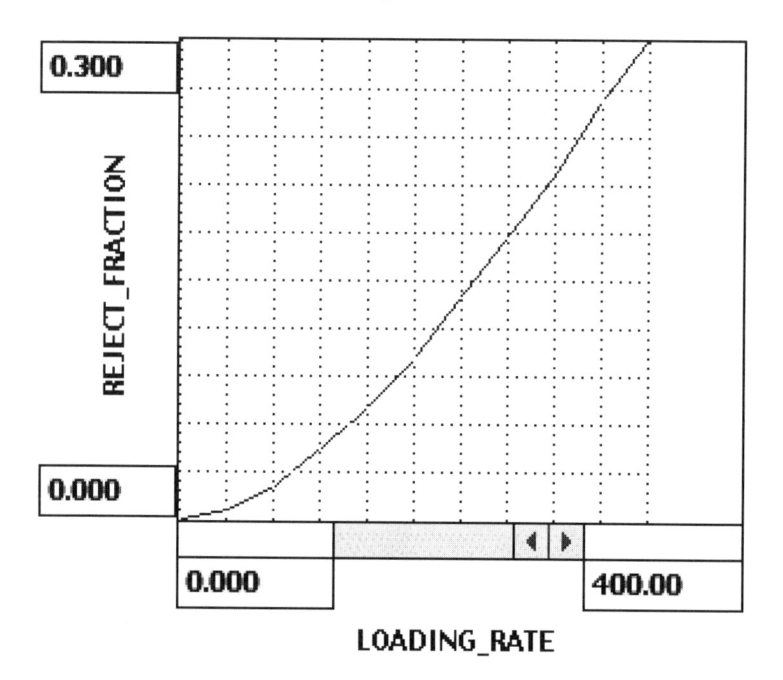

FIGURE 9.2. The empirical relationship between the LOADING RATE and the REJECT FRACTION

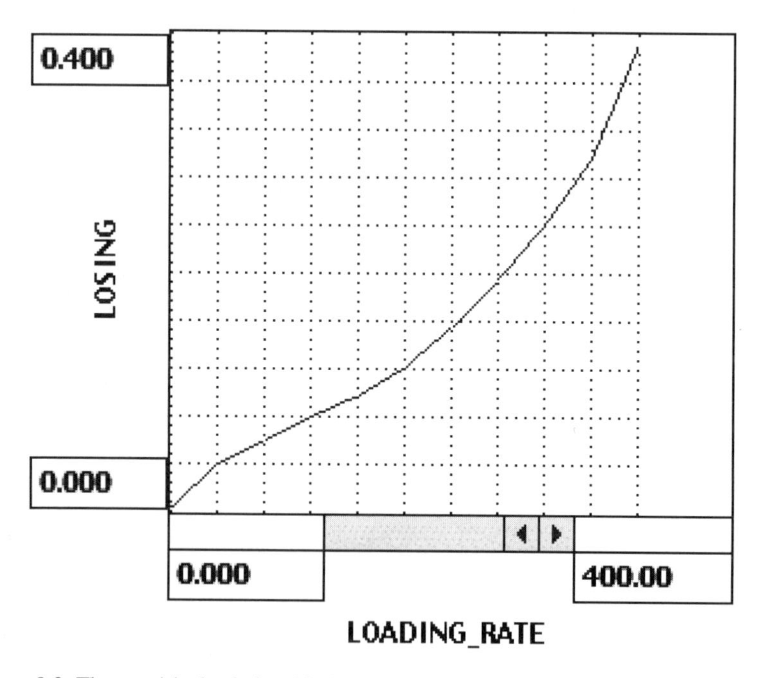

FIGURE 9.3. The empirical relationship between the LOADING RATE and LOSING, the rate at which rejected widgets are scrapped

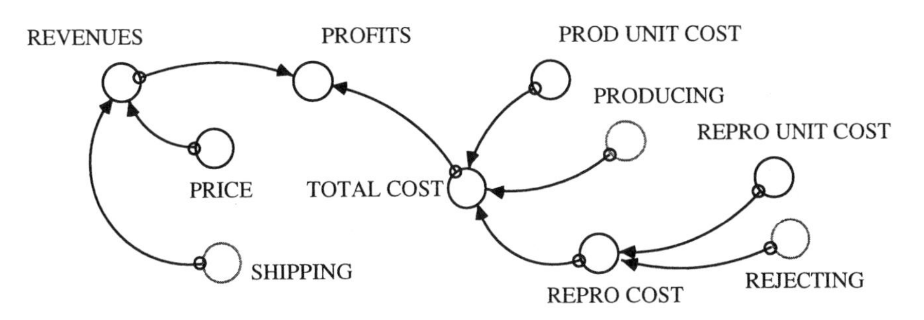

FIGURE 9.4. The scheme for calculating the cost and profits for the process, shown in figure 9.1

form TOTAL COST. The reprocessing cost is the product of the REJECTING rate and the unit reprocessing cost, REPRO UNIT COST. We also form the TOTAL UNIT COST by simply dividing TOTAL COST by SHIPPING. Finally, the PROFITS are found by removing the TOTAL COST from the REVENUES (PRICE times SHIPPING).

We now have our simple model constructed and described. We are ready to examine the way it functions and the tradeoffs between quality, speed, and cost.

9.3 The tradeoffs

We can vary the LOADING RATE and determine the impact on QUALITY, TOTAL COST, and PROFITS. Our expectation is that as the LOADING RATE increases, QUALITY will decline and TOTAL COST and PROFITS will increase. This is not quite the whole story. First, we examine the QUALITY as the LOADING RATE varies from 150 to 400 (see figure 9.5).

As expected, QUALITY declines as the LOADING RATE increases. At low LOADING RATES, the QUALITY was the highest. As we can see in figure 9.6, these rates are associated with negative profits.

The PROFITS (per hour) peak (~$6,900) when the LOADING RATE is about 320 widgets per hour.

However, the TOTAL COSTS keep climbing as the LOADING RATE increases (see figure 9.7).

The TOTAL COST consistently increases throughout the range of LOADING RATES. At the optimal profit rate, the TOTAL COST per hour was $17,400 and the QUALITY had declined to 95 percent.

> Learning Point: Tradeoffs between thruput rates (speed), quality, and cost are complex, but a clear optimum thruput for the goal of maximizing profit can be found.

We are left with several conclusions:

1. Reducing costs (by lowering the LOADING RATE) as a general strategy could lead us away from peak PROFITS. Only if our initial LOADING RATE

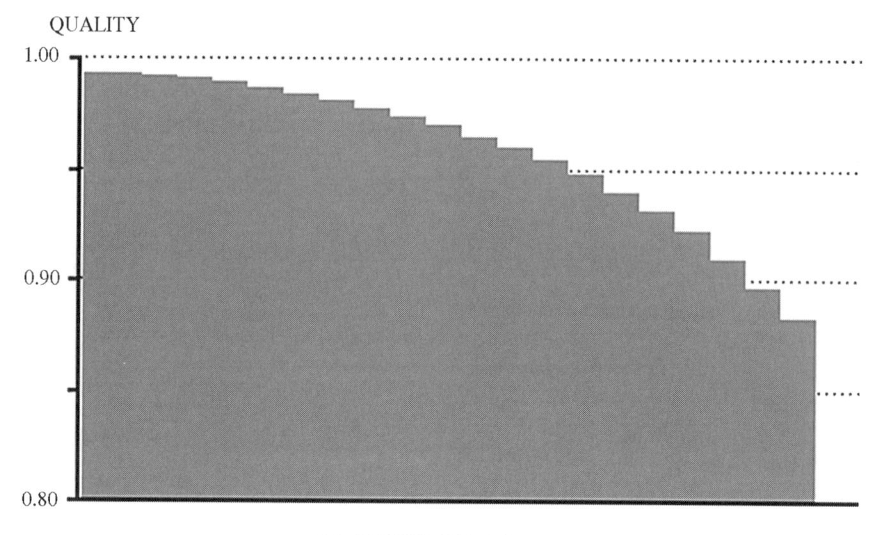

FIGURE 9.5. The variation of QUALITY as the LOADING RATE increases: Note the QUALITY range varies from 0.80 to 1.00

was greater than 320 widgets per hour would lowering costs lead to increased PROFITS. We must be careful to realize that our initial LOADING RATE could be higher or lower than the optimal LOADING RATE.

2. The LOADING RATE for peak PROFITS (320 widgets per hour) is clearly associated with a QUALITY less than perfect. This means that we will run the

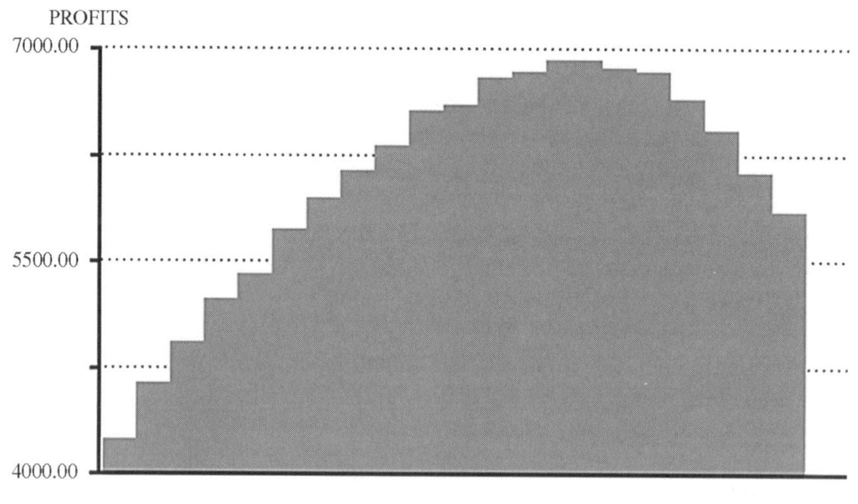

FIGURE 9.6. The variation in PROFITS as the LOADING RATE increases

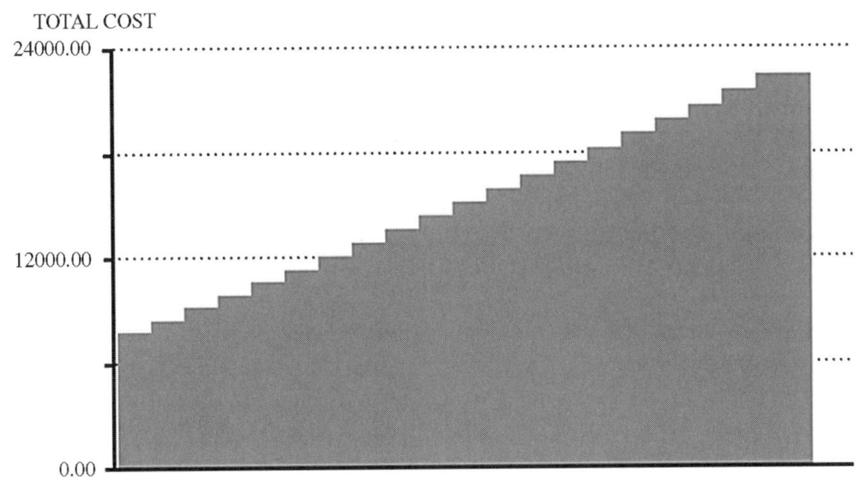

FIGURE 9.7. The variation in TOTAL COST as the LOADING RATE increases

production process so fast that we will not only produce imperfect widgets, but some them will be so imperfect that they will be scraped. At the optimal profit LOADING RATE, we are actually rejecting 69 widgets per hour and scraping 16 widgets per hour (5 percent).

3. It may well be that the reclaimed widgets were more fault-prone than those that passed INSPECTION in the first place. Consumers would then find that our quality is inconsistent. We will pay for this aspect of our product as some consumers cease to buy this product and possibly all other products made by this company. This could be a serious cost, which is not included in the model. Lowering the product price and attracting new customers to offset the lost ones might compensate for this sort of cost. Losses from inconsistent quality could be reduced eventually by a better reprocessing method, by slowing the production rate to produce fewer rejects, and by improving the production process so that fewer rejects are produced.

Note that all the monetary calculations are artifactual in this model. The effect of higher cost on the production process comes at the end of a scenario such as those presented in figures 9.5, 9.6, and 9.7. At that point, we modelers select the LOADING RATE that maximizes PROFITS. A more complex model would have linked the PROFITS to the physical production process and determined the profit-maximizing LOADING RATE as an output. We did not need to do this here because our goal is to elucidate the modeling process. If our modeling efforts were aimed at answering a specified real-world problem and intended for everyday use, it would be more complex. We strongly urge, however, that such a model be constructed out of simple models with the routine aid of those who will eventually use it.

9.4 Coping with uncertainty

Our model system so far has lag times that enable it to make the calculations of actual costs and profits and average quality that are unreachable with our mental models. We handle this complexity with a computer model and produce results that are easily understood. A real production system would also contain the added complexity caused by uncertainty. Production lines break down mechanically. Workers get sick. Input deliveries vary because of vendor problems. Whereas it is always appropriate to construct models of real systems first without uncertainty, we must now face the issue.

Let us imagine that the vagaries of machine and labor produce a delayed variation in the production time and lengthen it slightly. We choose a POISSON(10.5) rather than a constant 10 as the production time, and then limit it to only increases: production time = IF POISSON(10.5) > = 10 THEN POISSON(10,5) ELSE 10. This statement says that the mean production time would be 10.5 if we did not remove any shortening of that time. The POISSON function chooses only positive integers, such that the mean choice is 10.5. We accept only choices that are greater than the previous constant ideal production time of 10. Consequently, the new average production time is 11.5 rather than the ideal of 10, which is a 15 percent increase. We are assuming that the machine breakdowns and workforce sickness bring about these new production time delays.

We have to average the QUALITY, PROFITS, and TOTAL COST values because of their extreme variation. Typical of these variations is the one for PROFITS (see figure 9.8).

The PROFITS (after time = 20, allowing for the initial startup transients) are accumulated in CUM PROFITS and then divided by TIME – 20 to give the proper average. The QUALITY and TOTAL COST variables were averaged in the same way. The prescribed delays in production times lowered the average SPEED by 3.3 percent, lowered the profits by 4.1 percent, and lowered the QUALITY by 3.2 percent.

In a real way, this analysis is also a sensitivity test of the effect of production time increases on the important variables in the system. We could apply the same

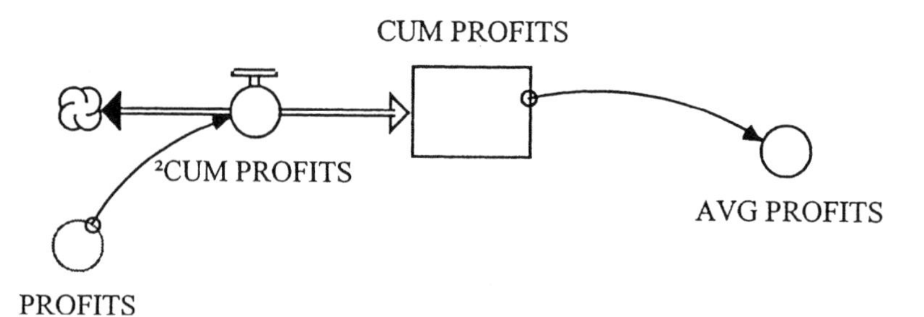

FIGURE 9.8. The calculation of the average profits when production times are randomly delayed

kind of randomness to the REPROCESSING cycle. We leave this as an exercise for the reader.

With this sort of analysis of the random variations in critical variables in the system, managers can determine where their best efforts should be directed. A manager may be interested mainly in the changes in the mean value produced by these random changes in the production times. These averages imply long-run effects. A more sophisticated manager may also be interested in the degree of variation of certain critical variables. (See chapter 7 for more discussion.) For example, the space allocated for production, inspection, reprocessing, and warehousing is finite. The manager must provide room in these places for the increased number of widgets caused by the randomness in production times. A still more sophisticated manager would be interested in whether such widget numbers are skewed toward the high or low side of the mean. If they were skewed toward the high side, the manager would be more likely to expand these spaces permanently to accommodate the peak flows. If the distribution were skewed toward the low side of the mean, the manager might tend to provide temporary space for the peak widget flow production.

In this sample model system, for example, we use the delay noted in production times and calculate (in a spreadsheet) the mean (3,555), standard deviation (310), and skewness (–0.51) of the number of widgets in the PRODUCTION space. With this information, the manager would determine the production space needed to handle more than 3,555 widgets only occasionally. The manager might choose to build the facility for the mean-plus-one standard deviation, reserving temporary facilities for overflow given that the space requirement is significantly skewed to the lower side of the mean. Exactly how much permanent space to provide depends on the cost of the permanent and temporary space and exactly how frequently the temporary space would be needed.

The reader can make this same sort of determination for the INSPECTION, REPROCESSNG, and WAREHOUSE spaces.

10

Modeling Supply Chains

Ability will never catch up with the demand for it.

—Malcolm Forbes

10.1 Introduction

Now we will expand our modeling capabilities to the entire supply chain. This chapter can be considered an extension to chapter 8, which focused on the complexity of the customer interface with a model of a single node. We will now consider the complexities caused by a supply chain rather than by a single node. We will use a simple customer interface. In this way, any complexities we see can be attributed to the multiple-node nature of a supply chain.

We define a supply chain as that entire set of activities of supply from start to finish. This definition covers the following:

- The physical structure from procurement of raw materials to delivery of finished products to the paying customer.
- The demand structure that covers the gathering of demand information and the creation of orders from that information.
- The logical structure that covers information flow and decision making throughout the supply chain.
- The financial structure that covers paying suppliers, gathering revenue from customers, and other financial activities that support building of facilities and other operations of the supply chain.

Generally, a supply chain is considered to be an assembly of nodes. These nodes can be factories, distribution centers, and so on. As the reader can anticipate, we still take the position that the construct of stocks and flows can model all the elements of a supply chain. (Although, again, we do not say that in specific cases it is the best approach to take—many vendors sell modeling capabilities that are specific to supply chains.)[1] However, this does not mean that the behav-

1. For example, Systems Dynamics, Inc.; see their Web site, <http://www.simulationdynamics.com>.

ior of supply chains is a simple extrapolation of the models we have seen so far. The structure of supply chains can lead to more complex behaviors than those seen up to now. In fact, supply chain evolution involves many of the complexity issues we discussed in chapter 2, section 2.1 (size, globalization, and so on). On top of this, "nodal identity" can make supply chains more complicated. By this we mean that in typical supply chains, people working at each node feel a great identity with the node. This is not difficult to understand. People can see the node, but the supply chain is so large and diverse that people have difficulty feeling a sense of belonging to the entire supply chain. The impact of this identification is that while a supply chain is a system and must be managed as a system, nodes tend to operate in a somewhat independent way. This shows up in many ways:

- Node-to-node communication is often poor.
- Even when information is available from outside the node, the node may not know how best to use it.
- Performance metrics are node focused.
- Decision making works well at the node level but is poor at the supply chain level.
- Generally, no one has authority for decision making at the supply chain level. We try to foster "team decisions" and any node will tend to resist decisions that appear to suboptimize the node locally.

Although a supply chain can have many suppliers, products, customers, nodes, and so on, we will take a simple view of a supply chain here. This view is shown in figure 10.1.

In our simple model, materials flow from the start node (node 1) to the final node (node N). Demands enter at node N and are translated into orders that flow down to the start node. Of course, translating demand into specific node orders can only be done using data about the capability of the nodes to produce and handle materials and products. For example, we must know manufacturing lead times so that when an order is created for a product we know when that product should

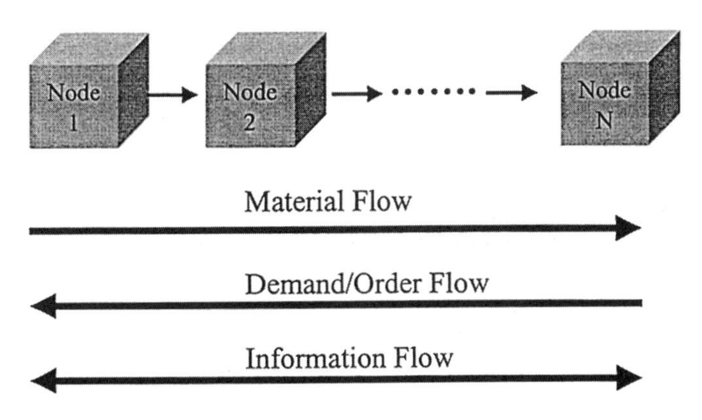

FIGURE 10.1. Simple supply chain model

be made. Systems that create node orders from demand and node performance data are called Manufacturing Resource Planning (MRP) systems (see Hopp and Spearman 1996). Information can flow in both directions in the supply chain. For example, suppose that node 2 runs into some production problems and must halt production for a time. One can imagine that knowledge of this would be useful to all other nodes in the chain and that failure to communicate such information by node 2 could prevent the other nodes from taking action that might minimize the impact of the production halt at node 2.

In real supply chains, three elements of complexity can be present:

- Dead time—it takes time for material at any node to reach a node downstream. Also, it can take time for orders to move from a node to a node upstream.
- Feedback—performance at any node will be impacted by other nodes. For example, node 3 cannot supply node 4 if it has not been supplied by node 2.
- Variability—production rates at nodes can fluctuate, demands can change, and so on.

Thus, we can appreciate why supply chains can demonstrate complex behavior, similar to what we saw in chapters 4, 5, and 6 when we looked at workflow processes. There is, however, another aspect of supply chains that increases the level of complexity in the supply chain environment, which essentially is related to the lack of overall coordination of the supply chain that we described earlier. Suppose that we have an error at any node in the chain. This error might be that the node has overproduced material or ordered too much material from the previous node. Then because each node is operating independently, a real possibility exists that this error will not be managed well overall. It may become even larger as it propagates to other nodes in the supply chain. In fact, such a phenomenon was observed in companies such as Procter & Gamble and Hewlett Packard.[2] In many cases, the demand for products by the end customer (say, the retailer) was fairly stable and appeared to be easily predictable over time. This might lead anyone to conclude that each node in the supply chains also would behave in a similar stable manner. At each node, they expected that production rates and order rates would be as stable as the overall customer demand rate (as measured by the standard deviation of the production, order, and demand rates, for example). Instead, what they found was completely the opposite! As they looked at the nodes in the supply chains, they found that the variability in the production and order rates at nodes was far greater than the variability in the customer demand rates. They also noted that the variability increased in the upstream supply chain. This phenomenon was termed the "bullwhip" effect because of its similarity to the action of a whip when it is flicked. The portion of the whip nearest the hand moves the slowest and the tip of the whip moves fastest. (It is the speed of the tip that causes the sound as you crack the whip!)

2. See Lee, Padmanabhan, and Whang 1995, 1997a,b.

> Learning Point: The bullwhip effect in supply chains is the tendency for production and order variability to increase as we move upstream in the supply chain from the customer demand signal.

Since these initial observations, this bullwhip phenomenon has been observed in many situations (Lee, Padmanabhan, and Whang 1997). It is considered a fundamental aspect of the potential behavior of real supply chains. Any introductory text on modeling must include a model that is both capable of reproducing the bullwhip effect and capable of illustrating ways that the bullwhip effect can be reduced.

Many years ago, a board game simulation that reproduces the bullwhip effect was developed at the Massachusetts Institute of Technology (MIT). Called the Beer Game, it has been played many times and is described fully in Senge 1990. Studies of the bullwhip effect often use the MIT Beer Game as the basis for the simulation. It is this approach that we intend to follow here. (We also should note that other types of modeling have been applied to supply chains. One example is agent based modeling; see Swaminathan, Smith, and Sadeh-Koniecpol 1997.)

10.2 Introduction to the Beer Game

The Beer Game (hereafter referred to simply as BG) is based on a simple supply chain consisting of four nodes as shown in figure 10.2.

The nodes are as follows:

- The Retailer who sells beer to the consumer.
- The Wholesaler who sells beer to the retailer.

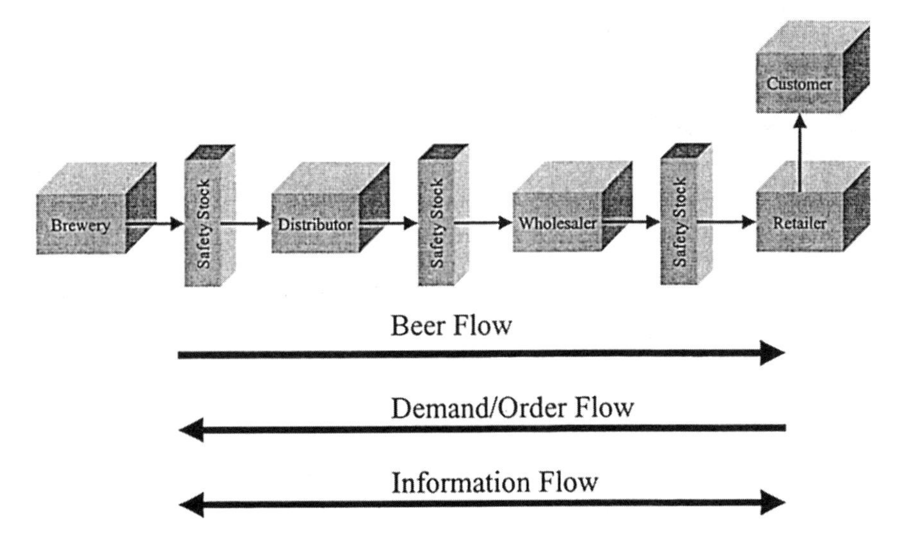

FIGURE 10.2. Beer Game supply chain

- The Distributor who sells beer to the wholesaler.
- The Factory that makes the beer and sells it to the distributor.

Only one type of beer is involved. The game is run on a weekly cycle with four players, one at each node. The beer is produced by the factory in cases and moves downstream to the retailer and the consumer. Each node tries to keep a safety stock of 12 cases of beer to hedge against uncertainty in supply and demand. The retailer receives demand from the consumer as the number of cases of beer consumed each week. If the retailer has the beer in stock, the demand is satisfied. If not, any unsatisfied order is backlogged. Based on the demand, current safety stock, and any other information, the retailer must also decide what orders to place to the wholesaler. Thus, the BG is essentially a make-to-stock supply chain. Once the retailer places an order, it takes two weeks for the order to reach the wholesaler. The wholesaler will try to satisfy orders from stock. Again, any unsatisfied orders are backlogged. Once the wholesaler sends the beer to the retailer, it takes two weeks for the beer to reach the retailer. The wholesaler must now also decide how many cases of beer to order from the distributor, and so on. Finally, the factory must decide how many cases of beer to manufacture. From this it can be seen that the only decision each player has to make is how many cases to order each week. All other actions are taken as prescribed by the procedure for satisfying orders as described. Note that failure to satisfy a consumer demand is not backlogged. We assume that if consumers cannot get the beer at the retailer, they will go elsewhere, and the sale is lost. Also, all of the nodes (including the brewery) are assumed to have infinite capacity. It is important to note here as well that when the game is played, a game board is used and all players can see the entire board. The only data that the players cannot see are the orders when they are placed. A player sees the orders only when they arrive at his or her node and the player does not show these arriving orders to the other players.

We must have a way of measuring the performance of players in the game. As it should be, this measurement is based on a financial measure of performance. It is also based on two competing pressures in a supply chain—one to satisfy the orders and one to minimize inventories. At the end of the game, we cost the number of backlogged cases of beer at $1 and cost every case of beer in inventory at $0.50. These numbers were chosen to reflect the fact that a lost sale is typically more costly that a unit of inventory. These two components are then used to calculate the cost of operating the supply chain. The objective of each of the four players is to minimize the cost of operation of the supply chain. We want to keep 12 cases of beer in stock at each node, and we always want to be able to satisfy orders (no backlogs). Consequently, if we run the game for N weeks, the benchmark cost is 4*12*$0.50*N = $24*N. This allows us to compare actual performance when the game is played. Typically, the overall actual cost of operation is many times this ideal benchmark cost. The reason for this is that the demand is a constant four cases per week initially, but after about five weeks, the demand rate is increased from four to eight cases per week. This change in demand rate leads to over- or underordering and over- or underproduction, and throws the supply chain into disarray. (In fact, in the authors' experience of running the game, the difficulties begin even before the demand rate has been changed at week 5.) Bullwhip type behavior is observed in the results. Even with such a simple change in the demand signal, the variability in orders and production rates up the chain can be huge, and it is difficult to see how such variability could be obtained from such a small change in consumer demand. Let us see if we can create a model that recreates this bullwhip behavior.

10.3 The Beer Game model

Two types of items are moved in the supply chain. The first item is the cases of beer that move from the brewery to the retailer and the consumer. The second item is the orders that move from the consumer and retailer to the brewery. Separate parts of the model handle these different items. Also note that the beer that leaves the brewery is the same as that which arrives at the consumer point. However, this is not true for the orders. Therefore, we will have one single line for the beer flow but separate lines for order processing at each node. Connectors will handle information transfer between nodes. With all the nodes and the logic included, the model becomes quite complex. Consequently, we will look at the model in sections. The easiest section is the customer section and is shown in figure 10.3. (Note that while we showed the supply chain with beer flow from left to right, the model was developed in the opposite direction because it was easier to start with the customer section and work to the right toward the brewery.)

As we said at the start of this chapter, the model in figure 10.3 uses a simpler form of the customer interface models used in chapter 8.

The Customer Ordering has the following equation:

Initial_Demand+STEP(Demand_Increment_Variable,5)

This allows us to set up an increase in demand at week 5. These orders flow into the stock Customer Order Backlog. The outflow Satisfying Customer Orders is given by the following equation:

INT(Customer_Order_Backlog)/DT

This means that we always try to satisfy all customer demand. The INT (Integer) built-in function is used throughout this model for order calculations simply to ensure that we place orders only for an integer number of cases of beer. The flow Selling To Customer comes from the Retailer Inventory stock, which is not shown here but will be shown later in figure 10.4. This flow is set by the Satisfying Customer Orders outflow. If inventory is available, the beer will flow from the Retailer Inventory stock. If Retailer Inventory does not have enough beer, what is available will flow from the stock. However, even if there is not enough beer to satisfy all customer demand, the stock Customer Order Backlog will be emptied for every update of the model. Thus, there can never be a build up of customer backlog. This reflects the idea that all unsatisfied customer demand is lost.

Figure 10.4 shows the retailer section of the model. This section of the model is more complex than the customer section in figure 10.3. On the beer supply por-

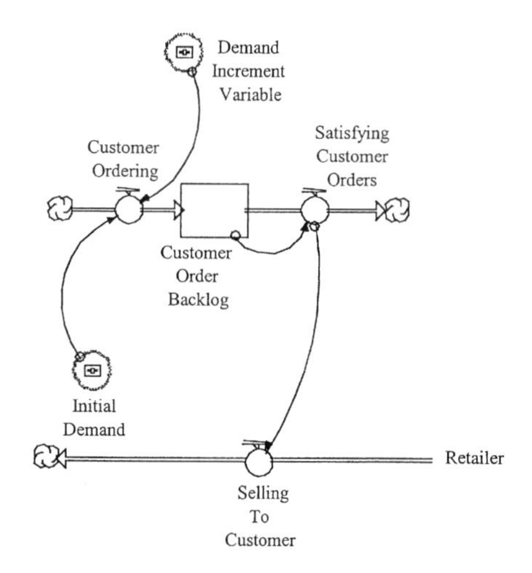

FIGURE 10.3. Customer section of the model

tion, we have two stocks: a conveyor (In Transit To Retailer) to represent the transportation of beer from the wholesaler to the retailer, and a stock (Retailer Inventory) to represent the inventory of beer at the retailer and to provide the inventory that the retailer uses to satisfy customer demand as we saw in figure 10.3. The converter Supply Transit Time is used to specify the two-week transit time from wholesaler to retailer. This part of figure 10.4 is simple. The complexity in figure 10.4 is related to the order portion of the diagram. More specifically, it is related to the calculation of the amount of orders that we must place for beer. A player in the BG can use three components to assess what orders should be placed.

1. The orders that the retailer receives from its customer. The retailer must place enough orders to replace what the customer is ordering from him. (More complicated approaches can be used where the retailer would use historical data to anticipate future requirements and use these projections to decide how much to order. The reason for this is that the retailer would realize that because it takes time for orders to translate into actual arrival of beer, the amount of orders should depend not only on current sales to the customer but also future sales. Because we cannot know the future, the best we can do is make some type of estimate based on historical trends, knowledge of the market place, and so on. If you think the customer sales are likely to increase, you would want to place more orders today. We looked at more complex ordering policies in chapter 8 and focus now on the supply chain aspects of the model. We leave it to readers to apply these alternate policies to the basic BG model.)
2. The difference between the target inventory and the actual inventory at the retailer. The bigger the difference between target and actual inventory, the more orders the retailer must place to the wholesaler. Note that sometimes a node

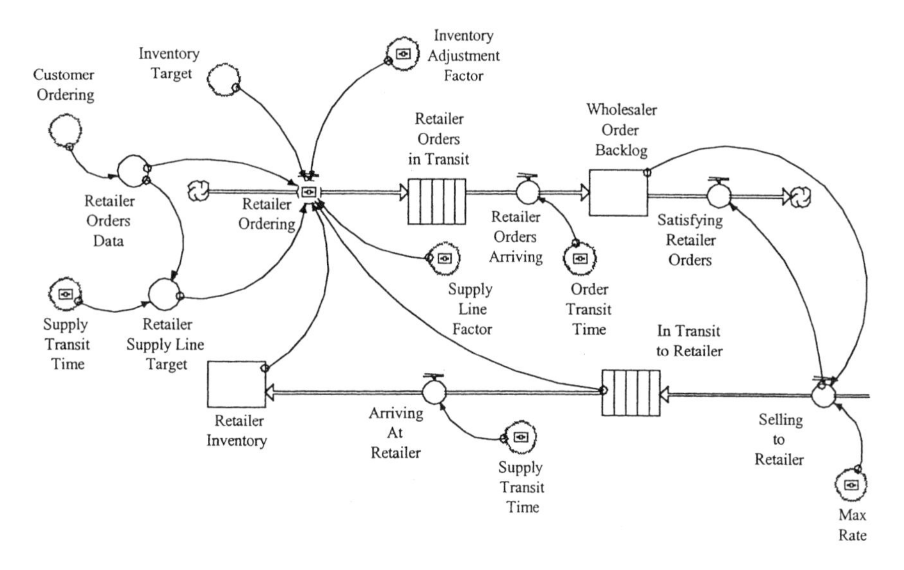

FIGURE 10.4. Retailer section of the model

has no actual inventory but, rather, a backlog. In these instances, we can think of the node as having an effective inventory equal to the actual inventory minus the backlog. Thus, when a backlog exists, the effective inventory will be negative. Consequently, we should look at the difference between the target inventory and the effective inventory. Note also that a backlog at a node refers to the backlog of orders from the node downstream of that node. By this definition, the retailer cannot have a backlog because all missed orders from the customer are not backlogged; they simply become lost sales.

3. The difference between the target amount of beer in transit to the supplier and the actual amount of beer in the supply line to the retailer. Again, the bigger this difference, the more orders the retailer needs to place.

We can look at these components separately. Component 1 is easy. We simply use the current customer orders, as contained in the converter Customer Ordering. For component 2, we use the following equation:

$$O = IAF * (IT_R - I_R + B_R) \tag{10.1}$$

where O is the order rate, IT_R is the retailer inventory target, I_R is the current retailer inventory level, and B_R is the current retailer order backlog (which we know is always 0). *IAF* (Inventory Adjustment Factor) is a factor that reflects how strong the linkage is between order rate and the inventory difference. If this value is small, say 0.1, then the linkage is weak. What this means is that the supply chain tends to react slowly to changes in inventory level. If this value is large, say close to 1, then the supply chain-reacts more aggressively to changes in inventory.

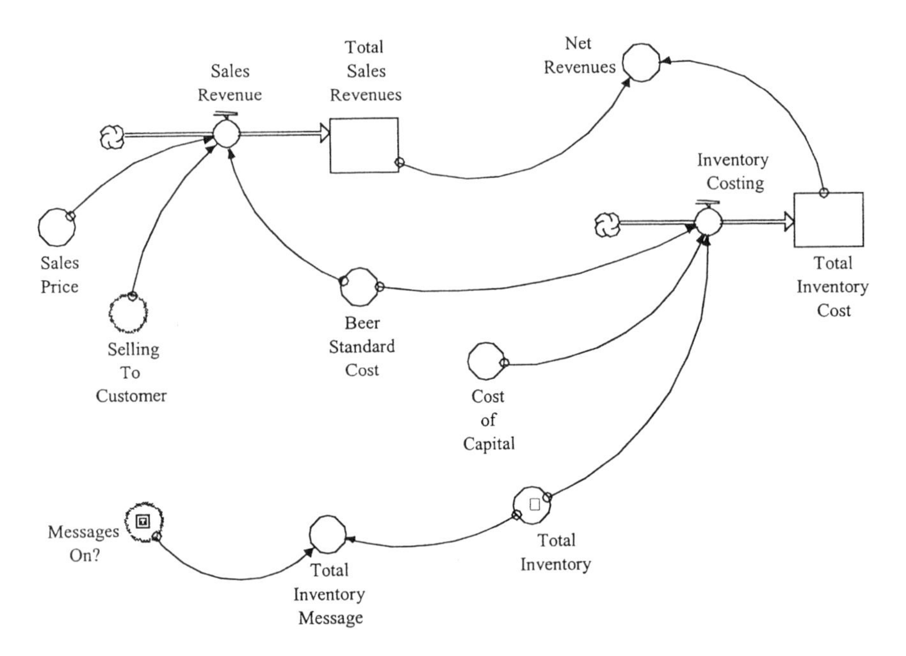

FIGURE 10.5. Financial section of the model

The inventory cost is then calculated as the total inventory level times the standard cost of the beer times the cost of capital for the beer. We assume the standard cost is $1 in the converter Beer Standard Cost and the cost of capital is 13 percent per year. In this case, the value in the converter Cost of Capital is 0.13/52.0. The equation for the inflow Inventory Costing is the following equation:

Total_Inventory*Cost_of_Capital*Beer_Standard_Cost/DT

The stock Total Inventory Cost accumulates the inventory costs over the simulation. The other part of the financial section calculates the sales revenues. The sales price for the beer is assumed to be $1.25, and this is represented by the converter Sales Price. The equation for the inflow Sales Revenue is the following equation:

(Sales_Price-Beer_Standard_Cost)*Selling_To_Customer

Therefore, in our financial calculations, we do not include any negative impact of backlogs at the wholesaler or distributor. (A backlog at the retailer is by definition a lost sale). The stock Total Sales Revenue accumulates the revenues over the entire simulation. (Note that this equation does not have a DT because Selling To Customer is a flow whereas Total Inventory is a stock). The converter Net Revenues contains the following equation:

Total_Sales_Revenues–Total_Inventory_Cost

This net revenue will be our performance measure for the entire BG supply chain. Thus, we can measure the extent of the bullwhip effect in the simulation by the value of the Net Revenues. The larger the net revenues, the smaller the extent of the bullwhip effect. As we look at the operation of the model, we will try to maximize this net revenue, as distinct from the BG itself, which tries to minimize operating costs.

The two converters that reference the word "message" in figure 10.5 will be used in section 10.5 when we discuss using the model in game mode.

The basic model we are creating here does not have any random variation included. This is consistent with the way the game is actually played. All transit times are fixed at two weeks; no losses of beer occur throughout the supply chain; and so on. The only source of random variation is the players. However, the authors' experience has shown that players typically try to develop a strategy when they play the game. They may change this strategy during the game when they determine that their current strategy is not working well. It appears that they do not change this strategy in a random fashion. Data from an actual BG session show that order and production patterns do not move around randomly. Instead, large cycles emerge in these patterns. Thus, we do not include random variation in the model, as it does not appear to be needed to reproduce the bullwhip behavior. We can surmise, however, that such random variation will make matters worse and increase the extent of the bullwhip effect. We leave it as an exercise for readers to modify the model to include random variation and to compare the performance to the present model.

> Learning Point: While random variation will increase the extent of the bullwhip effect in a supply chain, we do not need to include random variation to reproduce the bullwhip behavior.

Before we can run the model, we must specify the values of the inventory adjustment factor, IAF, and the supply line factor, SLF. What determines these values? As we said earlier, the IAF reflects how aggressively we respond to inventory changes. A less aggressive approach (low value of IAF) to making inventory changes results in small inventory/order fluctuations, but the risk of shorting the customer is high. On the opposite side, a more aggressive approach will lead to higher inventory/order fluctuations but a smaller risk of a stock-out. The ultimate determination may then depend on each player's perception of the relative costs of inventory versus stock-outs. For example, if a player's real job is with an organization where running out of product is the ultimate crime, then this player will tend to play with a larger value of IAF. The player may take the attitude that it is better to order too much and be sure of supply than to take any risks. Whether this perception provides the player with a good strategy for keeping the customer supplied will be obvious when we run the model. We will use an IAF value of 0.2 as a base case.

Learning Point: Determining the value of IAF depends on each player's perception of the relative costs of high inventory versus the cost of stock-outs.

Determining the value of SLF involves a different consideration. What we are saying with this factor is that the players will look at what is in transit to them from their supplier and then incorporate this into their order policy. While it is possible for the players to use the supply line adjustment, it is unlikely that they will do it in practice because of the computations involved, as shown by equations 10.2 and 10.3. In practice they are not likely to use this supply line adjustment much. Therefore, we will use an SLF value of 0.2 as our base case. Note that these values are close to those used by Joshi (2000).

Learning Point: In practical situations, the value of the SLF will tend to be low because players are unlikely to be able or willing to calculate an expected order rate based on the supply line.

In order to display the results of the base case, on figure 10.6 we show the order profiles for each player over the simulation time (52 weeks).

With all these graphs on the same scale, it is easy to see the bullwhip effect. So, even though the customer orders changed only from four to eight cases per week, we can see large fluctuations in the order rate switch in the supply chain. In fact, the fluctuations were so large that the brewery has had to shut down for nearly half a year! This is with an IAF value of 0.2—just imagine what it would be with a higher value of IAF. Note also that the order rate at the brewery does not fluctuate much more than the distributor. This is because the brewery is at the end of the chain and does not have a supplier. It can make whatever it needs. Therefore we should expect the situation at the brewery to be a little different than at the other nodes.

Figure 10.7 shows the effective inventory profiles. Figure 10.7 also indicates a bullwhip effect, although it is not as obvious as the one in figure 10.6. Certainly the inventory fluctuations at the retailer are much less than at the distributor/wholesaler. The brewery fluctuation is much smaller again because the brewery is at the end of the supply chain. Note that the cycle in the orders/inventory is of the order of 1/2 year = 26 weeks. This should not surprise us because the total delay time in the supply chain is 16 weeks (two-week delay for supply, two-week delay for orders, and four nodes altogether). But it begins to help us recognize the importance all this delay time has on the supply chain. It is a significant component of the difficulty with managing the supply chain and shows us how slow the supply chain response time can be. It means that when we see an issue, we cannot simply flip a switch and overnight everything will become satisfactory. This is also an issue because each node may not appreciate just how large the overall re-

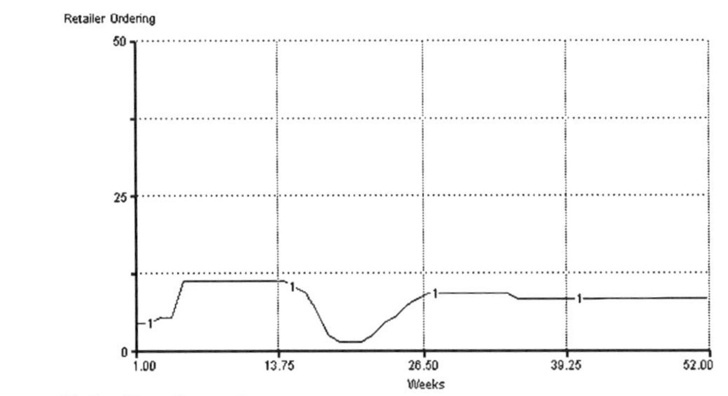

FIGURE 10.6a. Retailer orders

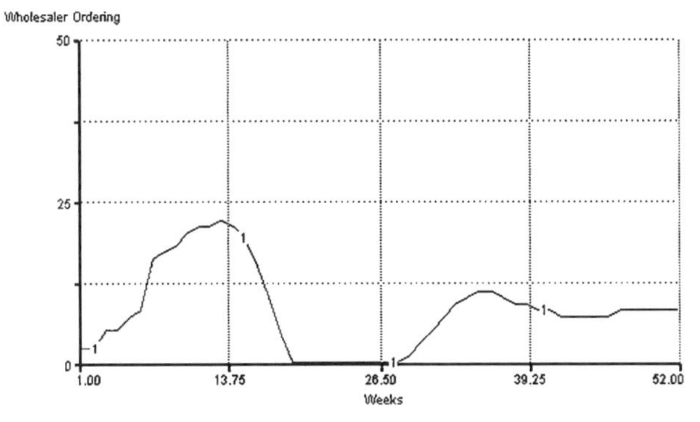

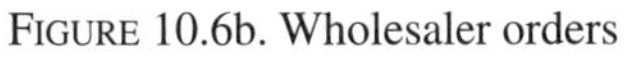
FIGURE 10.6b. Wholesaler orders

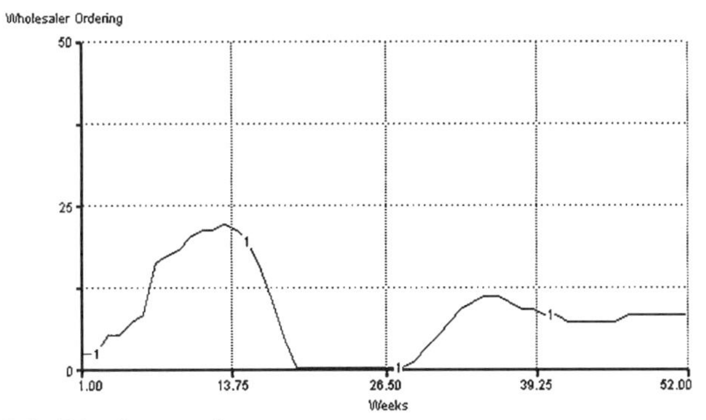

FIGURE 10.6c. Distributor orders

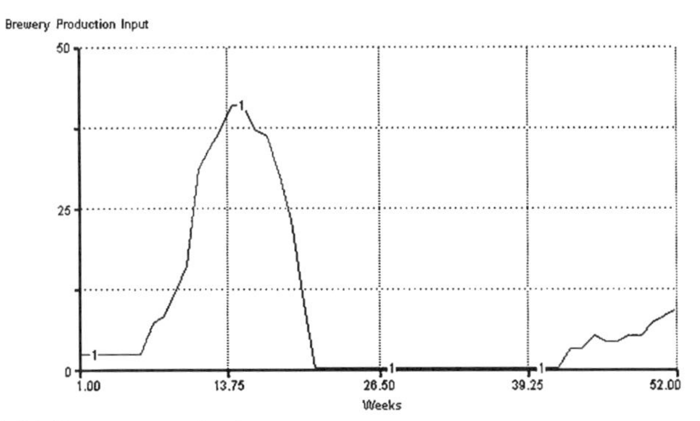

FIGURE 10.6d. Brewery production

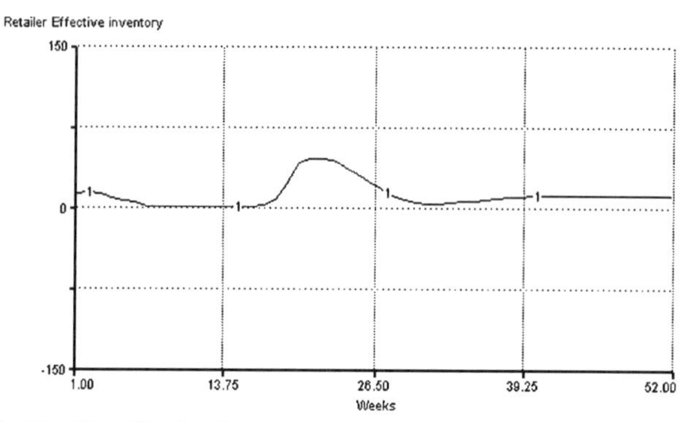

FIGURE 10.7a. Retailer effective inventory

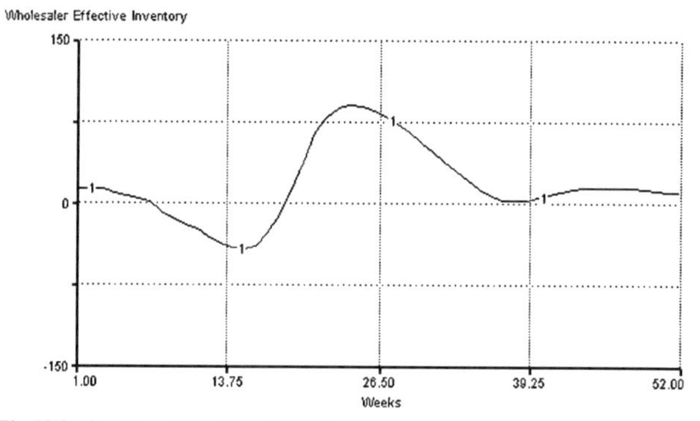

FIGURE 10.7b. Wholesaler effective inventory

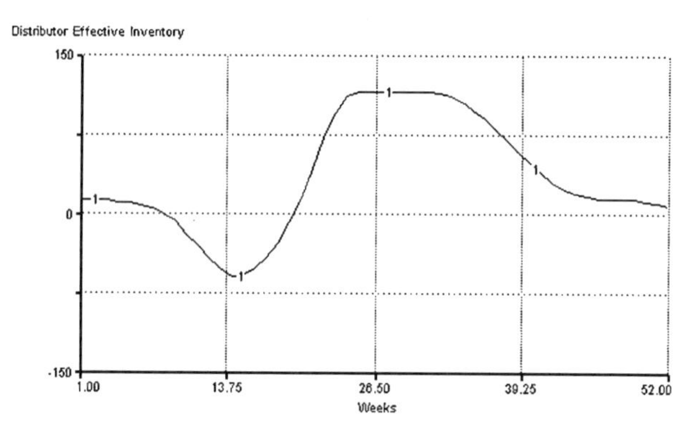

FIGURE 10.7c. Distributor effective inventory

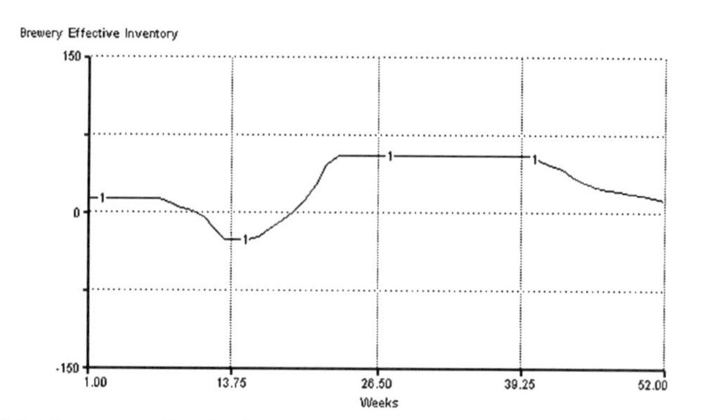

FIGURE 10.7d. Brewery effective inventory

sponse time of the supply chain is—they may understand only their own node, where the delay time is only four weeks.

A final point about the profiles in figure 10.7 is this: In practice, when the inventory gets really high, players will probably react as if their IAF value has changed. In other words, instead of the reaction of the players being linear as indicated by equation 10.1, a nonlinear response is likely to occur and have the following form:

$$O = IAF * (IT_R - I_R + B_R) \qquad (10.4)$$

If $B_R < N_1$, $IAF = IAF_1$
If $B_R < N_2$, $IAF = IAF_2$
If $B_R < N_3$, $IAF = IAF_3$, ...
where $IAF_1 < IAF_2 < IAF_3 <$ . . . . . .

where the Ns are numbers that will depend on the particular situation. What equation 10.4 is saying is this: When the backlog is low, take a smaller response than you might take; but if it gets large, increase your response even more. This type of response is a typical human response. When the gap is low, people can react as if it does not exist. When the gap is large, they respond in a way that is much larger than what we would expect by a simple linear extrapolation of their behavior when the gap was smaller. We leave it as an exercise for readers to modify the model so that it uses equation 10.4 instead of equation 10.1.

In the financial model we are using, we cost inventory based solely on a capital cost of 13 percent per year, and we assume that any lost sale will translate to a loss of $0.25. Consequently, we should operate the supply chain by favoring a scenario of higher inventories and lower risk of lost sales. This scenario would indicate that we should operate with a higher value of the Inventory Adjustment Factor, IAF. Figure 10.8 shows the variation of net revenues with IAF over the range 0.0 to 0.4, with the Supply Line Factor, SLF, kept at 0.2.

This shows that our efforts to over adjust based on inventory levels may have exactly the opposite effect that we hope for. The net revenue actually decreases with IAF, especially for higher IAF values, where the dropoff in net revenues is quite dramatic (and is driven by high inventory levels). This is because high IAF values cause such high fluctuations in the supply chain that it becomes too difficult to manage the inventory at reasonable levels, even if we are keeping the customer orders supplied. When keeping the customer supplied is critical, reacting more quickly to inventory fluctuations is not the simple answer we might have hoped for as a way to manage the supply chain. The inventory costs become so high that they drive the net revenue values. We may be keeping the customer supplied, but at what cost? This consideration is even more critical when we deal with supply chains for multiple products or where each node has many customers. If a customer thinks that supply is going to be an issue, then that customer may order too much, reasoning that "the squeaky wheel gets the grease." If the shortage does occur, that customer should get a bigger piece of the pie because they ordered more. And if it does not occur, they may simply cancel orders. Of course,

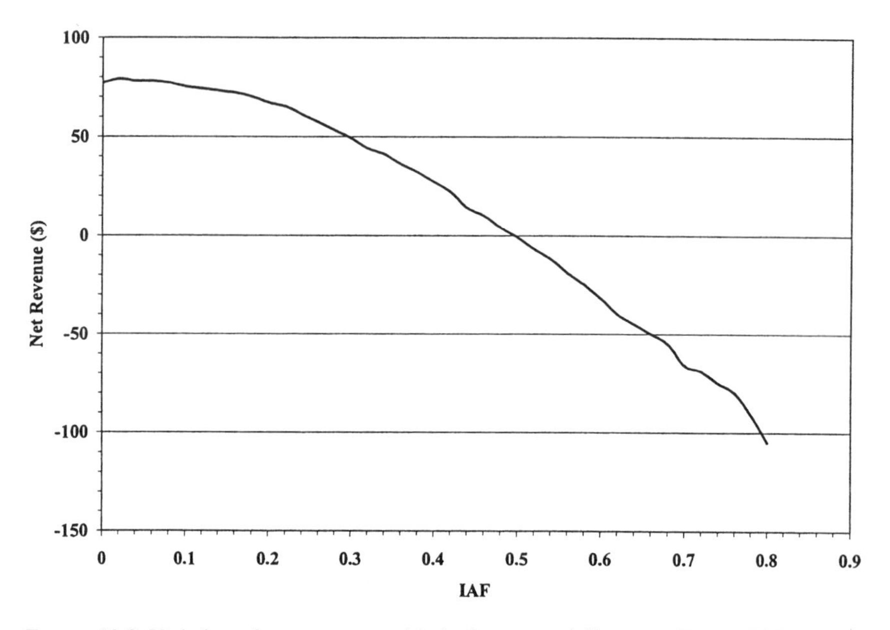

FIGURE 10.8. Variation of net revenues with the Inventory Adjustment Factor, IAF

the other customers may think the same, and we end up with a "Tragedy of the Commons" scenario as described by Senge (1990).

> Learning Point: Even though increasing the Inventory Adjustment Factor, IAF, value ensures that we react faster to inventory fluctuations when supply is critical, it is typically not a sound strategy for managing a supply chain.

What about the supply line adjustment? Figure 10.9 shows the variation of net revenues with SLF at various values of the IAF.

Figure 10.9 has a number of interesting aspects. First, for any given value of IAF, there is an optimum value of the SLF.[3] What this means in practice is that it is not enough to provide just the supply line information to the players. The players must know how to use this information. For example, suppose that the value of IAF is 0.3. Then the best value of SLF to use is about 0.5, for a net revenues value of $59. Using an SLF of 0.8 gives a net revenues value of $40, about the

3. From here on, when we refer to optimal conditions for the supply chain, we mean choosing the value of the Supply Chain Factor, SLF, that maximizes the net revenues, all other things being fixed. This is a simpler optimization to that which we considered in chapter 8.

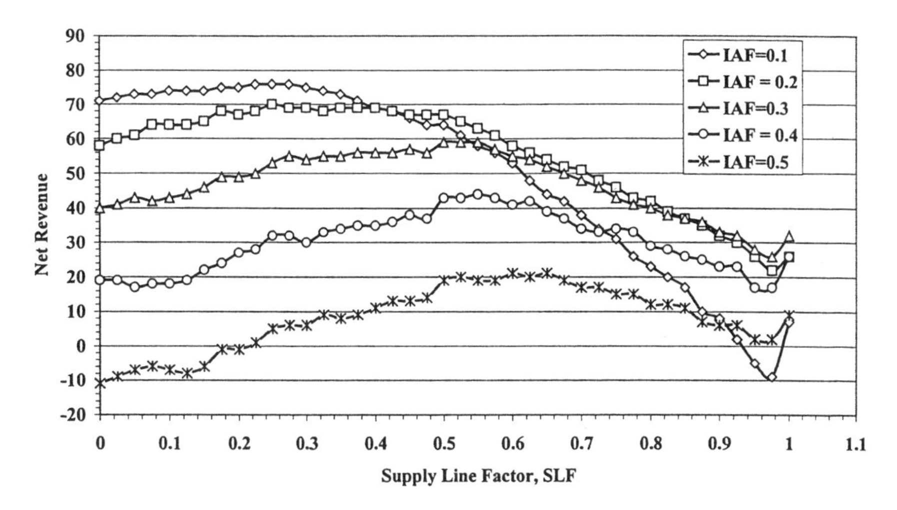

FIGURE 10.9. Variation of net revenues with the Supply Line Factor, SLF

same as if SLF = 0. Thus, choosing the right value of SLF is important. For our model, it could make a difference of 50 percent in net revenues. This issue is also more important because the curves in figure 10.9 cross over each other. For example, if we use an SLF value of 0.9, the net revenues for an IAF value of 0.4 are greater than for an IAF value of 0.1! In effect, it appears that the higher values of SLF are overcompensating for the values of IAF. This means that simply making supply information available to the supply chain is not enough. We also must look at how the information is used (that is, the algorithm that uses the information in making the ordering decisions). Using the information incorrectly can actually make the situation worse as measured by net revenues.

> Learning Point: In a supply chain, simply making information available to the nodes is not enough to guarantee improved supply chain performance. The algorithm to process the information must also be optimized for the information supplied and the supply chain being managed.

Information visibility and information usage are not the same. Note also that for the lower values of IAF, the net revenue/SLF curves tend to be flatter than for the higher values of IAF. The larger the value of IAF, the more critical it is to choose the optimum value of SLF. This again reflects the greater degree of variation we will see in the supply chain operation for larger IAF values. In addition, note that the larger the value of IAF, the lower the optimal net revenue value we get by varying SLF. Therefore, while choosing the optimal value of SLF can minimize the impact of IAF, it cannot eliminate it altogether and we will still pay the price for overreacting in the supply chain.

10.4 Further analysis of the Beer Game model

In this section, we will take a look at some ways that we can improve the performance of the BG model. Two of these relate to improving the actual supply chain properties and one is related to information visibility.

We made the point in the last section that the cyclical behavior of the supply chain response was driven by the delay times in the orders and supply line, which were both set to two weeks. Table 10.1 shows the impact of the delay times on the net revenues.

The column Total Effective Delay is simply the sum of the supply transit time and order delay time multiplied by four. The data in table 10.1 clearly show that the larger the delay times, the lower the net revenues. Also, the relationship is nonlinear. The impact of changing from a delay of 3 to 4 weeks on net revenues ($11 to minus $59) is much greater that changing from one to two weeks ($85 to $67). Again, typically a nonlinear effect such as this will not be obvious to the nodes. This again supports our position that it is both delay times and variability that degrade our ability to effectively manage these long chain processes.

The other interesting aspects to the data in table 10.1 are that the impact of delay time at the orders and the supply transit time are not the same. For example, adding an extra week of delay to orders when both the order and supply delay times are three weeks changes the net revenues from $11 to minus $14. Adding one week to the supply transit time changes the net revenues from $11 to minus $38. Thus, the impact of changing the delay times on the supply chain performance is quite complicated. We also have to remember that as we said earlier, once we change the supply chain structure or properties (by changing the delay times), the optimum value of SLF for a given IAF value will change. The last column in table 10.1 shows the results for the given delay times where the net revenues have been optimized by finding the best value of SLF. These data still show that supply transit times have a larger impact on net revenue than order delay times. This effect is present even when SLF = 0, so it cannot be due to the fact that the supply line has more impact resulting from the supply line adjustment portion

TABLE 10.1. Variation of net revenue with delay times.

Supply Transit Time (weeks)	Order Delay Time (weeks)	Total Effective Delay Time (weeks)	Net Revenues ($)	Net Revenues ($) (SLF optimized)
1	1	8	85	85
1	2	12	80	83
2	1	12	79	80
2	2	16	67	70
2	3	20	43	44
3	2	20	37	42
3	3	24	11	12
3	4	28	–14	–3
4	3	28	–38	–20
4	4	32	–59	–28

of the ordering equation. Instead, it appears more likely to occur because what drives net revenue is the inventory costs rather than lost sales. The larger supply transit time leads to larger fluctuations in the inventory levels in the chain.

A second supply chain attribute that may have an impact on the bullwhip effect is the fact that we have no capacity constraints in the system. This is true not only at the brewery, which has an infinite production capacity, but also at the other nodes, where we assume there are no transport capacity constraints. Therefore, it might be of interest to see what impact capacity constraints have on the net revenues. A converter Max Rate is defined and the flows that specify the selling of beer to the downstream nodes are set so that if the calculated selling rate is greater than Max Rate, then Max Rate is used. We saw this earlier where we used the MIN built-in function when specifying the brewery production rate. For the base case, the maximum selling rate is below 50 cases per week. In figure 10.10 we show the impact of Max Rate on net revenues over the range 1 to 50.

Above a Max Rate value of about 15, the net revenues are fairly consistent in the range of \$67 to \$70. As we reduce the Max Rate below 15, the net revenues line begins to rise until it peaks at about \$82. Thereafter, it drops until at a Max Rate value of 1, the net revenues value is \$9. It should not surprise us that the capacity value that gives the maximum net revenues is 8, the value of the customer demand after the step change in demand at week 5. In fact, as readers can verify, the maximum net revenues always occur when the Max Rate is set to the value of the customer demand after the step change has occurred. Essentially, the capacity constraint caps the amount of orders we can satisfy, which prevents us from over-

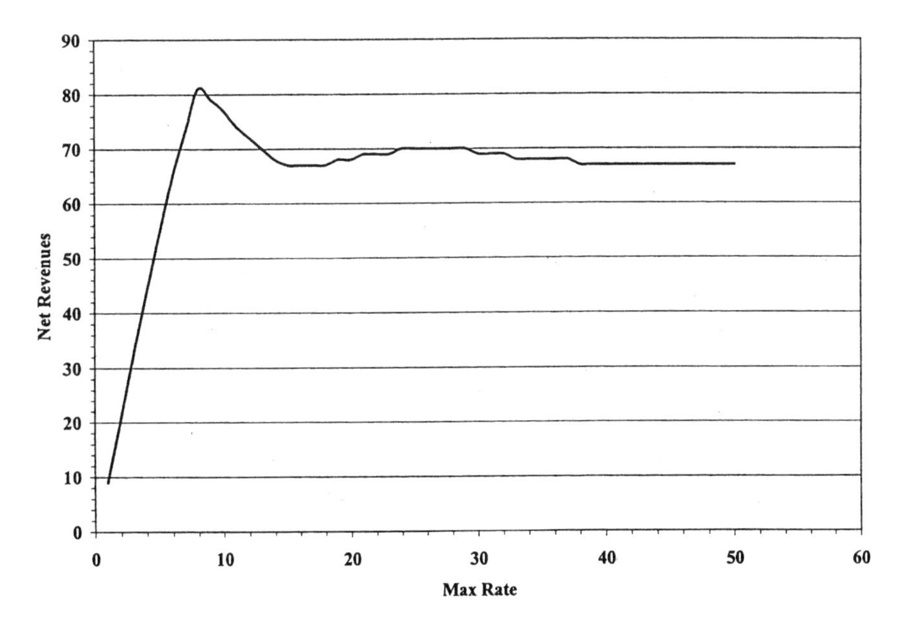

FIGURE 10.10. Impact of capacity/transportation constraints on net revenues

producing beer even when the supply chain control mechanism indicates that more beer should be supplied. This is an interesting aspect of process management and control that we should recognize. Many times we think that the more constraints we have in a system, the less optimal a solution we can find to a problem. This is true in a steady-state problem where we are looking for a single, optimal solution to a problem.

For example, consider the problem of scheduling a production run of several different products through a factory. The constraints could be resource availability and order due dates. The problem is to decide the sequence for making the products so that some objective is optimized. Suppose we find an optimal schedule for a given set of constraints. If we impose even more constraints, then the new optimal schedule cannot be any better than the original best schedule. In a dynamic situation, however, the result is different. In this case, adding a constraint may actually improve the behavior of the system because the constraint prevents the system from moving into regions of poor performance. This is exactly what the capacity constraints appear to be doing.

> Learning Point: In a dynamic system, constraints of the process may actually improve process performance over time. In the BG, this is shown by the way a capacity constraint can actually lead to improved net revenues.

Of course, capacity constraints do not lead to a practical control mechanism for the supply chain. Such capacity constraints are simply attributes of the supply chain. If we set them incorrectly, we will reduce net revenues. What the earlier results indicate, however, is that it is critical to consider such constraints when planning how the supply chain is to be managed. From a management perspective, having too much capacity will create a set of issues, just as having too little capacity will be problematic. When we have too much capacity, we must find ways to reduce the tendency to overproduce just "to be sure."

The third modification to the basic model involves improving our information visibility in the chain. If we look back at the equation a node uses to decide how many orders to make, the first component was the orders data. We assumed that because a node could only see order data from the next node downstream, the orders data would be specified by looking at the ordering data from the next node downstream. However, this approach means that any errors in ordering a node are propagated *upstream* in the chain. What would happen if, instead of this approach, the customer ordering information were to be made available to all the nodes in the supply chain? We would hope to see an improvement in performance and higher net revenues. To test this hypothesis, we modified the orders data by introducing a new parameter, the Demand Information Factor (DIF), and modifying the orders equations to the following:

Wholesaler: Customer_Ordering*Demand_Information_Factor
+ (1-Demand_Information_Factor)*Retailer_Orders_Arriving

Distributor: Customer_Ordering*Demand_Information_Factor
+ (1-Demand_Information_Factor)*Wholesaler_Orders_Arriving

Brewery: Customer_Ordering*Demand_Information_Factor
+ (1-Demand_Information_Factor)*Distributor_Orders_Arriving

In effect, if DIF = 0, then we have the base case model. If DIF = 1, then a node completely ignores the orders arriving from the downstream node and uses the customer orders only as its indicator of the orders that must be replenished. Thus, DIF = 0 represents the worst case in demand information visibility, and DIF = 1 represents the maximum in demand information visibility. Figure 10.11 shows a comparison of the distributor ordering profile for the case where DIF = 0 (profile 1) and DIF = 1 (profile 2).

The improvement gained from sharing the end-customer demand information is clear. The amplitude of the cycle in profile 2 is much less than profile 1. Figure 10.12 shows the variation in net revenue as a function of DIF when IAF = 0.2 and we have optimized the value of SLF.

Again the positive impact of sharing customer demand information throughout the supply chain is clear. The data in figure 10.12 also indicate that we have a law of diminishing return. In other words, the impact of increasing the DIF when the DIF is small is greater than an equivalent change in the DIF when the DIF is large. Thus, the initial information sharing gives the biggest benefit as measured by its impact on net revenues. This sharing of customer demand by supply chains has been widely reported in the literature (Lee, Padmanabhan, and Whang 1995, 1997a,b). In many instances, the retailer may allow the supplier to view their in-

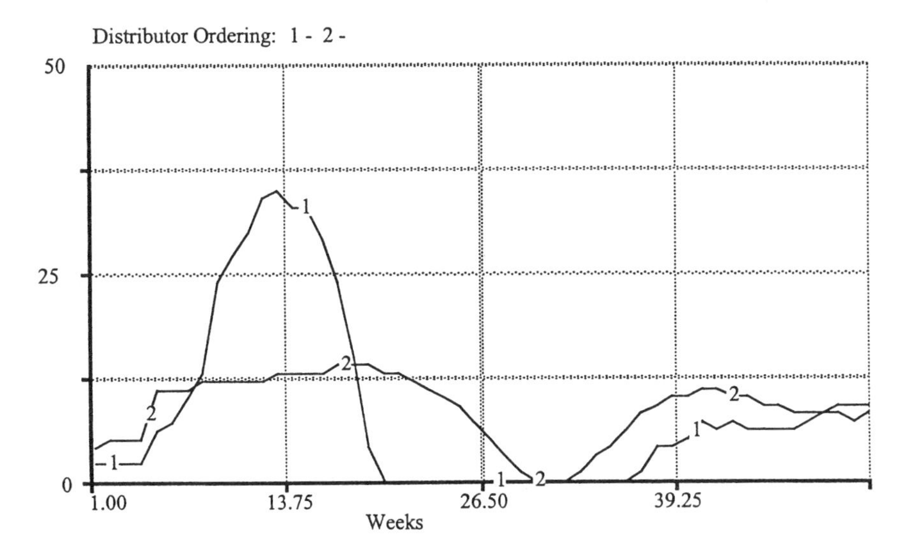

FIGURE 10.11. Distributor ordering profile for DIF = 0 and 1

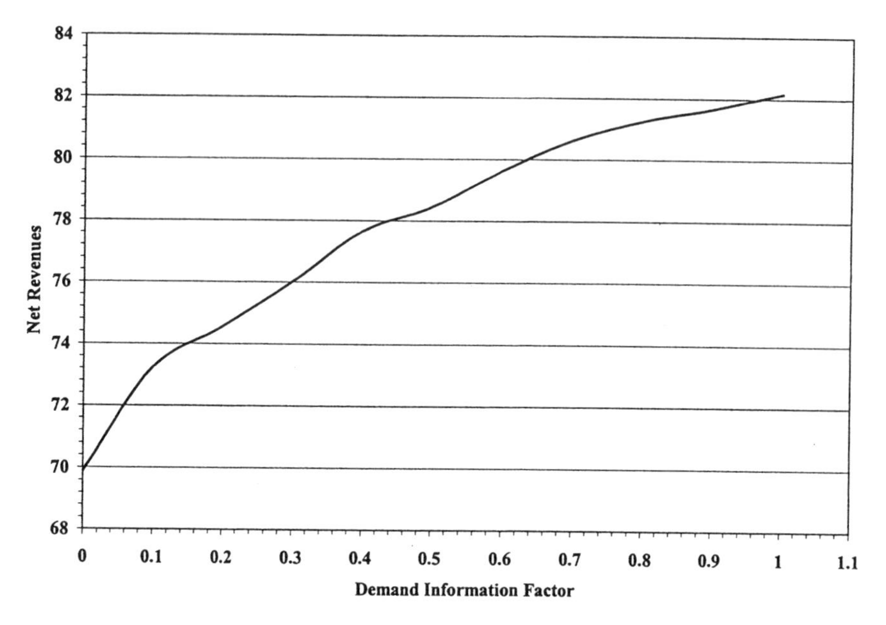

FIGURE 10.12. Variation of net revenues with demand information factor, DIF

formation systems directly so that the suppliers can see the customer demand data and determine how to keep the retailer stocked adequately. This kind of cooperation demands a high level of trust between the retailer and the suppliers. We should also note that, as in the case of supply-line data sharing, it is not enough to simply share the data between nodes. The algorithm that uses the data must also be considered. If the nodes do not agree on a common algorithm or if they all use a poor algorithm, the demand data sharing could make matters worse.

> Learning Point: Sharing demand data among the nodes of a supply chain can improve the operation of the supply chain (provided that the data is accurate and is used properly).

In the simple supply chain representing the BG, using the demand data is fairly simple. However, in more complex supply chains where each node has multiple products, multiple suppliers, and multiple customers, deciding how to best use the demand data will be more difficult.

10.5 Modifications to the basic model

We leave it as an exercise to readers to consider the following modifications to the base model provided.

- Adding random variation to the order delay times and supply transit times.
- Adding random variation to the customer demand signal.

- Using the nonlinear response equation 10.4.
- Using different values of IAF, SLF, and DIF at each node to see if a relationship exists between position in the chain and the impact of these parameters.
- Using different order delay times and supply transit times at each node to see if a relationship exists between position in the chain and the impact of these parameters.
- Alternative financial models of performance for the supply chain.

10.6 Using the Beer Game model in game mode

It is possible to run the model in a game mode that simulates closely the actual conditions under which the game is played. Figure 10.13 shows the main interface provided for the model. Before running the model in game mode, the parameters of the model can be set up here. The base case parameters give a reasonable set of parameters with which to play the game.

In order to run the model in game mode, click the Down button at the lower right-hand corner of the interface. This will move the interface down and reveal the interface shown in figure 10.14.

The data shown in the data table are approximately what a player can see in the actual game. (Of course, readers can add and subtract to this table as they see fit

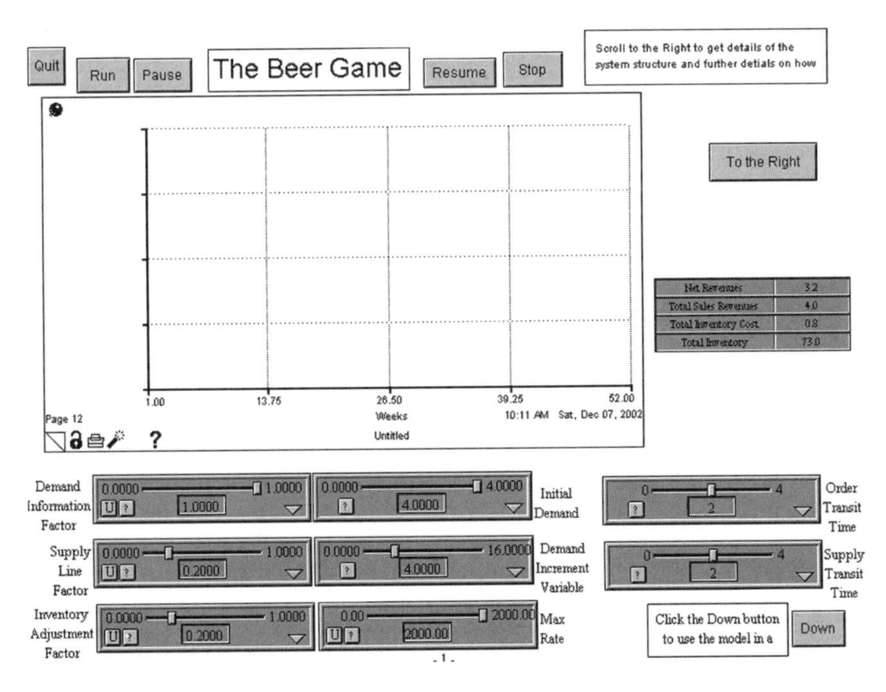

FIGURE 10.13. Main user interface to Beer Game model

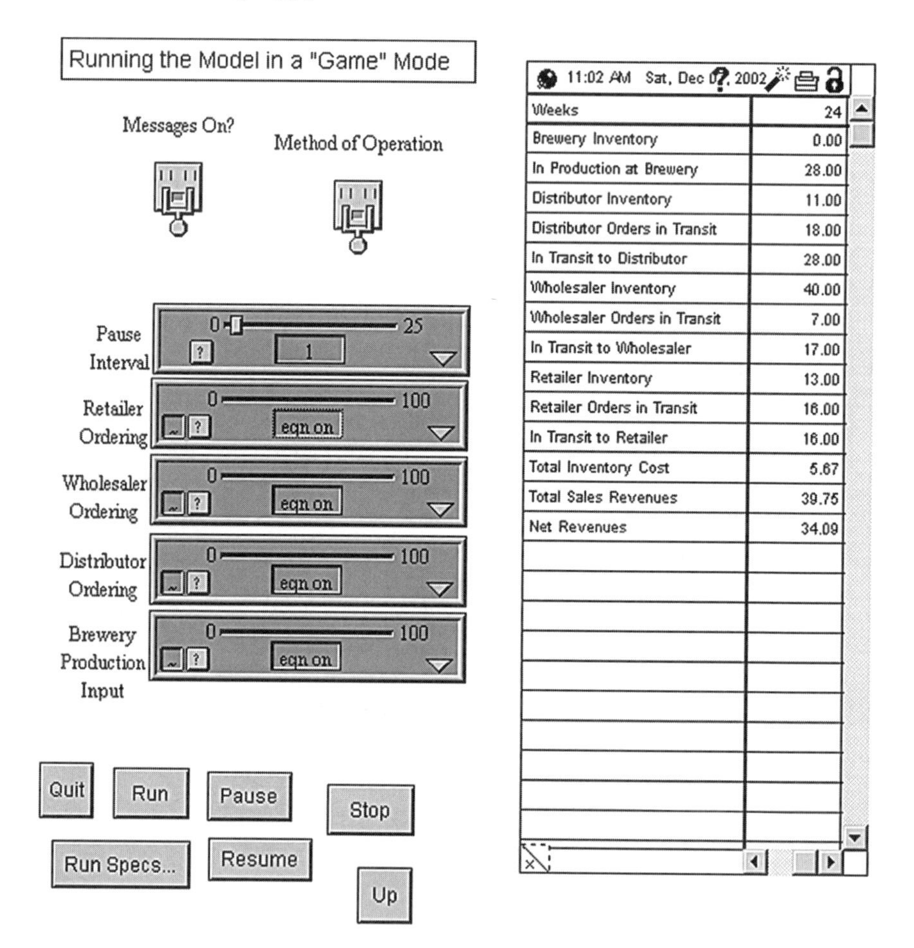

FIGURE 10.14. Model interface for game mode

to modify the game play). Now the reader clicks the Method of Operation switch to turn on game mode. In effect, this pauses the model after every time step to give the players time to make decisions on what to order. To start the game, click the Run button. When the players have entered their order numbers, they can click the Resume button to move the model to the next time step (each time step is still one week). In order for players to enter orders, we must modify the sliders for the position they are playing. By default, the sliders are set in eqn on mode, meaning that the orders are calculated from the normal model equations. Suppose that someone wants to play the retailer role. Then they click the ~ button on the Retailer Ordering slider; the eqn on text disappears; and a number appears instead. Now the player can enter an actual number of orders. The model still takes care of all the calculations that must be done to move orders and beer through the chain (in the actual game, the players must do all this). Thus, a player can simply play one role and pit his or her skills against the automatic ordering algorithms used by the other nodes for the values of IAF, SLF, and DIF that were set on the main model interface.

Of course, if the game is played with four players, all ordering sliders will have the eqn on turned off and the IAF, SLF, and DIF parameters will not be used at all. The Messages On? switch can be used to turn on some warning messages that will appear if inventory is too high or if the retailer is out of beer. This can serve as a wakeup call for players if they are not managing the supply chain well. One of the interesting aspects of the BG that can be illustrated with the game mode of play will occur if you have players who are familiar with the game and have played before. If we then set up the IAF, SLF, and DIF parameters for poor performance (say, IAF = 0.5, SLF = DIF = 0.0), then even though experienced players know what to do, they will still not be able to maintain good performance in the chain because the other nodes are tuned for poor performance. The chain is only as good as its weakest link!

Learning Point: Even if some of the players know how to behave so as to maintain good performance for the supply chain, that is not enough to guarantee good performance overall. **All** nodes in the chain must exhibit the right behaviors in order to guarantee success.

11

The Dynamics of Management Strategy

AN ECOLOGICAL METAPHOR

> Actually we shall call these sets of rules the "strategies" of the game.
> —J. Von Neumann and O. Morgenstern, *Theory of Games*

11.1 Introduction

When a corporation is composed of many plants, how should it control the operation of these plants in the least intrusive way; that is, what is the minimal set of appropriate constraints the corporate level should apply? These constraints include specific operating conditions and require specific operational information. An analogy with ecology is presented, along with an opportunity to develop a dynamic model of the optimal expansion of the corporation. Consider two identical firms with the same owner. The only difference between these firms is geographic location. It is the nature of managers in each firm to optimize their own operations, whereas it is the duty of the manager at the next higher level in the ownership hierarchy to run the firms such that their combined operation is optimized. As seen from the level of the firms, individual firm optimization is a tactical process, while combined optimization is a strategic process. Tactical decisions are made at your level in the management hierarchy, and strategic decisions are made at the next highest level. (The decision to optimize the operation of the two firms is strategic and is made at the next level higher than that of the firms themselves. The decision as to how many firms this company should own is also a strategic one, as seen from the firm level.)

This chapter is intended to shown how, in an ideal world, the tactical and strategic decisions about company expansion are most appropriately distributed across various levels of the business environment. We model the dynamic process of industry expansion until the market is saturated and economic profits are zero. In the example that follows, the corporate level takes signals (constraints) from the market, including such signals as interest rates and market prices for the products being produced and the input unit costs. These corporate managers also determine the scheduling of new plants and instruct those at each of the plants to pick the best technology, build the plants at the desired rate, and expand the production of each plant to maximize the profits for that period. Decisions about how to meet daily and weekly production goals are set at the plant level and achieved by those in the plant operations division. In this example model, the market is the highest level in the hierarchy, the corporate level is next, the plant level is the next most detailed, and the plant operations division is the lowest.

11.2 Hierarchy in nature

We can find considerable guidance on the subject of hierarchical dynamic systems in the history of ecosystem modeling. McMahon et al. (1978) lay out a biological system of hierarchical description. They show three descriptions that span from the highest possible level to the organism. At the highest level are phylogenetic (Kingdom, Order, Species, Organism), coevolutionary (Community, Population, Breeding Population, Organism), and matter-energy exchange (Biosphere, Ecosystem, Organism). They lead to a single hierarchical description for the organism (Organ, Cell, Molecule, Subatomic Particle). The particular branch one chooses depends on one's point of view in modeling. Allen and Starr (1982) had a critical observation on how to get results in modeling biological systems:

> The limiting factor is not computational power; of that we have plenty. And it is not our ability to translate field experience into mathematical forms. A new and powerful mathematics is not what is needed; rather we are limited by a lack of familiarity with the consequences of coupling subsystems with different cycle times.

Allen and Wyleto (1983) point out that unified models inevitably contain contradictions that destroy their credibility. Models representing multiple levels of the system resolve these conflicts and can reveal emergent properties of the system.

Some observers will focus on the structural aspects of the system and claim it is unchanging. Others will examine the flows in a system and claim that nothing in the system is constant. Lavorel, Gardener, and O'Neill (1993) demonstrate that complex systems *will* develop hierarchical structures. To us, the most useful and insightful view of hierarchy in systems comes from Johnson (1993). His claim is that hierarchy theory implies the idea that strategy and tactic are relative. One must choose the level in the hierarchy and discuss the definitions from that point of view. The dynamical behavior of the agents at a specified level in the hierarchy are determined by interactions with the next lowest level. The specified level gets the constraints on its actions from the level just above it. We say, therefore, that strategic decisions appear to those at the specified level as the result of decisions from the level above, and its own tactics determine how its own dynamics will develop. From these dynamics come the constraints for the next lowest level. Johnson goes on to point out how Stommel (1963) defined the relationship between cycle times and levels in a hierarchical system. A forest, for example, can be thought of as organized by size from Landscape, to Patch, to Canopy, to Tree, to Branch, to Leaf, and so on. As one moves down this hierarchical description, the cycle times become significantly shorter. Plotting these sizes versus their corresponding cycle times yields a straight line on log-log paper, indicating a power law relationship. In fact, Hannon and Ruth (1997) showed that the change in the total biomass of any species at the Landscape level is essentially zero. Barring earthquakes and meteors, the Landscape level is defined as that level where each of the individual biomass components is steady. As one proceeds toward the small-scale, the turnover times shorten dramatically and the predictability is diminished.

If we adapt this ecological view of hierarchical systems organization, then the process of proper constraint determination is a crucial one. The constraints must not be too general nor too specific, so as not to usurp the widest possible range of functioning of the next level down in the hierarchy. In the business world, we would say that the constraints must be set to maximize the creative response of those at the next level in the hierarchy and to engender ownership of their part of the process. The typical set of business constraints is likely to be too conflicting, or too indefinite and therefore maximizing the possibility of unfulfilled mission.

To those at the lower or receiving level, these constraints appear to be strategic decisions. Those at the higher level, however, see the same constraints as tactical decisions. The dynamics of the higher level are the result of its interaction with the next lowest level. This view of the system is repeated down through the organization until it ceases to be useful.

We shall now develop a model of two interacting levels in a hierarchy as a way to crystallize the meaning of the proceeding rather general discussion.

11.3 A model demonstrating the hierarchical nature of an expanding business

Here is an example model for a hierarchical business system, a multiplant company in a competitive market. Level 0 is the market environment. Level 1 represents the corporate management. Level 1 constrains several of its own plants at Level 2, the plant management level. The rest of the production capacity for this product lies in the hands of competitors, but it is assumed they use exactly the same production technology and that all corporate managers know the expansion schedule for the rest of the industry. Our corporation will own several of the plants, but not enough for them to act differently from a competitive enterprise. In the model, Level 2 is the management of one plant as it turns a single INPUT into OUTPUT. INPUT is a stock. We must increment this stock consistently to find the maximum profit output level of the firm. The time constant for Level 1 is 1 year, while the time constant for Level 2 is 1 month. Consequently, the meaning of time = 1 for the model is 1 month. To improve the model accuracy, the model time step, DT, is shorter (1/8). Level 3 is the collection of production line employees in a plant. Our model focuses on the dynamics and interactions of Levels 1 and 2.

For simplicity with no loss of usefulness, the costs of units of all the inputs are summed into a single variable, UNIT COST. This cost includes the cost of the input, labor and energy resources, all taxes, capital (averaged), and depreciation and maintenance. Level 1 management has negotiated with the outside environment (another hierarchy) for the unit cost of this single input. From their analysis, they have also discovered the demand curve for their product (PRICE = 10–0.004*PLANTS*OUTPUT). Level 1 has also received the necessary discount rate from the market environment (Level 0). The discount rate is used by Level 1 to compare the RETURN RATES of any actual scheduling of the plants. These are the constraints on Level 1 from Level 0.

At the beginning of each year, the corporate managers specify the number of new PLANTS they expect to see developed that year, both of their own and of their competitors. The entire industry begins with one plant and assumes new plants will be added each year as predicted by the SCHEDULE. As we shall see in the example, this schedule is a parameter that can be changed to meet specified goals. Level 1 management surveys the industry building plans and determines how many identical plants the competition is building in the coming year. They add their own building plans to that of the competitors to conform to the SCHEDULE.

Level 1 management uses the demand curve and the industry-wide building schedule to determine the PRICE. Level 1 chooses to require profit maximization on the part of their plant managers. So PRICE, UNIT COST, number of new plants to be built, and the maximum PROFIT goal are the constraints Level 1 management gives to Level 2. Level 1 management is also responsible for modeling the market and the plant simultaneously, as shown in figure 11.1.

Level 2 management is responsible for choosing the proper technology, building and running the plants, and generating a PROFIT that each PLANT monitors constantly. PROFIT is simply revenues (PRICE*OUTPUT) less COST (UNIT COST*INPUT). The OUTPUT is determined by a production function: OUTPUT = A*INPUT^2–B*INPUT^3, with A (50) and B (15) determined by statistical analysis of OUTPUT and INPUT data from the historical use the use of this technology. A and B represent the particular production technology chosen by the Level 2 management. This form of the production is necessary because we must be able to find the output level where the marginal and average costs are equal. This equality signals the equilibrium level for the industry as a whole. The variable MONTH simply counts the months, beginning with month 1 and ending with month 12, and then starts over. During each year, the plant(s) receive the PRICE signal for the year from Level 1. They are CHANGING INPUT to find the new level of OUTPUT that maximizes PROFIT, after RESETTING the INPUT to a low level at the end of each year.

The technique for finding the maximum profit is quite a simple one: plant output expansion grows until the profit starts to decline. The Level 2 plant managers start their scheduled new plants at the beginning of each year with near-zero input (STARTING INPUT) and increase it slowly (EXPANSION RATE) to find their maximum profit OUTPUT value. The plants scheduled for startup that year are brought fully on line to their profit maximizing level of output in about six months and are held at that level. Because the market price drops each year as

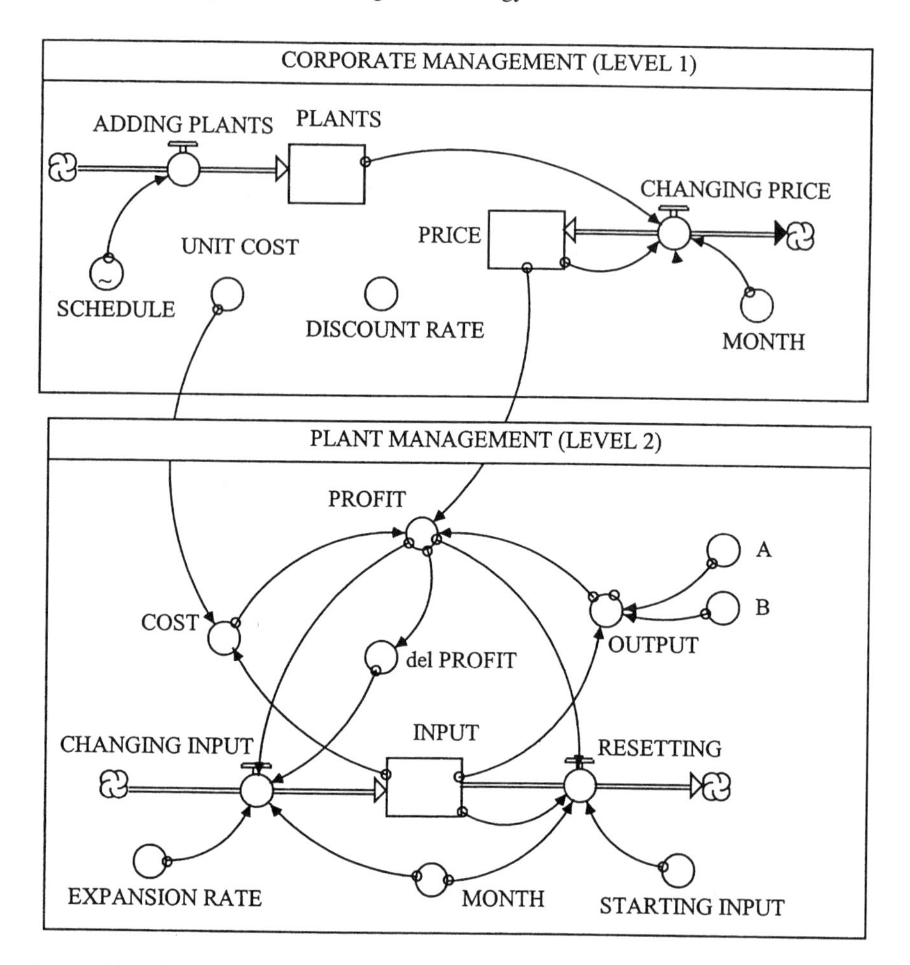

FIGURE 11.1. The model of Level 1 and Level 2 dynamical interaction: For clarity, some of the secondary control variables are omitted in this figure; all of the equations are presented at the end of the chapter

plant expansion continues, new plants must seek a new (lower) optimizing level of output each year.

In this example, the plants are scheduled as follows (see SCHEDULE in the model): one plant for year 1; five plants per year for year 2; four plants per year for years 3 and 4; and two plants per year thereafter.

Corporate management would experiment to find the optimal schedule that will maximize the cumulative discounted profit. In such calculations, management would not smooth out the capital costs but would assume they occurred as spikes. The calculations they would have to make can be seen in figure 11.2.

In the model, plants start up each year and continue to run for the rest of the model at the optimum output rate for the price in that year (see figure 11.3).

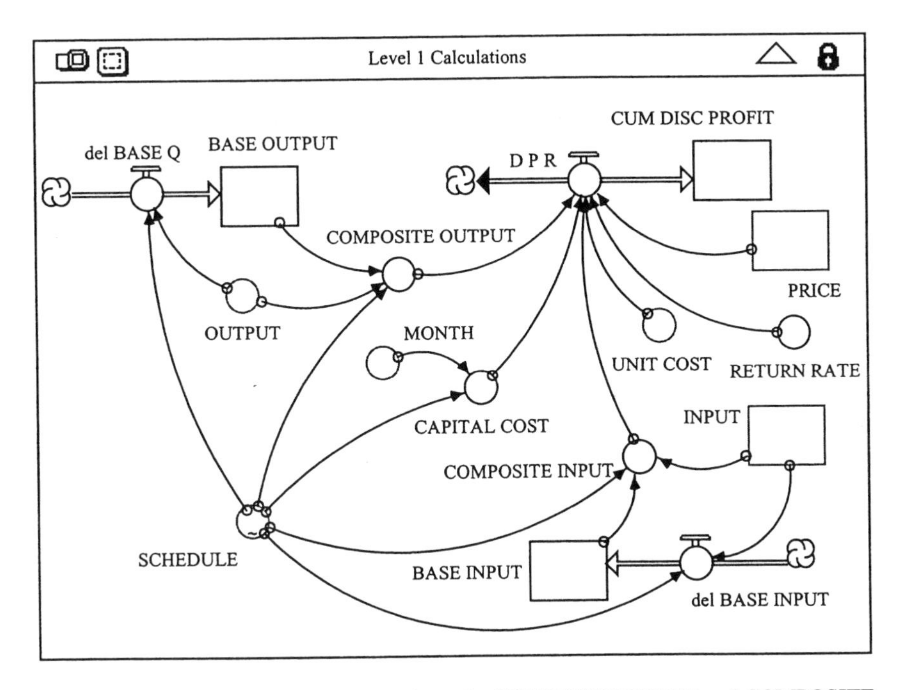

FIGURE 11.2. The portion of the model where the COMPOSITE INPUT and COMPOSITE OUTPUT are made: These values, along with the periodic CAPITAL COST for initial startup and for each year's new plants, allow corporate managers to calculate the CUMULATIVE DISCOUNTED PROFIT (cumulative discounted EVA)

The peak profit per plant declines as more and more plants are added. Eventually, after 120 months, the total operation has 26 plants. The addition of any more plants to this market would provide maximized but negative profits. The two levels have interacted to reveal the pattern of industry-wide profits for the particular new plant growth pattern forecast by the management in Level 1. From this point on, the production of all 26 plants is assumed to continue at the rate achieved in the last year of expansion.

Because of the complex nature of the startup process, it takes Level 2 slightly more than six months to get their new plants to the profit maximizing output level. This output is then fixed for the rest of that year's model run. Figure 11.4 shows only the first year's output of a single plant. Once each plant reaches the optimal output level, that plant continues to run for the rest of time. Each plant produces about 82 units per month.

The total output is 2,140 units per year at 0 economic profit. At this point, we have reached the market equilibrium. The company is paying all its bills, labor, taxes, and stockholders, but nothing is left to invest in further plant construction.

We have imagined these plants built by a single corporation, and yet our process of price taking by the plants is a description of a competitive environment. Market information about the construction of plants belonging to other compa-

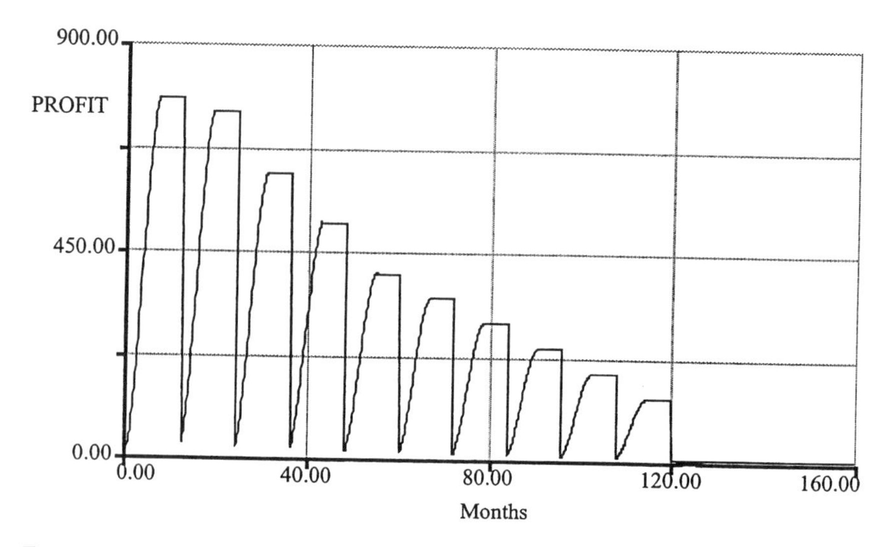

FIGURE 11.3. The annual additions to profit of the new plants starting up at the first of each year: Each year's new additions (see SCHEDULE) continue to run from then on. At the end of year 10, additional plants cease to add any profit to the collection as market equilibrium has been reached and no further plants are built. After 120 months, all 26 plants continue with constant output (82.13 units per plant, per year) that maximize the collective (0) profit while selling their output at the equilibrium price ($1.46).

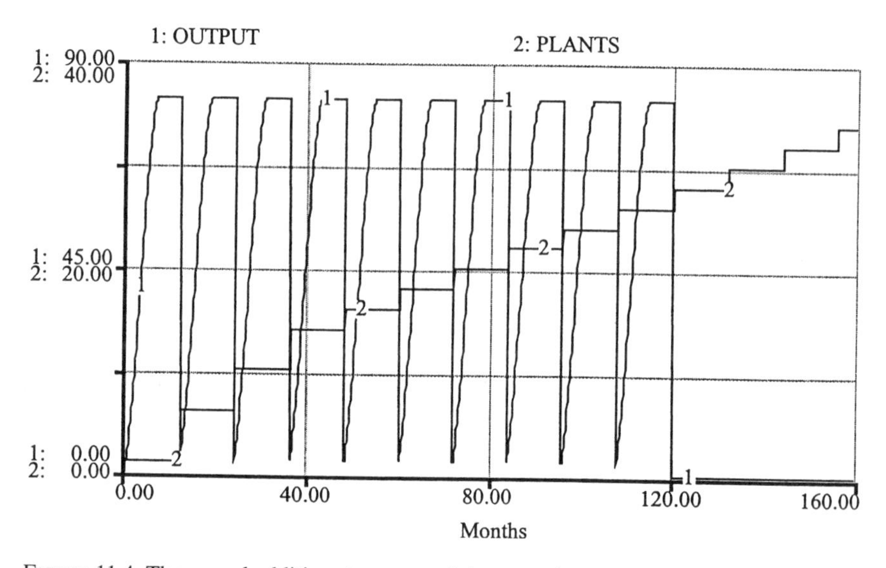

FIGURE 11.4. The annual additions to output of the new plants starting up and the cumulative number of plants in operation: Each year's new additions continue to run for the rest of time. Note how little the optimum output changes for the new plants even though the profit declined continuously. This is because of the particular choice of production function.

nies must be included as a part of this schedule. The industry-leader corporation tries to be the first to construct new plants with the hope of causing other companies to postpone or reduce their plans. Under such circumstances, it is easy to see how an industry could become overbuilt in the sense that the optimal building schedule is exceeded. Although our corporation may not be interested in maximizing profits for the entire industry, the process—a kind of cooperative competition—described in this model may not be too far from reality in industries where there are no exclusionary rights such as patent protection.

It is important to understand how the optimizations at the two levels proceed. The UNIT COST contains all the costs, including capital costs (20 percent). For each unit of input, these capital costs are smoothed out and added to payable taxes and operation and maintenance costs to form the UNIT COST. Each year, Level 2 managers bring the new plants on line and up to the output level that maximizes profits, given the UNIT COST and the PRICE for that year. The Level 1 managers take a broader view. For each proposed SCHEDULE, they determine the RETURN RATE that brings the CUMULATIVE DISCOUNTED PROFIT of the entire set of plants up to 0 just as the market reaches equilibrium. For the example, that RETURN RATE for the given plant scheduling is 19 percent, and the graph of the CUMULATIVE DISCOUNTED PROFIT is shown in figure 11.5.

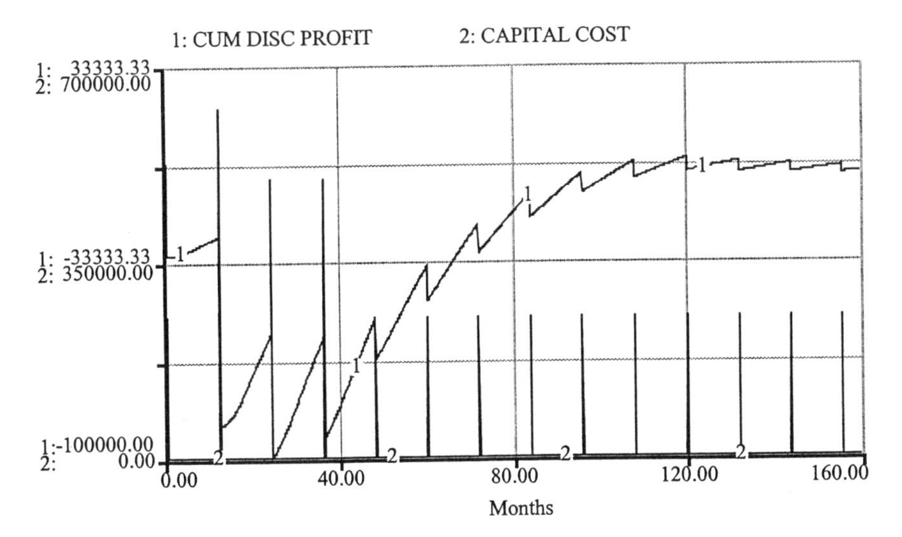

FIGURE 11.5. The CUMULATIVE DISCOUNTED PROFIT for the company is brought to 0 by adjusting the RETURN RATE, just as the series of plant building has stopped (time = 120 months), for a particular plant scheduling sequence: The sequence is varied to determine the largest RETURN RATE achievable. If this rate is greater than the given DISCOUNT RATE, the plan is feasible. If this rate is the greatest of all the investment options for this firm, then the option is desirable. The CAPTIAL COSTS are shown as they occur at the beginning of each period.

The RETURN RATE for this scenario is 19 percent, which is above the given market rate of 17 percent. Therefore, the scenario is feasible. If investment funds are available after all projects with higher return rates are satisfied, then the scenario is desirable.

According to economic theory, marginal cost should equal price when profits are maximized. Theory also indicates that when profits are maximized at 0, the average and marginal cost should equal price. This occurs when the time = 120 months for this particular plant scheduling sequence. The sequence would next be varied to determine the largest RETURN RATE achievable (see figure 11.6).

After 120 months, the average cost switches from below to above the price. At the optimal point, price, marginal cost, and average cost would be equal. Because the plants appear as discrete units, this equality is not precisely achievable.

In the real world, Level 1 might over- or underconstrain Level 2. Level 1 management could direct the plant to buy a certain technology and then insist that it be run according to a given schedule to produce the requisite profits. This could be a disaster if the corporate managers were not really good technologists or were not familiar with the real history of the technology required. It is much better to put these decisions in the hands of those who will be responsible for efficient operation of the plants. They are more likely to have the appropriate level of knowledge required for such purposes, and they have added incentive to make sure their own decisions turn out to be good ones.

Level 1 could be guilty of constraining too loosely. For example, without the proper signal to stop building plants, the industry becomes overbuilt. Some plants will operate in the red.

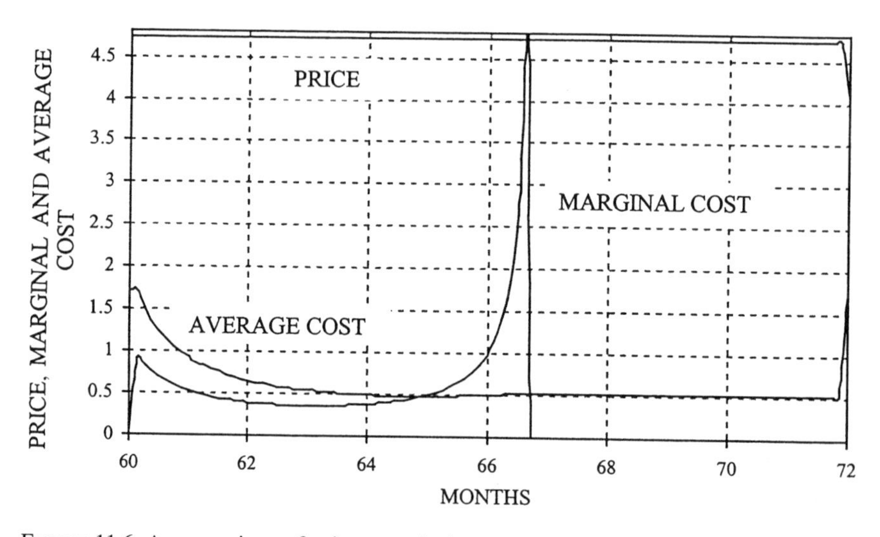

FIGURE 11.6. A comparison of price, marginal cost, and average cost in a portion of the year 9 of expansion: Maximum profit was achieved in about 5 to 6 months. The exactness of the equivalency of price and marginal cost depends on the choice of DT. AC and MC are equal when X = A/2B.

Level 1 managers may be late in giving the signal on when to start the new plants, causing a delay in the construction schedule and ultimately in the most profitable production schedule. In fact, our model has an unnecessary delay in plant startup timing in the early part of each year. This was purposely done to convey the nature of the process. However, the plants could be started in the last six months of the preceding year so that they would be fully operational at the beginning of the year they were scheduled to start.

Level 1 could be delayed in the gathering of information about plans of the competition. In addition, the information they gather could be wrong, wrongly interpreted, or largely ignored. They might direct the construction of new plants in an effort to get the "jump" on the competition. We are assuming that the competition is using a model to determine the optimal construction schedule in this industry. The question of optimal path becomes a game as to whose account the profits will actually accrue. Assuming that the profit rates are controlling the rate of plant expansion, the company with the highest rate of profit expands the fastest, constrained by the availability of investment funds. We assume that whoever this is will realize that an optimal expansion path exists. This constraint on the rate of building new plants is not now in the model. It comes from Level 0, the (lending) market. The reader is encouraged to add it.

The Level 1 managers are faced with uncertainty, particularly with regard to the true nature of the demand curve and the intentions of the competition. What policies can they develop to cope with such uncertainty? One answer is extensive sensitivity testing with such a model to determine its response to this uncertainty. For example, what is the variation in the ultimate number of plants if the shutoff price (10) in the demand curve was really 9? The answer, with the same building schedule, is that such a change reduces the total amount of plants to 24 but lowers the return rate to only 4 percent, clearly well less than investment in the external market. This particular scenario is not feasible.

What about the effects of varying several such parameters at once in the hope of finding some peculiar conjunction that seriously affects the key results? Try varying the shutoff price and the slope of the demand curve simultaneously. Such sensitivity analysis can be performed efficiently using either ithink or BERKELEY MADONNA™.

Surely the technology will improve as we build more plants. This could be handled in the model by representing A and B, the technology parameters, as graphical functions. A would rise to an asymptote and B would decline toward 1, with the rate of rise/decline based on cumulative output. This is the typical "learning curve" approach to modeling dynamic technical change.

Further modeling complications might take the form of firm behavior under conditions of a limited number of production companies (oligopoly), or a limited number of buyers (monopsony), or monopolistic collusion, as described by Ruth and Hannon (1997).

The deeper details of the programming techniques can be found in the model documentation found inside the various model variable symbols.

The entire process can be described in hierarchical terms. Level 1 receives from the market (Level 0) what they view as strategic constraints. These appear in the form of the necessary discount rate and the unit cost from the suppliers of input. Level 1 must drive down this unit cost to a point that allows them to at least make some initial level of economic profit.

Level 1 constrains Level 2 to produce at maximum profit. Level 1 also issues PRICE and UNIT COST constraints and makes the decision on how many new plants will be built and at what time. They require Level 2 to behave in such a way as to maximize profits under these constraints. These are tactical constraints in the eyes of corporate management at Level 1, but they are strategic constraints when seen by plant management at level 2. The distinction between strategic and tactical is in the eyes of the beholder. Level 2 issues tactical constraints, such as what technology to use and when to stop increasing output, to their production managers (Level 3). The production managers see these as strategic constraints and issue their interpretations of them as detailed instructions for plant operation. The dynamics at Level 1 are determined by their interactions with Level 2. Level 2 has chosen the technology of actual production, the startup level and rate for the plants, and they report the timing and levels of production to Level 1. The dynamics at Level 2 are determined from their interaction with those who actually run the production machinery—those who face the day-to-day breakdowns and shortages that spoil their best-laid plans.

Finally, Level 2 could be two or more different types of plants in the same chain; that is, one of these two receives its input from the other one. We could elaborate the optimal production rate for each, separately considered. We also could optimize the output rate of the last plant, considering the three of them as a single plant. These are modeling ideas for the reader.

The firms depicted in this model do not overshoot their maximum profit production goals because they have perfect information about the market: the market prices are not made random and neither inventory nor shipping delays occur. Our model is in contrast with the production models in chapter 20 of John Sterman's *Business Dynamics* (2000). Such variation could be added to our model but it would obscure the point of this chapter: the concept of hierarchy in corporate decision making.

11.4 Equations for the complete model

[Note that some of these equations contain DT. It is used in Delay functions to cause the smallest possible delay. DT also is used in the denominator (see CHANGING PRICE equation) to make the equation independent of the DT choice.]

```
A = 50 {Units of Q per Units of X}
B = 15
INIT PRICE = 20
STARTING_INPUT = .5
INIT INPUT = STARTING_INPUT
OUTPUT = if time < = STOP_TIME then A*INPUT^2 – B*INPUT^3 else 0
```

NC_UNIT_COST = GRAPH(time)
(0.00, 60.0), (30.0, 60.0), (60.0, 60.0), (90.0, 60.0), (120, 60.0), (150, 60.0), (180, 60.0), (210, 60.0), (240, 60.0), (270, 60.0), (300, 60.0)
CONSTRUCTION_COST = 19000
DISCOUNT_RATE = (1 + (.17))^(1/12) – 1
PLANT_LIFE = 140
EXPECTED_INPUT = 2
CAPITAL_UNIT_COST = CONSTRUCTION_COST*DISCOUNT_RATE/((1 + DISCOUNT_RATE)^PLANT_LIFE – 1)/EXPECTED_INPUT
UNIT_COST = NC_UNIT_COST + CAPITAL_UNIT_COST
COST = UNIT_COST*INPUT
PROFIT = PRICE*OUTPUT – COST
Profit_Lag = PROFIT – delay(PROFIT,DT)
Profit_Rise = IF Profit_Lag > 0 THEN 1 ELSE IF Profit_Lag = 0 THEN 2 ELSE 0
MONTH = MOD(TIME,12) + 1
EXPANSION_RATE = .03
CHANGING_INPUT = IF Profit_Rise = 1 OR MONTH = 1
THEN EXPANSION_RATE/DT
ELSE 0
del_OUTPUT = OUTPUT – DELAY(OUTPUT,DT)
del_COST = COST -DELAY(COST,DT)
MARGINAL_COST = IF (del_OUTPUT > 0) AND (OUTPUT > = .2) THEN del_COST/del_OUTPUT ELSE 0
INIT PLANTS = 0
SCHEDULE = GRAPH(TIME)
(0.00, 1.00), (3.00, 1.00), (6.00, 1.00), (9.00, 1.00), (12.0, 3.00), (15.0, 3.00), (18.0, 3.00), (21.0, 3.00), (24.0, 3.00), (27.0, 3.00), (30.0, 3.00), (33.0, 3.00), (36.0, 3.00), (39.0, 3.00), (42.0, 3.00), (45.0, 3.00), (48.0, 3.00), (51.0, 3.00), (54.0, 3.00), (57.0, 3.00), (60.0, 3.00), (63.0, 3.00), (66.0, 3.00), (69.0, 3.00), (72.0, 3.00), (75.0, 3.00), (78.0, 3.00), (81.0, 3.00), (84.0, 3.00), (87.0, 3.00), (90.0, 3.00), (93.0, 3.00), (96.0, 3.00), (99.0, 3.00), (102, 3.00), (105, 3.00), (108, 3.00), (111, 3.00), (114, 3.00), (117, 3.00), (120, 3.00), (123, 3.00), (126, 3.00), (129, 3.00), (132, 3.00), (135, 3.00), (138, 3.00), (141, 3.00), (144, 3.00), (147, 3.00), (150, 3.00), (153, 3.00), (156, 3.00), (159, 3.00), (162, 3.00), (165, 3.00), (168, 3.00), (171, 3.00), (174, 3.00), (177, 3.00), (180, 3.00), (183, 3.00), (186, 3.00), (189, 3.00), (192, 3.00), (195, 3.00), (198, 3.00), (201, 3.00), (204, 3.00), (207, 3.00), (210, 3.00), (213, 3.00), (216, 3.00), (219, 3.00), (222, 3.00), (225, 3.00), (228, 3.00), (231, 3.00), (234, 3.00), (237, 3.00), (240, 3.00), (243, 3.00), (246, 3.00), (249, 3.00), (252, 3.00), (255, 3.00), (258, 3.00), (261, 3.00), (264, 3.00)
ADDING_PLANTS = if Time < = STOP_TIME then (PULSE(SCHEDULE,0,1000) + PULSE(SCHEDULE,12,12)) else 0
INIT BASE_OUTPUT = 0
COMPOSITE_OUTPUT = BASE_OUTPUT + OUTPUT*SCHEDULE

```
CHANGING_PRICE = IF MONTH = 13 – 1*DT and time < STOP_TIME
    THEN (20*exp(-.001*COMPOSITE_OUTPUT) – PRICE)/DT
        ELSE 0
Positive_Profit = If PROFIT < = 0 then 0 else 1
RESETTING = IF MONTH = 13 – DT
    THEN (INPUT—STARTING_INPUT)/DT
        ELSE IF Profit_Rise = 2 AND Positive_Profit = 0
            THEN INPUT/DT
                ELSE 0
AVERAGE_COST = If OUTPUT > .2 then COST/OUTPUT else 0
INIT BASE_INPUT = 0
del_BASE_INPUT = If MONTH = 13- DT then (PLANTS*INPUT—
BASE_INPUT)/DT ELSE 0
del_BASE_Q = IF MONTH = 13 – DT
    THEN (PLANTS*OUTPUT – BASE_OUTPUT)/DT
ELSE 0
INIT CUM_DISC_PROFIT = 0
CAPITAL_COST = IF MONTH = 4 and PROFIT > 0 and TIME < STOP_TIME
    THEN CONSTRUCTION_COST*SCHEDULE
        ELSE 0
COMPOSITE_INPUT = BASE_INPUT + INPUT*SCHEDULE
RETURN_RATE = (1 + (.21))^(1/12) – 1
D_P_R = (PRICE*COMPOSITE_OUTPUT – CAPITAL_COST/DT –
NC_UNIT_COST*COMPOSITE_INPUT)*EXP(-RETURN_RATE*TIME)
PRICE(t) = PRICE(t – dt) + (CHANGING_PRICE) * dt
INPUT(t) = INPUT(t – dt) + (CHANGING_INPUT – RESETTING) * dt
PLANTS(t) = PLANTS(t – dt) + (ADDING_PLANTS) * dt
BASE_OUTPUT(t) = BASE_OUTPUT(t – dt) + (del_BASE_Q) * dt
BASE_INPUT(t) = BASE_INPUT(t – dt) + (del_BASE_INPUT) * dt
CUM_DISC_PROFIT(t) = CUM_DISC_PROFIT(t – dt) + (D_P_R) * dt
OUTPUT = if time < = STOP_TIME then A*INPUT^2 – B*INPUT^3 else 0
NC_UNIT_COST = GRAPH(time)
(0.00, 60.0), (30.0, 60.0), (60.0, 60.0), (90.0, 60.0), (120, 60.0), (150, 60.0),
(180, 60.0), (210, 60.0), (240, 60.0), (270, 60.0), (300, 60.0)
CAPITAL_UNIT_COST = CONSTRUCTION_COST*DISCOUNT_RATE/((1
+ DISCOUNT_RATE)^PLANT_LIFE – 1)/EXPECTED_INPUT
UNIT_COST = NC_UNIT_COST + CAPITAL_UNIT_COST
COST = UNIT_COST*INPUT
PROFIT = PRICE*OUTPUT – COST
Profit_Lag = PROFIT – delay(PROFIT,DT)
Profit_Rise = IF Profit_Lag > 0 THEN 1 ELSE IF Profit_Lag = 0 THEN 2
ELSE 0
MONTH = MOD(TIME,12) + 1
CHANGING_INPUT = IF Profit_Rise = 1 OR MONTH = 1
    THEN EXPANSION_RATE/DT
        ELSE 0
```

```
del_OUTPUT = OUTPUT – DELAY(OUTPUT,DT)
del_COST = COST – DELAY(COST,DT)
MARGINAL_COST = IF (del_OUTPUT > 0) AND (OUTPUT > = .2) THEN
del_COST/del_OUTPUT ELSE 0
SCHEDULE = GRAPH(TIME)
(0.00, 1.00), (3.00, 1.00), (6.00, 1.00), (9.00, 1.00), (12.0, 3.00), (15.0, 3.00),
(18.0, 3.00), (21.0, 3.00), (24.0, 3.00), (27.0, 3.00), (30.0, 3.00), (33.0, 3.00),
(36.0, 3.00), (39.0, 3.00), (42.0, 3.00), (45.0, 3.00), (48.0, 3.00), (51.0, 3.00),
(54.0, 3.00), (57.0, 3.00), (60.0, 3.00), (63.0, 3.00), (66.0, 3.00), (69.0, 3.00),
(72.0, 3.00), (75.0, 3.00), (78.0, 3.00), (81.0, 3.00), (84.0, 3.00), (87.0, 3.00),
(90.0, 3.00), (93.0, 3.00), (96.0, 3.00), (99.0, 3.00), (102, 3.00), (105, 3.00),
(108, 3.00), (111, 3.00), (114, 3.00), (117, 3.00), (120, 3.00), (123, 3.00), (126,
3.00), (129, 3.00), (132, 3.00), (135, 3.00), (138, 3.00), (141, 3.00), (144, 3.00),
(147, 3.00), (150, 3.00), (153, 3.00), (156, 3.00), (159, 3.00), (162, 3.00), (165,
3.00), (168, 3.00), (171, 3.00), (174, 3.00), (177, 3.00), (180, 3.00), (183, 3.00),
(186, 3.00), (189, 3.00), (192, 3.00), (195, 3.00), (198, 3.00), (201, 3.00), (204,
3.00), (207, 3.00), (210, 3.00), (213, 3.00), (216, 3.00), (219, 3.00), (222, 3.00),
(225, 3.00), (228, 3.00), (231, 3.00), (234, 3.00), (237, 3.00), (240, 3.00), (243,
3.00), (246, 3.00), (249, 3.00), (252, 3.00), (255, 3.00), (258, 3.00), (261, 3.00),
(264, 3.00)
ADDING_PLANTS = if Time < = STOP_TIME then
(PULSE(SCHEDULE,0,1000) + PULSE(SCHEDULE,12,12)) else 0
COMPOSITE_OUTPUT = BASE_OUTPUT + OUTPUT*SCHEDULE
CHANGING_PRICE = IF MONTH = 13 – 1*DT and time < STOP_TIME
      THEN (20*exp(-.001*COMPOSITE_OUTPUT) – PRICE)/DT
            ELSE 0
Positive_Profit = If PROFIT < = 0 then 0 else 1
RESETTING = IF MONTH = 13 – DT
      THEN (INPUT—STARTING_INPUT)/DT
            ELSE IF Profit_Rise = 2 AND Positive_Profit = 0
                  THEN INPUT/DT
                        ELSE 0
AVERAGE_COST = If OUTPUT > .2 then COST/OUTPUT else 0
del_BASE_INPUT = If MONTH = 13 – DT then (PLANTS*INPUT
–BASE_INPUT)/DT ELSE 0
del_BASE_Q = IF MONTH = 13 – DT
      THEN (PLANTS*OUTPUT – BASE_OUTPUT)/DT
            ELSE 0
CAPITAL_COST = IF MONTH = 4 and PROFIT > 0 and TIME < STOP_TIME
      THEN CONSTRUCTION_COST*SCHEDULE
            ELSE 0
COMPOSITE_INPUT = BASE_INPUT + INPUT*SCHEDULE
D_P_R = (PRICE*COMPOSITE_OUTPUT – CAPITAL_COST/DT –
NC_UNIT_COST*COMPOSITE_INPUT)*EXP(-RETURN_RATE*TIME)
```

12

Modeling Improvement Processes

There's always room for improvement, you know—it's the biggest room in the house.

—Louise Leber, on being chosen Mother of the Year
New York Post, 14 May 1961

12.1 Introduction

In this final chapter, we look at modeling processes that are used to improve processes. This assumes, of course, that we realize that improving processes is a process in itself. It implies that we can analyze such improvement processes so that we can "improve the process by which we improve processes." Organizations that do not recognize improvement as a process do not invest energy trying to determine how to improve the improvement process. Approaches such as Six Sigma have definitely helped in this regard by placing emphasis on the process that is used to create improvement. For example, the Six Sigma approach uses an improvement process called DMAIC (Define, Measure, Analyze, Improve, and Control). Like any process, the improvement process can be analyzed to see where the bottlenecks are and to help identify ways that the improvement process can be improved.

Deming (1986) identified a generic model for an improvement process. Called the PDCA cycle, it is shown in figure 12.1.

PDCA stands for Plan-Do-Check-Act.[1] This cycle represents a basic form of improvement. First we PLAN how we want the system to operate. Then we operate the system (DO). As the system operates, we gather performance data. This performance data is then compared with the PLAN to see what gaps exist between plan and actual performance (CHECK). Once these gaps have been identified, they are analyzed; improvements are identified; and improvements are implemented (ACT). This means that the rate at which we improve depends on the output of the PDCA process, which in turn depends on the number of times the

1. Sometimes the process is referred to as the PDSA cycle, which stands for Plan/Do/Study/Act.

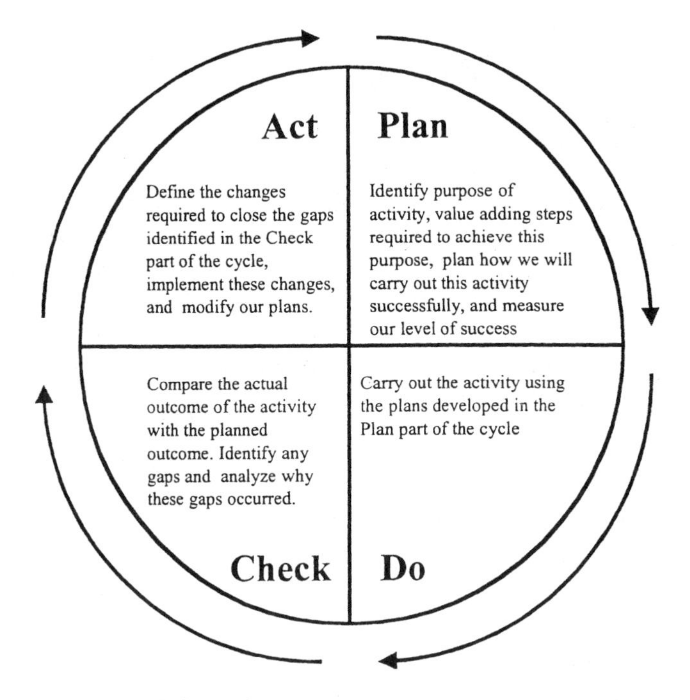

FIGURE 12.1. Deming's PDCA cycle

PDCA process is run and the improvements that occur every time the PDCA process is run. In principle, this is no different than saying that the total output of a production process is the output per production run multiplied by the number of production runs.

In any particular situation, the rate at which the PDCA cycle runs will depend on many factors. At any one time, one of these factors will be a bottleneck to the rate of improvement. In order to speed up the improvement process, this bottleneck must be identified and its performance improved. For example, it may turn out that the bottleneck to improvement is the amount of time people can devote to improvement versus the time they must spend dealing with production hiccups. Unless the organization figures out a way to divert more of people's time on improvement, this bottleneck will always exist. The PDCA process provides a generic starting point for the development of improvement models in any given situation. Studies on the rate of improvement for a process naturally lead to the study of learning curves.

12.2 Learning curves

Once we can model an improvement process in an organization, we must consider what the output of such a model will be. It must be possible to measure the process improvement; otherwise we will never be able to verify that improvement has occurred. We must be able to identify which measurements of the process represent the key outputs that we want to improve. Then we must look at the trajectory of these measurements over time and see how they behave. If they are getting better, then we could contend that improvement is taking place. If not, then there is no evidence of improvement. Referring back to the discussions in chapter 2, these measurements are the outputs of the process that represents compliance, effectiveness, and efficiency. A graph of the trajectory of a measurement that is used to show improvement is generally referred to as a *learning curve.* Such improvement curves are indicative of learning within an organization if the improvement can be sustained over a significant period of time. (It is not unusual for organizations to be able to show short spurts of improvement that cannot be sustained. Such short spurts would not be considered as evidence of organizational learning.) Learning curves were first described by Wright (1936). Originally, these curves were referred to as *experience curves.* The reason for this is that, in his original work, Wright studied how the manufacture of aircraft wings improved as the workers' experience increased. In particular, he studied the amount of labor hours required to make an aircraft wing. The idea was that as the workers became more experienced in manufacturing a wing, the amount of time it would take them to perform this task would decrease. Thus, a plot of labor hours per wing versus time would show improvement, where improvement means a reduction in the labor hours required. The learning curve might look like figure 12.2.

The more negative the slope of the learning curve, the faster the labor hours per wing is decreasing; that is, the faster the learning is occurring. Of course, labor hours per manufactured wing is an efficiency measure, a measure of productivity. It is possible to use this approach for any organizational measure. Also, some

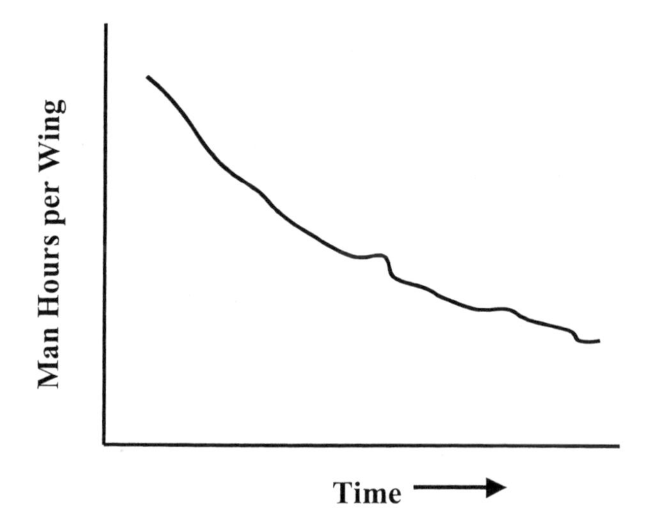

FIGURE 12.2. Typical learning curve

measures may be bigger-is-better measures, but the idea of a learning curve still applies. Thus we could construct a learning curve of customer service for an organization. Garvin (1998) gives an example of this. Many articles have been written on the use of learning curves that go beyond their use for measuring improvement due to increasing worker experience. Examples of such work can be found in Levy 1965, Nadler and Smith 1963, and Steedman 1970. It is because of this extension to other drivers for learning that we use the term *learning curve* rather than *experience curve* as a more general descriptor of these curves.

At this point, let us examine an important element of the way learning curves are set up. In figure 12.2, time was used as the x-axis. Analysis of what is actually going on in an improvement environment shows that, in general, time is not always the best variable to use for the x-axis. Looking back at figure 12.1, we can see that improvement can potentially occur every time the PDCA cycle is run. Each cycle represents an opportunity to improve. Then the more opportunities there are to improve, the more likely improvement might occur. Therefore, the x-axis in figure 12.2 should be the cumulative opportunities to improve that have been created as the PDCA cycle is run. In Wright's study (1936), this means that the x-axis should be the cumulative number of wings manufactured. If the system we are studying is the production of a product made by a batch manufacturing process, the x-axis is the cumulative number of batches made of a product. If the process is an order handling process, then the cumulative number of orders processed might be the best choice for the x-axis. If the system is a continuous process, then the best choice for the x-axis could be the total operational time for the process. Sometimes the opportunity to learn may depend on more than one variable. For example, consider safety in a manufacturing area. The measure of performance might be the accident rate. Suppose that the main people who have accidents are the production operators and the maintenance operators. For the production operators, the appropriate measure of learning opportunity might be production batches. For the maintenance operators, the appropriate measure of learning opportunity might be the number of maintenance jobs they perform. This may or may not be strongly correlated to number of production batches manufactured. Thus, in general, the x-axis of a learning curve should be the appropriate opportunity for improvement.

In his original work, Wright (1936) also noticed that the trajectory for worker productivity followed a definite pattern. We will study this pattern in more detail in the next section when we model an improvement process. This was the real breakthrough in the study of improvement. If there is no apparent pattern to the trajectory, then it is of limited use. However, if patterns do exist, these curves become useful. In particular, we can do the following:

- Characterize the rate of improvement by a limited number of parameters. Most learning curves characterize the learning rate by one single parameter. This characterization is useful in distinguishing between incremental and breakthrough improvements. Improvements that continue along the current learning curve can be considered as incremental improvement. We are operating a system that is fundamentally the same over the whole of the learning curve. Breakthrough improvement, by contrast, is characterized by a significant change in the learning rate.

> Learning Point: Incremental improvement can be considered as improvement along the existing learning curve. Breakthrough improvement, on the other hand, involves improvement that creates a significant change in the learning curve.

This is illustrated in figure 12.3.

- Compare a new data point to the existing learning curve to see if the new data point is consistent with the current rate of improvement or if a shift in the learning rate occurs. If this approach uses statistically based decision rules, this is equivalent to using the learning curve as a control chart with a mean that is constantly changing.

- Use the learning curve to decide if further investment in the improvement of a measure is value adding. Intuitively, we can accept that as the amount of improvement increases, the less improvement opportunity remains in a particular measure and the harder it will be to create this improvement. Continuing to invest in a product that is far along a learning curve may not be a good business decision.

- Use the learning curve as part of the target-setting process. In most organizations, management carries out a target-setting process on a regular basis, usually every year. It is not unusual to hear employees complain that the targets management adopts "appear to be pulled out of thin air." The targets may be so aggres-

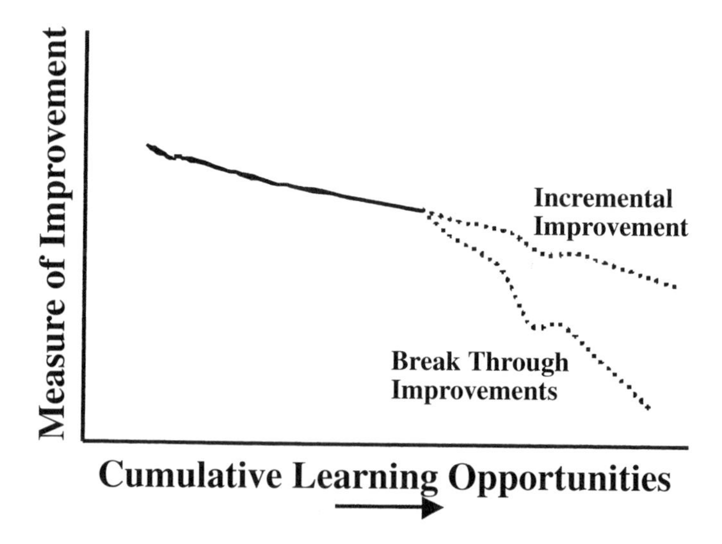

FIGURE 12.3. Incremental improvement versus breakthrough improvement

sive that there is no way that the organization can achieve them. They may be so easy that the organization has to do hardly anything to achieve them. However, if the organization has a learning curve, then this learning curve can be extrapolated into the future. Choosing a target that is on the extrapolated learning curve implies that it can be reached using the same incremental improvement techniques that the organization uses today. This is a nonaggressive improvement target. Choosing a target that is more aggressive than what is indicated by the extrapolated learning curve indicates that some breakthrough type improvement activities will be required to reach these targets. This is an aggressive improvement target. This approach to target setting is shown in figure 12.4.

• Compare the improvement rates in different systems. This comparison can help pinpoint opportunities for replication of improvement strategies from system to system, but such a comparison should be done with caution. Many times, an approach that works in one part of an organization will not work at all in another part of the same organization. This is because certain assumptions that allow the approach to work in one area are not valid in another area. Consequently, the approach fails when it is extended to another area. For example, introducing information technology to an area can have major benefits if the fundamental business processes are working well. In contrast, if information technology is introduced into an area that has poor business processes, the main effect may be to simply produce poor results faster.

The notion of a learning curve has implications for the way an organization carries out product development work. Figure 12.5 illustrates this idea.

Suppose that we have a new product and that we are ready to launch at a point where some attribute of the product, say unit cost, is at point A in figure 12.5. The cumulative number of units manufactured represents the opportunity for improve-

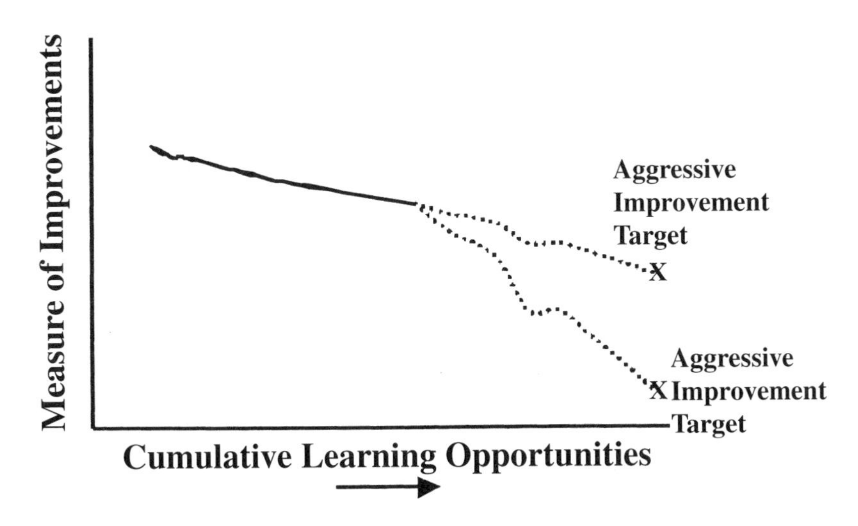

FIGURE 12.4. Using a learning curve as part of a target-setting process

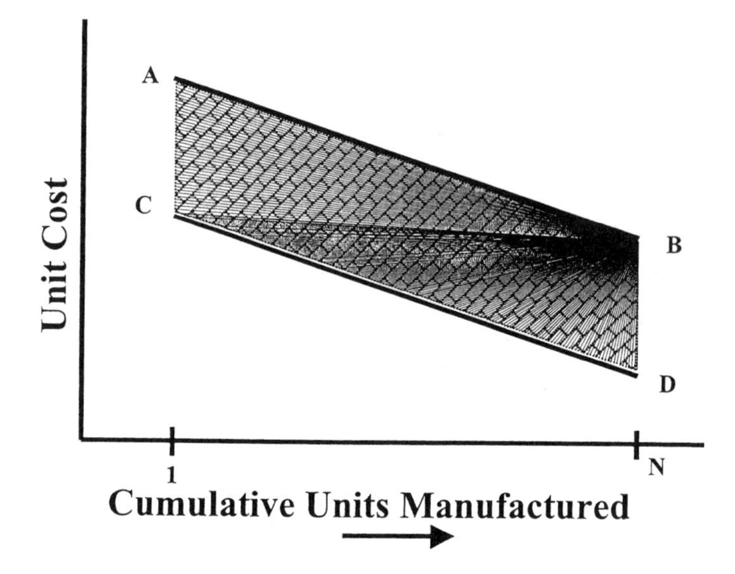

FIGURE 12.5. Impact of improvement on product development

ment. Over the life cycle of the product, we make N units of the product and we generate the learning curve AB in figure 12.5. The area underneath the curve AB, represented by ABN1, represents the total manufacturing cost for all units over the life cycle of the product. Now suppose that we put more upfront development work into the product. Then at the launch time, the product unit cost is now at point C. If we can create the same improvement rate as before, the new learning curve becomes CD. Now the total manufacturing cost for the product over its life cycle is given by CDN1. The impact of this strategy is the difference of these manufacturing costs and is given by the area ABCD. While this strategy might appear attractive, it has some unrealistic assumptions. Some of these are as follows:

- If more time is spent on upfront development work, then it might be impossible to maintain the same improvement rate. Many times the rate of improvement is dependent on the amount of development resources applied to a product. If new products are being developed, for example, there might be pressure to apply available resources to these new products. This leaves fewer resources for creating improvement in existing products. Hence, we get a lower learning rate.

- It cannot be assumed that just because we put more effort into upfront development work that the initial unit cost will be reduced sufficiently to make this development work beneficial. Most developmental work must be done at small scale. There is a great deal of learning that can occur only when you are actually manufacturing the product at production scale.

- The analysis assumes that the upfront development work will not impact the launch date. In many cases, particularly in highly competitive businesses, getting a new product into the marketplace is crucial. The benefits of being in the marketplace first may completely overshadow incremental unit cost reductions in the early phase of the product life cycle. Of course, as the demand for the product grows and the product moves into the mature phase of its life cycle, cost reduction becomes crucially important.

12.3 Modeling an improvement process

Based on the previous sections, we can see that in order to develop a model of an improvement process, we must follow four steps, which we record as a learning point.

> Learning Point: Modeling improvement in an organization involves a four-step process:
>
> Step 1: Identifying improvement measures that can be plotted in a learning curve format.
> Step 2: Identifying the opportunity for the improvement measure.
> Step 3: Identify the bottleneck in the PDCA cycle.
> Step 4: Creating a model of the PDCA cycle that is operating in the organization by linking the improvement measures in step 1 to the opportunity measure in step 2.

In this section, we will create a model for a productivity improvement process. We will consider a production organization in which a certain product is being produced. We assume that the cost of the first unit is known. As we begin to produce units, improvement takes place. Following the approach outlined above, we first identify the measure that we want to improve. This measure will be the unit cost. Next we ask what variable represents the opportunity for learning. In our example, this will be the cumulative number of units manufactured. Next we look at the PDCA process we have in place. We will consider the following process. At any given time, we have a list of projects that can be implemented to improve the process and reduce the unit cost. These projects can be sorted in terms of their impact on unit cost. Each project must be implemented one at a time so that the impact of the improvement can be judged independently from other improvements. This means that the bottleneck to the improvement process is the rate at which the projects can be implemented. It will take a certain amount of production after the project has been implemented before we can decide whether the project has been successful or not. Thus, the model will have three parts:

- A section modeling the production of units.
- A section modeling the unit cost.
- A section modeling the implementation of improvement projects.

The model is shown in figure 12.6.

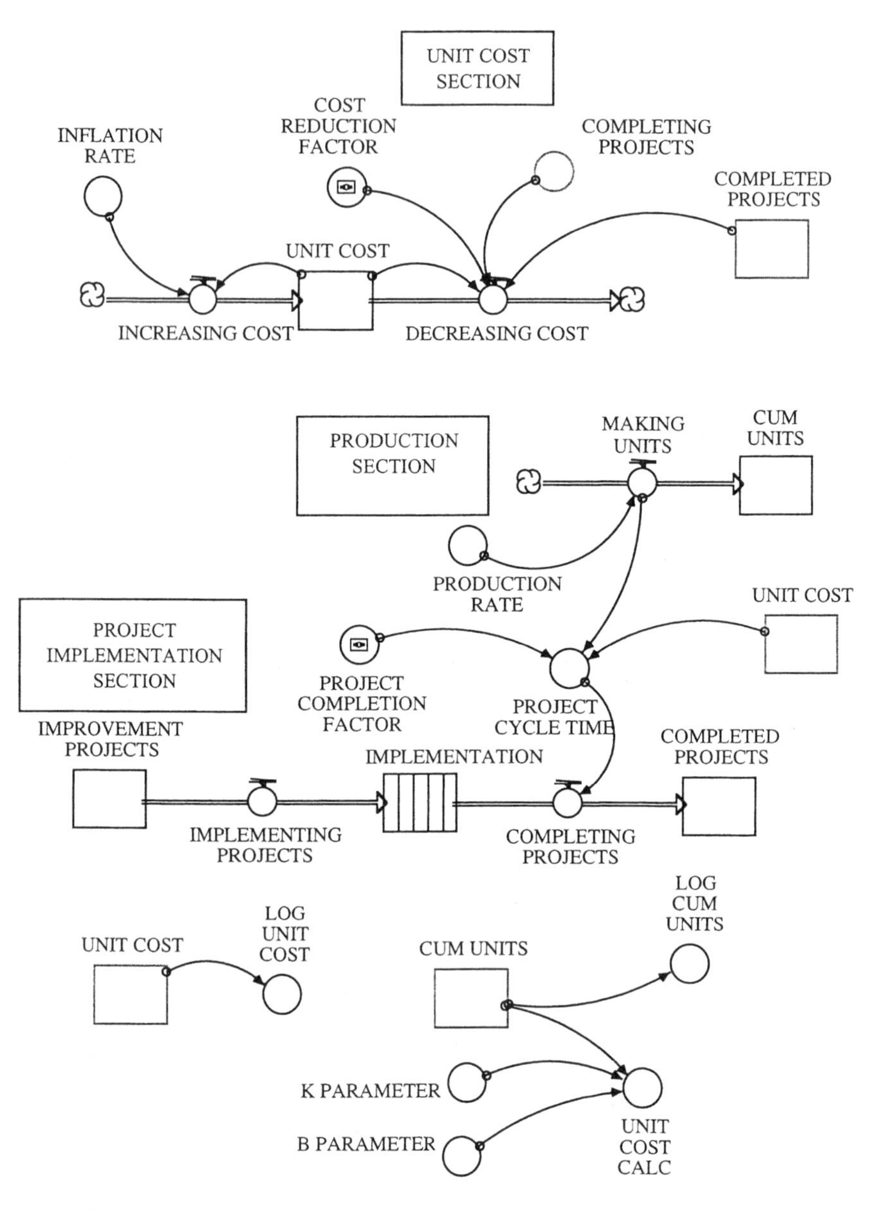

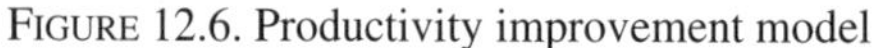

FIGURE 12.6. Productivity improvement model

The unit cost is modeled as a stock with two competing influences. There is a tendency for unit costs to increase because of inflation. In the model, this inflationary tendency is represented by the converter INFLATION RATE which causes the UNIT COST stock to increase via the inflow INCREASING COST. The equation for this inflow is this:

INFLATION_RATE*UNIT_COST

Now, while inflation is a real phenomenon, learning curve analysis factors out the impact of inflation when measuring the rate of improvement. The rational for this is that inflation is an external influence over which the organization has no control. Therefore, inflation should not be counted against the rate of improvement. So for all the results presented here, INFLATION RATE will be set to 0. It is included in the model for completeness, and readers can run the models themselves with inflation included to see what impact it has.

What about the tendency to decrease costs? As the unit cost decreases, one can expect that further cost reductions will become more difficult to produce. Also, it is reasonable to assume that the first projects that the organization recognizes and implements are the ones with the biggest impact on the unit cost. Therefore we can say that DECREASING COST is proportional to UNIT COST and inversely proportional to COMPLETED PROJECTS. Therefore, the equation used for the outflow DECREASING COSTS is this:

```
IF (COMPLETED_PROJECTS>0) AND (COMPLETING_PROJECTS > 0)
        THEN (COST_REDUCTION_FACTOR*UNIT_
        COST)/(DT*COMPLETED_PROJECTS) ELSE 0
```

The converter COST REDUCTION FACTOR is a constant of proportionality in the relationship between cost reduction rate and (UNIT COST, COMPLETED PROJECTS). The larger this number, the more impact each project has on the rate at which the unit costs are reduced. This obviously means that the rate of improvement can be increased by identifying and implementing improvement projects that have more impact on the unit cost. Note that the reason for the (COMPLETING_PROJECTS > 0) portion of the IF statement is to ensure that the cost reduction only occurs when the project is complete (that is, there is a flow in the outflow COMPLETING PROJECTS).

The production section is modeled simply by using an inflow that accumulates the production in a stock. The converter PRODUCTION RATE represents the production of units per unit time. We will assume that PRODUCTION RATE is constant even though in real situations the production rate tends to start off small and then (we hope!) ramp up as the market for the product grows. The equation for MAKING UNITS is this:

```
PRODUCTION_RATE/365
```

Of course, the factor 365 is because the production rate is an annual production rate and the simulation time unit is days.

The project implementation section is modeled by using a conveyor to represent the activity of implementing a project. This conveyor is fed from the stock IMPROVEMENT PROJECTS that represents the identified improvement projects. The initial value of this stock is set high so that there are always projects in this stock. Note that, in reality, it is unlikely that all future improvement projects will have been identified at time 0. Improvement projects are being identified all the time. However, the model assumes that projects are always available so that the availability of projects is not the bottleneck to improvement. It is the speed of implementation of each project that we are assuming is the bottleneck. If project availability were an issue, it would be easy to modify the model to show an inflow to the stock IMPROVEMENT PROJECTS. (We leave these modifications as an exercise for readers.) Once completed, projects are accumulated in the stock COMPLETED PROJECTS. So the only remaining parameter we need to consider is the cycle time for the conveyor IMPLEMENTATION.

In order to understand how the cycle time can be specified, consider a model of how projects might be implemented in an organization. A project is chosen for implementation and is then implemented. Before the next project is implemented, we must decide if the first project has been successful. To do this, we look at the behavior of the process after the improvement has been implemented and compare the behavior to what is was before the project was implemented. Once we can state with a high enough degree of confidence that the project has been successful (or not), we can move on to the next project. Of course, if our production process were such that no unit-to-unit variability occurred, it would be easy to see if the project had been successful. Simply compare the last unit prior to project implementation with the first unit produced after project implementation. Because variability is a fact in the real world, it will take more than one unit to discern if the project has been successful or not. Intuitively, it is not difficult to accept that the time to discern success depends on the production rate, P, and the amount of the change in the variable we are trying to discern, Δ. Therefore Δ represents the amount by which the unit cost is expected to change when the project is implemented. Denoting the time required to discern success by T, we can write the following equation:

$$T \propto \frac{1}{P\Delta} \tag{12.1}$$

As the production rate goes up, the time to discern success goes down. For large values of Δ, we expect the value of T to be small. It is easier to see big changes than small changes. Using the statistical theory of confidence intervals,[2] it is reasonable to write this equation:

$$\Delta \propto \frac{\sigma}{\sqrt{P}} \tag{12.2}$$

where σ is the standard deviation of the unit cost. Also, as improvement proceeds, it is reasonable to expect that the variation in the improvement measure (unit cost) decreases. Therefore σ can be modeled as proportional to 1/(Initial Unit Cost—Unit Cost). Combining all this leads to the following equation for T:

$$T \propto \frac{1}{P\Delta} = \frac{1}{P\left(\frac{\sigma}{\sqrt{P}}\right)} = \frac{1}{\sigma\sqrt{P}} = \frac{1}{\left(\frac{1}{(U_i - U)}\right)\sqrt{P}} = \frac{(U_i - U)}{\sqrt{P}} \tag{12.3}$$

2. See, for example, Wadsworth 1997.

where U represents the initial unit cost and U_i represents the initial unit cost. Equation 12.3 leads to the following:

$$T = K_P \frac{(U_i - U)}{\sqrt{P}} \tag{12.4}$$

K_P is a constant of proportionality that will be referred to as the project completion factor. This is the conveyor cycle time equation that will be implemented in the model. The equation for the cycle time, which is in the outflow COMPLETING PROJECTS, is thus:

PROJECT_COMPLETION_FACTOR*(INIT(UNIT_COST)-UNIT_COST)/(SQRT(MAKING_UNITS))

The model is now complete.

12.4 Model results

We will run the model using the parameter values shown in table 12.1. The results for these parameter values are shown in figure 12.7.

As we would expect, the decrease in unit cost is most pronounced in the early part of the curve and the rate of decrease slows down as production of units proceeds. Figure 12.8 shows a comparison of unit cost and project cycle time over the same simulation run. This figure clearly shows how the unit cost and the project cycle time are mirror images of each other, consistent with the idea that the bottleneck to improvement is the cycle time of the improvement projects.

Recall that what makes learning curves really useful is the ability to identify a pattern in the curves. What pattern we can see in figure 12.7? It turns out that if we plot the log of each variable (UNIT COST and CUM UNITS) we get the graph in figure 12.9. (Converters for LOG UNIT COST and LOG CUM UNITS were added to the model, and the log values were calculated using the Log10 built-in function in STELLA.)

TABLE 12.1. Base model parameters.

Time Unit	Days
DT	0.0625
Model time	1,825 days (5 years)
UNIT COST (initial)	1,000
CUM UNITS (initial)	0
IMPROVEMENT PROJECTS (initial)	1,000
IMPLEMENTATION (initial)	0
COMPLETED PROJECTS (initial)	0
INFLATION RATE	0
PRODUCTION RATE	1,000 units/year
COST REDUCTION FACTOR	0.3
PROJECT COMPLETION FACTOR	0.05

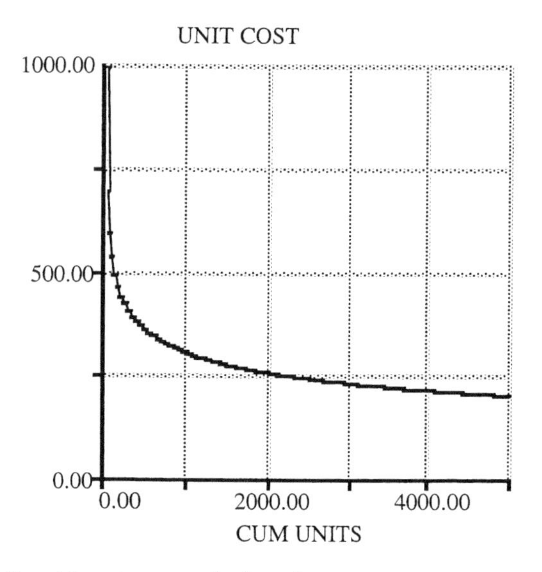

FIGURE 12.7. Unit cost learning curve for base data

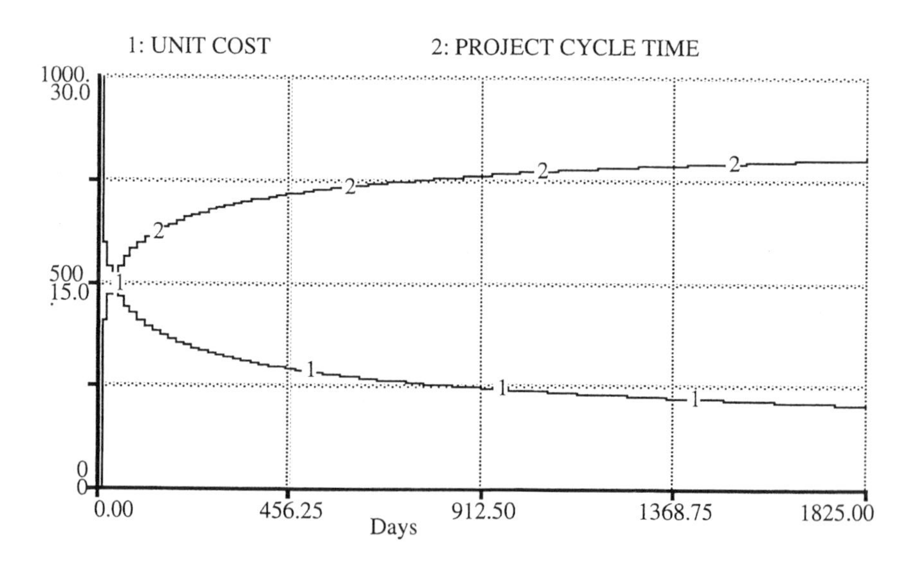

FIGURE 12.8. UNIT COST and PROJECT CYCLE TIME profiles

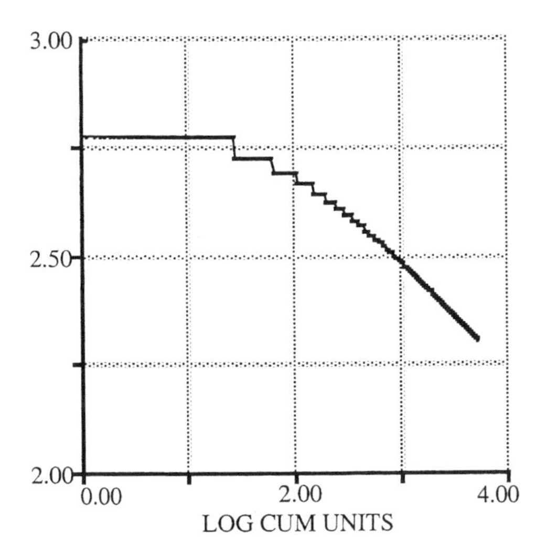

FIGURE 12.9. Learning curve on a log-log basis

For the majority of the learning curve, the variation of UNIT COST with CUM UNITS is linear on a log/log basis. This means that if we denote UNIT COST by U and CUM UNITS by C, the following equation holds for this linear portion.

$$\mathrm{Log}(U) = a + b\mathrm{Log}\,(C) \tag{12.5}$$

where $b < 0$ because U decreases as C increases. This can be rewritten as the following:

$$U = kC^{b} \qquad k = 10^{a} \tag{12.6}$$

Thus, U and C are related by a power law function. This is exactly what Wright (1936) found in his original analysis. Equation 12.6 has become the starting point for any learning curve analysis since then. We have been able to reproduce the behavior of learning curves by assuming that the bottleneck to improvement is because of the rate at which improvement projects can be implemented! (Of course, we cannot say that the relationship in equation 12.6 could not be produced by making other assumptions about the improvement bottleneck.)

> Learning Point: It is possible to model the basic form of the learning curve by assuming that the bottleneck to improvement is the rate at which improvement projects can be implemented.

In order to calculate the values of k and b for our base case model, the data were copied into Microsoft Excel and the data fit to a power law curve. The results obtained are k = 1530 and b = –0.236. A converter, UNIT COST CALC, was added to the model so that the results of the calculated relationship could be compared to the actual UNIT COST data calculated from the model. The results are shown in figure 12.10.

The comparison between the UNIT COST values from the model and those from the power law relationship are close, especially after the initial region where the rate of decrease is high. This confirms that our model is capable of reproducing the traditional form of the learning curve relationship, equation 12.6. We should note that in equation 12.6, the parameter k represents the unit cost at C = 1, that is, the unit cost of the first unit. So this parameter contains the information about initial conditions. The improvement rate information is therefore contained in the parameter b. To see what influences improvement rate, we should look at what influences the parameter b. For this reason, parameter b is referred to as the learning rate parameter.

By looking at the value of the parameter b, we can see what the rate of improvement is like. So if the measure is "smaller is better" (as it is for unit cost), b will be negative if improvement is taking place. The more negative the value of b, the higher the improvement rate. However, telling people that your learning rate is –0.25 does not convey much information because they will find it difficult to interpret what such a value of b implies. For example, they will know that a value of b = –0.25 is better than a value of b = –0.2, but they will be unable to infer how much better. This is because the impact of b on the unit cost is through the nonlinear log function. Most people cannot think in "logs." Because of this, it has become traditional to express the learning rate parameter in a different format. Sup-

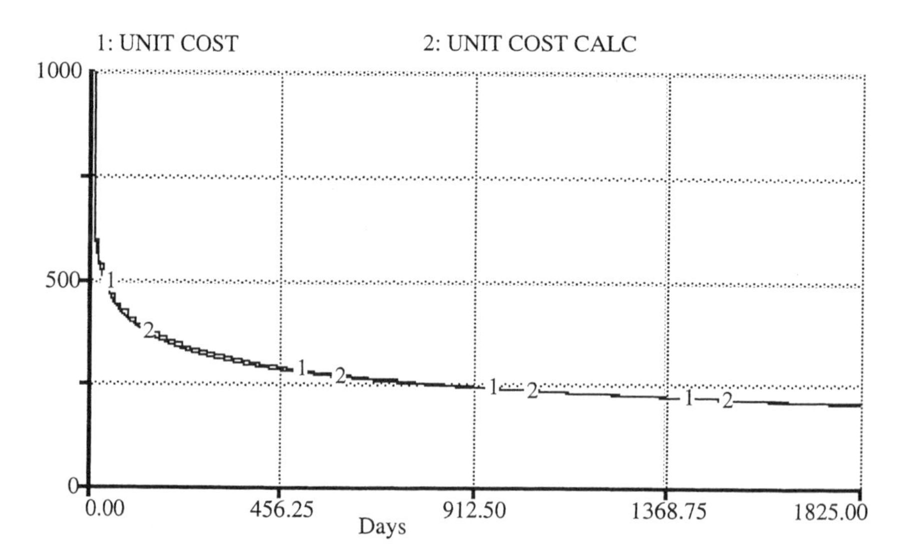

FIGURE 12.10. Comparison of UNIT COST versus UNIT COST CALC

pose the unit cost and cumulative units at some time t_1 are U_1 and CU_1 and at time t_2 are U_2 and CU_2 respectively. Then, we can say this:

$$U_1 = kCU_1^b \qquad U_2 = kCU_2^b \qquad \Rightarrow \qquad \frac{U_1}{U_2} = \frac{kCU_1^b}{kCU_2^b} = \left(\frac{CU_1}{CU_2}\right)^b \tag{12.7}$$

Now suppose that we pick the times such that CU_2 is twice CU_1. Then equation 12.7 says that the ratio of the unit cost values is 2^b. Therefore, we can characterize the learning rate by asking what change in unit cost occurs when the cumulative units double. This ratio depends only on the learning rate parameter b. Consequently, it completely characterizes the rate of improvement when equation 12.6 describes the learning curve. By convention, the ratio of the unit costs when the cumulative units double is referred to as the learning rate. It is defined as a ratio < 1 and is expressed as a percentage. So for a unit cost measure, where smaller is better, the ratio is defined as 2^b, because $b < 0$. If the measure is changed to "bigger is better" the learning rate is defined as 2^{-b}, because $b > 0$. So for the value –0.236 that we found for b, the learning rate is $2^{-0.236} = 0.85 = 85\%$. Therefore, we can say that every time the cumulative units double, the unit cost is reduced by 85 percent. This is a more intuitive way to characterize the improvement rate than saying that the learning rate parameter is –0.236.

So now we can ask this question: What impacts the learning rate? In our model, we had two parameters, COST REDUCTION FACTOR and PROJECT COMPLETION FACTOR. We can now vary these numbers and see what impact they have on the learning rate. Figure 12.11 shows the variation in learning rate with

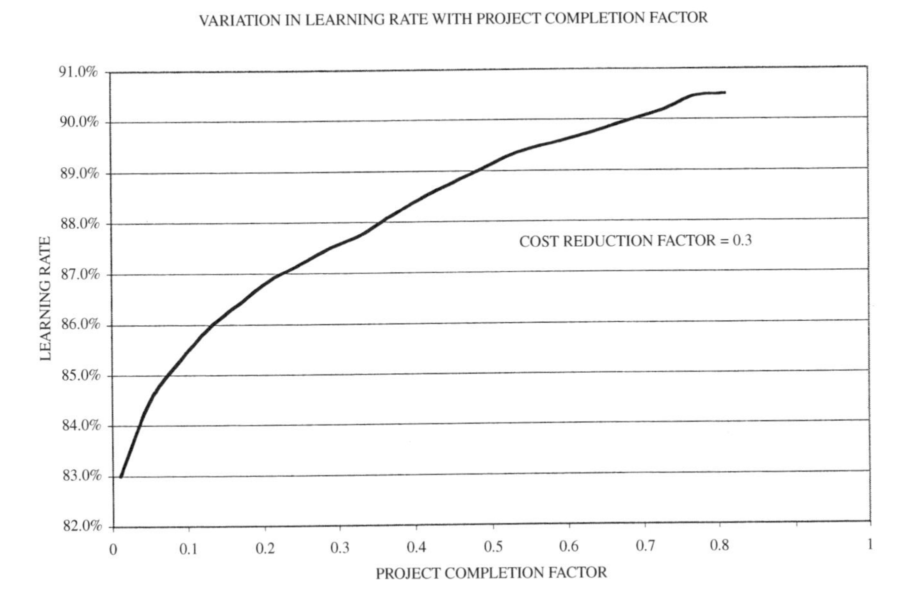

FIGURE 12.11. Variation in learning rate with PROJECT COMPLETION FACTOR

PROJECT COMPLETION FACTOR and figure 12.12 shows the variation with COST REDUCTION FACTOR.

Of course, a higher value of PROJECT COMPLETION FACTOR means that the project completion cycle times are longer, so the improvement rate is smaller, and the learning rate is larger. The opposite is true for COST REDUCTION FACTOR. Both graphs show that the variation is quite linear, which indicates that the nonlinearities in the learning curve behavior have been absorbed into the learning rate value. The graphs also show that the learning rate is more sensitive to COST REDUCTION FACTOR than to PROJECT COMPLETION FACTOR. Graphs such as figures 12.11 and 12.12 allow us to determine the impact that improving the improvement process can have on the improvement rate.

The base case results show that at the end of five years, the unit cost has been reduced from $1,000 to $202. Using this base case as a starting point, we leave it to readers to answer the following questions:

1. Marketing can launch the product in a new market when the unit cost drops below $400. When should they plan the launch?
2. If the COST REDUCTION FACTOR is increased to 0.4, what will be the impact on the launch date?
3. Suppose that the initial unit cost was to drop from $1,000 to $500 and the COST REDUCTION FACTOR was decreased from 0.3 to 0.2. Will this have a positive or negative impact on production costs over the five-year period?
4. At the end of year 5, management has set the following targets for the unit cost at the end of the next three years: $180, $165, and $155. Can these targets

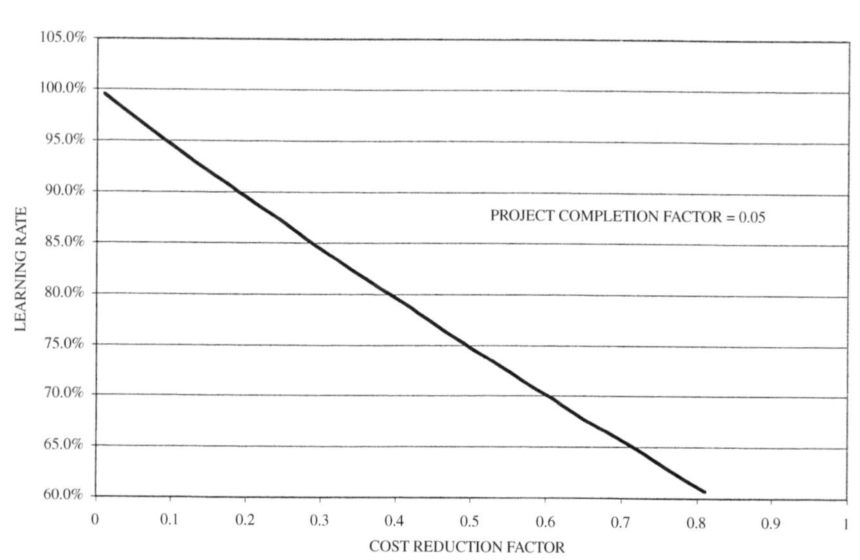

FIGURE 12.12. Variation in learning rate with COST REDUCTION FACTOR

be accomplished if we remain on the current learning curve (incremental improvement) or will we need to change the learning rate (breakthrough improvement)?

Finally, it is possible to improve the current model in a number of ways. Some ideas for improving the model follow:

1. Model the effect of production rates on the learning rate. Use intuition first and then use the model to confirm this intuition.
2. Introduce random variation into the model by using random functions to specify COST REDUCTION FACTOR, PROJECT IMPLEMENTATION FACTOR, and PRODUCTION RATE.
3. Add in the effect of the time value of money to the benefits section of the investment optimization model.
4. Add in the impact of inflation to unit cost.
5. Modify the model so that the organization can work on more than one improvement project simultaneously.
6. Add in more detail to the model to explain PROJECT COMPLETION FACTOR.
7. Add in more detail to the model to explain COST REDUCTION FACTOR.

12.5 Other types of learning curves

The productivity learning curve presented here represents only one possible type of learning curve. Although we do not explore them here, learning curve models have been developed for the following:

- Event type measures, such as accidents, in which the bottleneck to improvement is the ability to learn from each accident. When an accident occurs, we should perform a root-cause analysis to learn why the accident occurred. Then we use the new knowledge to prevent this and similar types of accidents from occurring again. Also, we can gather safety data not directly related to accidents and use this data to reduce the accident rate. An example of this is the data collected for behavior-based safety programs. The idea here is that unsafe behaviors have a strong causal relationship to the accident rate. If we can observe such unsafe behaviors and reduce their frequency, the accident rate will be reduced as well. The bottleneck to improvement is our ability to gather and analyze the data and create improvements from this analysis. Such models lead to the half-life learning curves discussed by Schneiderman (1986, 1988) and Stata (1989).
- Reliability improvement in product development. For example, suppose that we are a design engineering team, and our project is to design a new product for our organization. One of the key selling points for this product will be that it is far more reliable than competitors' products already on the market. Therefore, a goal has been set for the mean time between failures (MTBF) for this product. At the start of the project, we develop a prototype for the product.

Then the product is tested to see what the MTBF is. Typically, the MTBF will be much shorter than the design criterion. In order to improve the MTBF, we analyze the failure that occurred. Once the design flaw has been found, the product is modified and retested. Now the MTBF has increased, but it is still not at the design criterion. Further failure analysis is carried out. More design flaws are identified and further improvements are made to the MTBF. Our ability to improve the MTBF depends on our ability to expose the design flaws and fix them. Such models lead to the Duane reliability growth curves described by O'Connor (1991).

Appendix A

Modeling Random Variation in Business Systems

A.1 Introduction

In chapter 2, we said that the presence of uncertainty was one of the main reasons that modeling of business processes is so complex. In general, all of the variables in any process can vary randomly as the process runs or are unknown: these are random variables. Some of the key features of random variables and how they are modeled are described here. The reader can consult any of a number of texts on probability or statistics for details (Breyfogle 1999).

Random variables come in two basic types, discrete and continuous. Discrete random variables can take on a fixed set of values only. For example, in describing the number of telephone calls received by a customer service department over some time period, you can have only a discrete number of calls. You cannot have half a call. In contrast, continuous random variables can take all values. For example, the time it takes to complete a task can be, in theory, described to as many decimal places as possible.[1]

A random variable x is completely described by its probability distribution (density) function, P(x). For continuous random variables, this function is defined such that the probability of x being in the range x_1 to x_2 is given by the following:

$$\operatorname{Prob}\left(x_1 < x \leq x_2\right) = \int_{x_1}^{x_2} P(x)dx \tag{A.1}$$

For discrete random variables, P(x) is defined such that the probability of x being equal to x_1 is simply given by $P(x_1)$. We can also define the cumulative distribution function, C(x), which is the probability of the random variable being less than or equal to x. For a continuous random variable, we can thus say the following:

$$C(x_1) = \operatorname{Prob}\left(x \leq x_1\right) = \int_{-\infty}^{x_1} P(x)dx \tag{A.2}$$

For a discrete random variable, C(x), we can say the following:

1. Of course, in practice, the number of decimal places quoted will depend on the error associated with measuring the time.

$$C(x_1) = \Pr ob(x \le x_1) = \sum_{all\, x \le x_1} P(x) \tag{A.3}$$

Of course, $C(-\infty) = 0$ and $C(\infty) = 1$. Also, $P(-\infty) = 0$ and $P(\infty) = 0$. A typical graph of these two functions is shown in figure A.1.

In theory, we need the entire information in the probability density function to fully describe the behavior of x. In practice, however, we tend to describe a random variable in terms of two key summary variables: the mean (average), μ, and the standard deviation, σ.

The mean is calculated from the following formula:

$$\text{C} \quad \mu = \int_{-\infty}^{+\infty} xP(x)dx \qquad \text{D} \quad \mu = \sum_{all\, x} xP(x) \tag{A.4}$$

where C denotes the formula for a continuous random variable and D denotes the formula for a discrete variable.

The mean is measure of central tendency. That is, it measures where the values of x tend to cluster.

The standard deviation is calculated from the following formula:

$$\text{C} \quad \sigma^2 = \int_{-\infty}^{+\infty} (x-\mu)^2 P(x)dx \qquad \text{D} \quad \sigma^2 = \sum_{all\, x} (x-\mu)^2 P(x) \tag{A.5}$$

The standard deviation measures how the data are spread out from the mean to form the measure of dispersion. Notice from equation A.5 that the calculation of σ depends on the value of μ. This presents a difficulty if we want to calculate σ at every time step in a dynamic model. Suppose that we have run the model to time t and we know μ at time t. Now we calculate the model at time (t+dt). In general, μ will be different at time (t+dt). This means that all the calculations for σ at time (t+dt) are based on the value of μ at time t and not time (t+dt), as it should be. To get around this issue, a different but equivalent form of equation A.5 is used. Equation A.5 can be written in the following form (Harris et al. 1998):

$$\text{C} \quad \sigma^2 = \left[\int_{-\infty}^{+\infty} x^2 P(x)dx\right] - \mu^2 \qquad \text{D} \quad \sigma^2 = \left[\sum_{all\, x} x^2 P(x)\right] - \mu^2 \tag{A.6}$$

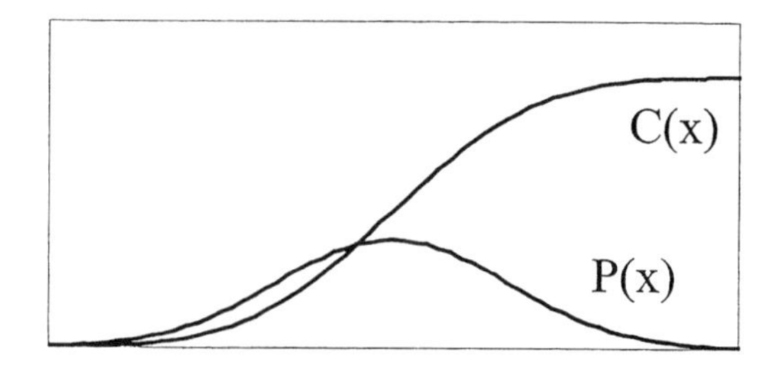

FIGURE A.1. The functions P(x) and C(x)

Figure A.2 shows a model fragment that implements these formulas to calculate the mean and standard deviation of a variable.[2]

VARIABLE represents any general variable in a model. ADDING V is a biflow that accumulates the value of the variable into the stock SUM V. The equation of this biflow is simply VARIABLE/DT. The DT is required so that when the flow is integrated, it is multiplied by DT, and the DTs cancel each other out. This ensures that the value of VARIABLE is added each time step irrespective of the value of DT. ADDING V2 is an inflow that accumulates the square of VARIABLE. Its equation is (VARIABLE*VARIABLE)/DT. Note that because this is a square function, a biflow is unnecessary here. The mean is given in the converter MEAN V, which has the following equation:

IF TIME > 0 THEN SUM_V/(TIME/DT) ELSE 0

The IF TIME > 0 is required to avoid dividing by 0 errors when the model initializes. The standard deviation is given in the converter STD DEV V, which has the following equation:

IF TIME > 0 THEN SQRT((SUM_V2/(TIME/DT))—MEAN_V*MEAN_V)
ELSE 0

Readers can try various distributions in this model to check that the correct mean and standard deviations are calculated.

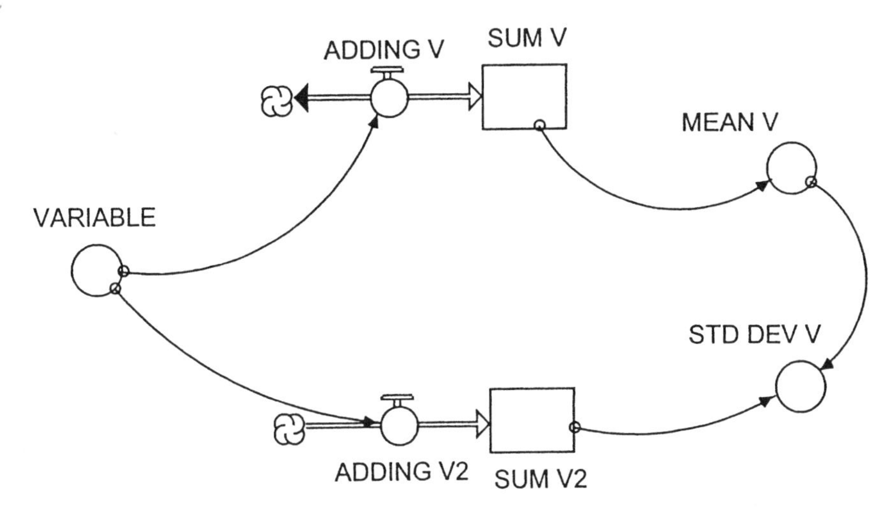

FIGURE A.2. Calculating the mean and standard deviation in a model

2. Readers can run the model (MODEL A.2.ITM) specifying different random functions for VARIABLE and verify the calculations.

We now look at the most common distributions that occur in dynamic modeling of business processes as well as the formulas for their mean and standard deviation. The equations for the distributions have been included even though the reader does not need to implement them in a modeling package such as ithink. The reason for this is to encourage the reader to become familiar with the details of the distributions. Otherwise, there is a real danger of misunderstanding their usage and misapplying them.

A.2 The uniform distribution

The uniform distribution[3] is typically used to model a situation where the only known information about a random variable is that it is in some range with minimum value a and maximum value b. (Without any other information, we generally assume that any value between a and b is equally likely.) Typically, the parameter can take on all values between a and b, so the distribution is continuous. The probability distribution is constant within the range a to b and 0 elsewhere. A graph of the uniform probability distribution is shown in figure A.3.

The equation for the distribution is this:

$$P(x) = \frac{1}{(b-a)} \qquad \text{if } a \le x \le b \qquad P(x) = 0, \text{ otherwise} \qquad \text{(A.7)}$$

The mean and standard deviation of this distribution are given by this:

$$\mu = \frac{(a+b)}{2}, \qquad \sigma^2 = \frac{(b-a)^2}{12} \qquad \text{(A.8)}$$

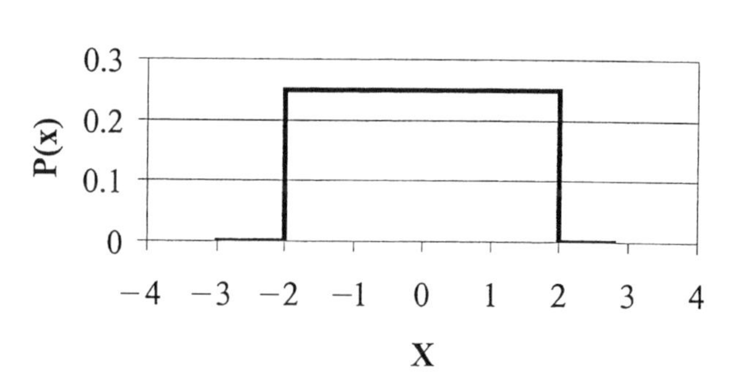

FIGURE A.3. The uniform probability distribution: For this graph, b = 2, and a = −2

3. Also referred to as the *rectangular distribution.*

In ithink, the uniform distribution is implemented using the function Random(a,b).

A.3 The triangular distribution

The triangular distribution is like the uniform distribution except that we have another piece of information. As well as knowing the range, we also know the most likely value for the random variable. The triangular distribution thus has three parameters. The triangular distribution is also typically a continuous distribution. A graph of this distribution is shown in figure A.4. The most likely value is m, the lower bound is a, and the upper bound is b. Note that if $m = (a + b)/2$ then the distribution is symmetrical. Otherwise it is nonsymmetrical.

The equation for the distribution is

$$P(x) = \frac{2(x-a)}{(m-a)(b-a)} \quad \text{if } a \le x \le m \quad P(x) = \frac{2(b-x)}{(b-m)(b-a)} \quad \text{if } m < x \le c$$

$$P(x) = 0, \text{otherwise} \tag{A.9}$$

The mean and standard deviation of this distribution are given by

$$\mu = \frac{(a+m+b)}{3}, \quad \sigma^2 = \frac{\left[a^2 + m^2 + b^2 - mb - a(m+b)\right]}{18} \tag{A.10}$$

One advantage of using the triangular distribution is that it is easy to get people to give you the parameters. Just ask for the minimum, maximum, and most likely value for the random variable. Care should be taken, however, that people do not

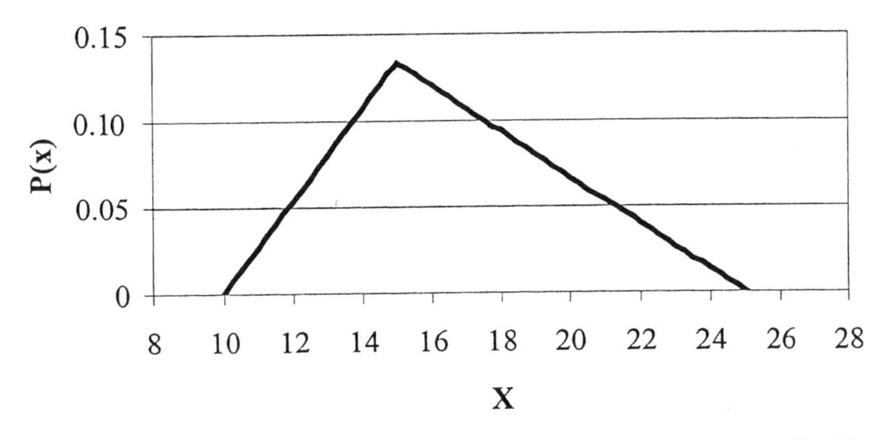

FIGURE A.4. The triangular probability distribution: For this graph, a = 10, m = 15, and b = 25

confuse average with most likely value and give the average when you ask for the most likely. The two are only the same when the distribution is symmetrical.

ithink does not have a built-in function for a triangular distribution, so we must create it ourselves. For many standard distributions, it is possible to generate the distribution if we can generate uniform random numbers in the range (0,1). Suppose r denotes such a uniformly distributed random variable in the range (0,1). Then it can be shown that the random variable t, given by the following equations, has a triangular distribution with parameters (a,m,b) (Devroye 1986).

$$\textit{If } r \le \left(\frac{m-a}{b-a}\right), \quad \text{then} \quad t = a + \sqrt{(b-a)(m-a)r},$$

$$\text{else} \quad t = b - \sqrt{(b-a)(b-m)(1-r)} \tag{A.11}$$

The following model illustrates the use of these equations in a simple process where the processing time varies with a triangular distribution with parameters a = 10, m = 15, and b = 25. This is the same data as used in figure A.4.

R has the equation RANDOM(0,1) and the equation for PROCESS TIME is this:

```
IF R<= (M-A)/(B-A) THEN A+SQRT((B-A)*(M-A)*(R))
        ELSE B-SQRT((B-A)*(B-M)*(1-R))
```

This model was run with a DT = 1 and a total run time of 10,000 time steps. The values for PROCESS TIME were written to a table and copied to a spreadsheet. This gives the histogram shown in figure A.6.

Comparing figure A.6 with A.4 shows that the model does, indeed, generate the expected triangular distribution.

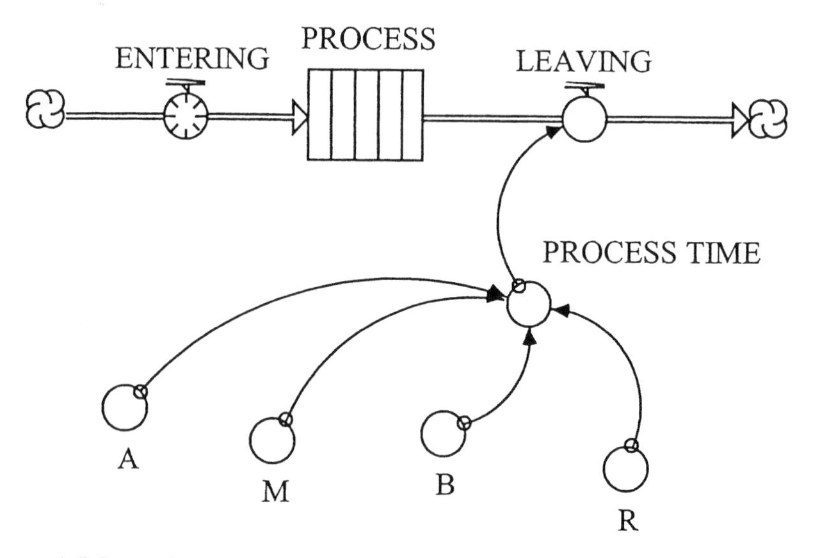

FIGURE A.5. Processing time with a triangular distribution

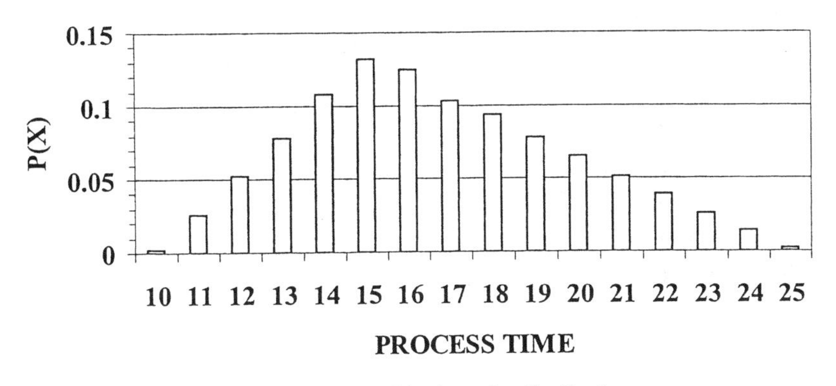

FIGURE A.6. Histogram of process time with triangular distribution

A.4 The normal distribution: Common cause process variation

In his original work on quality control, Shewart[4] distinguished between two different types of variation in a process. If the performance of a process is stable and predictable over time, then the only sources of variation affecting the process are called common cause sources of variation. These sources of variation are constant over time, leading to a constant level of performance in the process. This means that when only common cause sources of variation are present, the probability distribution of a process performance measure is constant over time. This type of variation is present in every process. Sometimes, as well as these common cause sources of variation, we have variation that is unpredictable. The probability distribution of the process performance measure is no longer constant over time.

It turns out that the normal probability distribution is used frequently to describe common-cause variation. One reason for this is that common-cause variation results from the combined effects of a large number of different sources of variation. Typically, each of these sources individually does not contribute significantly to the overall variation. In such cases, it can be shown that the resultant distribution will tend to follow a normal probability distribution[5] even if the individual sources of variation do not. This distribution has the well-known bell shape and is shown

4. See Shewart 1980, 1986; Deming 1980; or Wheeler 1992 for a complete discussion of the Shewart model of variation.
5. For those readers with an interest in statistics, this result comes from the Lyapunov limit theorem. See Rozanov 1977 or Harris 1998.

in figure A.7.[6] Because it arises naturally as a limiting case, it is considered the basic distribution for modeling many situations where a random variable is continuous and tends to cluster around a mean value. For example, measurement errors are frequently represented by the normal distribution, consistent with the fact that in many situations, there are many individual sources of error contributing to the overall measurement error. However, the reader should not conclude that the name "normal" indicates that all other distributions are abnormal, and that describing a random variable by a nonnormal distribution is unusual. The normal distribution is a limiting distribution as discussed earlier, and its utility is as a useful approximation to continuous distributions that occur in practice.

The equation for this distribution is given by the following:

$$P(x) = \frac{1}{\sqrt{2\pi}\sigma} e^{-(x-\mu)^2/2\sigma^2} \tag{A.12}$$

The mean, μ, and standard deviation, σ, appear explicitly in the equation for the distribution.

The normal distribution also arises quite naturally in processes where the process is monitored by picking a random sample from a stream of product. For example, suppose we are making packets of breakfast cereal, and we want to monitor the weight of the packets to ensure that the weight is consistent over time. Rather than weigh each box, we can take a random sample of a number of boxes and just weigh the boxes in the sample. The average of the box weights in the sample is calculated and monitored, typically using a control chart. It can be shown that if we can assume the individual box weights in the sample all have the same distribution, then the distribution of the average is normal.[7] This holds irrespective of the distribution of the individual box weights. The sample size does

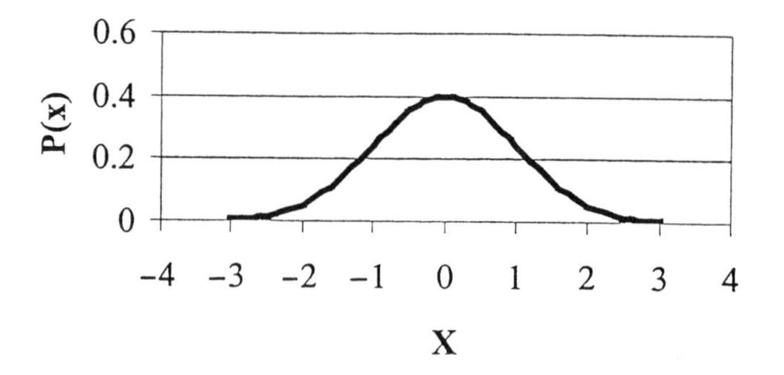

FIGURE A.7. The normal distribution: For this graph, the mean is 0, and the standard deviation is 1

6. It is also referred to as the Gaussian distribution, after the eminent mathematician Karl Friedrich Gauss.

7. This is called the Central Limit Theorem. See Harris 1998.

not have to be large for approximation of a normal distribution to be good. In fact, sample sizes of four or greater are usually considered adequate.

The normal distribution is symmetrical about the mean. A related distribution that can be used for skewed distribution is the lognormal distribution. Like the triangular distribution, the lognormal distribution is not a standard distribution in ithink. A random variable with a lognormal distribution can be derived from a random variable with a normal distribution. A model for the lognormal distribution is provided (Model A.7). Readers are invited to run this model to familiarize themselves with this distribution.

A.5 The exponential distribution: Equipment failure times

Suppose you have a piece of equipment that can fail in a random fashion. The time it takes for the equipment to fail can be modeled using the exponential distribution.[8] There are two ways that the exponential distribution arises when modeling the distribution of such random failure times. First, suppose that the failure can occur when the stress placed on the equipment is greater than some given value, usually called the strength of the equipment. As the equipment is used, the stress will vary. Now suppose that the stress varies randomly with a normal distribution. Then it can be shown that the equipment failure time is exponential. Second, suppose that there are a large number of ways (modes) that the equipment can fail. In addition, the failure of anyone of these individual modes means a failure of the overall equipment.[9] Then, irrespective of the distribution of the failure time for the individual modes, the overall equipment failure time has an exponential distribution. In this sense, the exponential distribution plays the same role for failure times as the normal distribution does for common cause variation. The exponential distribution is shown in figure A.8.

The equation for the exponential distribution is this:

$$P(x) = \frac{1}{\lambda} e^{-\frac{(x-a)}{\lambda}}, \qquad x \geq a \tag{A.13}$$

The distribution is only defined for $x \geq a$, as one would expect if it is used to model failure times. Parameter a is called the location parameter. It impacts the mean but not the standard deviation of the distribution. In many applications, $a = 0$, so that the distribution has only a single parameter. The mean and standard deviation are given by the following:

$$\mu = a + \lambda, \qquad \sigma^2 = \lambda^2 \tag{A.14}$$

so that the mean and standard deviation when $a = 0$ are the same. Thus, λ represents the mean failure time for the equipment.

8. See Dodson 1995 for more details on the exponential distribution.
9. We say that this is a serial failure system.

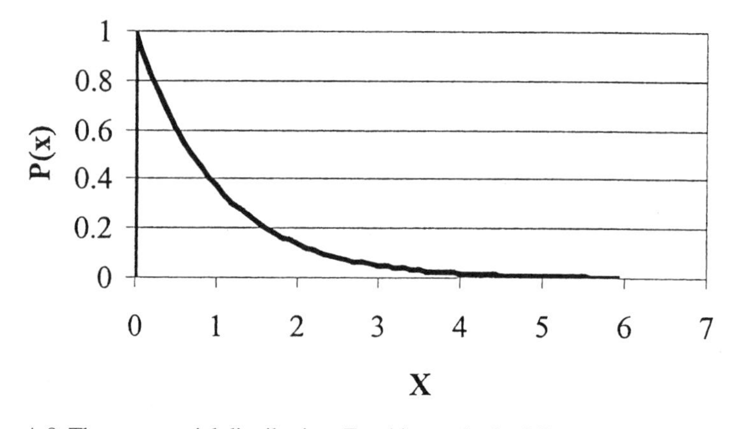

FIGURE A.8. The exponential distribution: For this graph, the failure rate is 1, and the location parameter is 0

It should be noted that sometimes equipment failure is not a function of time but instead depends on cycles. For example, consider an airplane and its potential failure modes. Some of these failures depend not so much on time but on the number of take off and landing cycles. Consequently, we may be more interested in the number of cycles to failure than the time to failure. Another example is a photocopier. Here, the number of copies made could represent cycles for the photocopier. Maintenance programs must take these cycles, as well as clock time, into account.

The exponential distribution has a unique property called the "memoryless" property. Suppose that you are responsible for the reliability of a piece of equipment. The equipment has been running for a time t without failure. Someone comes to you and tells you that they need to know what the probability of failure is for the equipment in the next time period. The question is this: How does the probability of failure in the next time period depend on the fact that the equipment has not failed up to the current time?[10] There are three possibilities:

1. The fact that it has already run for a time t indicates that it is more likely to fail in the future. We are getting closer and closer to the ultimate failure.
2. The fact that it has already run for a time t indicates that it is less likely to fail in the future. The lack of a failure up to time t gives us confidence about the performance of the equipment in the future.
3. The fact that it has already run for a time t has no bearing on its probability of failure in the next time period.

Possibility (c) looks at first like a strange situation—the probability of failing in time (t+dt) does not depend on t. It is as if the equipment has no "memory" of the fact that it has run for a time t without failure: hence, the idea of describing this as

10. For readers who know statistics, this is a problem in conditional probability. We want the probability of one event given that another event has (or has not) occurred.

a "memoryless" property. It turns out that the exponential distribution is the only probability distribution with this property.[11]

This property is at the heart of the famous paradox. Suppose you travel to work every day using a bus. You get the bus at a nearby stop. You know that on average, a bus arrives every 30 minutes. The buses arrive randomly, so you reason that on average you will wait 15 minutes for a bus. However, if the bus arrival times have an exponential distribution, then on average you must wait 30 minutes. When you arrive at the stop, the length of time that has elapsed since the arrival of the last bus has no impact on the length of time you must wait for the next bus. This appears to defy logic, and it has been called the "Waiting Time Paradox."

The memoryless property has implications for our ability to obtain analytical[12] solutions to mathematical models of processes. If we assume that the time between events is exponential, the equation for the number of events in a time interval t to (t+dt) is independent of t. Essentially, we do not have to keep track of previous event times. This greatly simplifies the mathematics. An example of this type of analysis would be the arrival of service calls at a service center. If the time between calls is an exponential random variable, then modeling the system (to calculate such quantities as average time the service center is busy) is greatly simplified. That is why much of the introductory literature on modeling of stochastic processes assumes that random variables have the exponential distribution.

A.6 The Poisson distribution: Modeling defects in products

Many times the quality of a product is measured by counting the number of defects in the product. The criteria for determining a defect are defined using the customer quality requirements. For example, suppose that we were producing wallpaper. We could define a defect as a blemish in the finish of the design on the wallpaper. Our definition of a defect would include not only the types of blemishes but also the magnitude that they would need to be before we would consider it a defect. For example, a blemish might have to cover an area greater than 1 mm^2 before it is considered a defect. As well as the definition of a defect, we also must specify the area of opportunity for such defects. In the case of the wallpaper, the area of the wallpaper that we examine when counting the defects would be a valid area of opportunity. The area of opportunity is important because we would expect that the number of defects would increase with the size of the area of opportunity. Therefore, the average number of defects per unit area of opportunity is a fundamental parameter describing the occurrence of such defects. The area of opportunity does not have to be an actual area. (For example, if we were counting the number of accidents in a manufacturing organization, a typical area of oppor-

11. It is possible to generalize the exponential distribution to include cases (a) and (b). This distribution is called the Weibull distribution. See Dodson 1995.

12. By "analytical solutions," we mean solutions that can be obtained in equation form without the need for dynamic simulations.

tunity is the number of years worked. The number of accidents is expected to increase as the number of years an individual has worked increases, all else being equal.) Suppose the number of defects is random and satisfies the following assumptions:

- The number of defects depends only on the size of the area of opportunity.
- The number of defects in one area of opportunity is independent of the number of defects in any other area of opportunity.

In this situation, the number of defects is described by the Poisson distribution. A graph of a typical Poisson distribution is shown in figure A.9.

Note that the earlier assumptions mean that the probability of a defect is constant across any given area of opportunity. This is important if we want to model the occurrence of defects in a manufacturing area where multiple processes are being run. If the processes are batch processes, then the number of batches produced in a given time period will represent the area of opportunity. If the probability of success is different for each process, then the overall defect rate will not satisfy the first assumption because the overall probability of failure will depend on the mix of processes in the time period as well as the total number of batches processed in the time period. In this case, it will be important to model each process separately and then add up the defects from each process to get the total number of defects.

The equation for this distribution is the following:

$$P(x) = \frac{a^x}{x!} e^{-a}, \qquad a > 0, \quad x = 0, 1, 2, 3, 4, 5, \ldots \qquad \text{(A.15)}$$

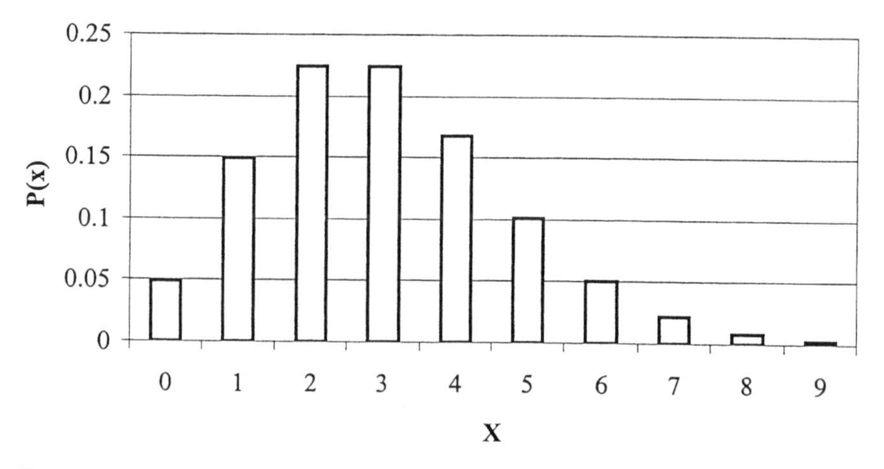

FIGURE A.9. The Poisson distribution: For this graph, the average number of defects per unit area of opportunity is 3

Here, $P(x)$ is the probability that the number of defects in a given area of opportunity will be x. The Poisson distribution has a single parameter, a, where a is the average number of defects for a given area of opportunity. This parameter is thus a dimensionless number. If d represents the average number of defects per unit area of opportunity, then for an area of opportunity A, we can say that $a = d^*A$. The mean and standard deviation of this distribution are given by the following:

$$\mu = a \qquad \sigma^2 = a \tag{A.16}$$

The Poisson distribution is related to the exponential distribution in this way: if the number of defects within an area of opportunity has a Poisson distribution, then the area of opportunity between defects has an exponential distribution. For example, suppose that the number of calls arriving at a customer service center per hour has a Poisson distribution with an average of six calls per hour. Then the time between calls has an exponential distribution with a mean interarrival time of 60/6 = 10 minutes.

A.7 The pass/fail and binomial distribution: Product failures

After a process is completed, a quality check is often performed to test whether or not the quality of the product is acceptable. If the product quality is acceptable, then this product is made available for further processing or for shipment to a customer. If the product fails, then it must be further processed or reworked before it can be shipped to the customer. This is shown schematically in figure A.10.

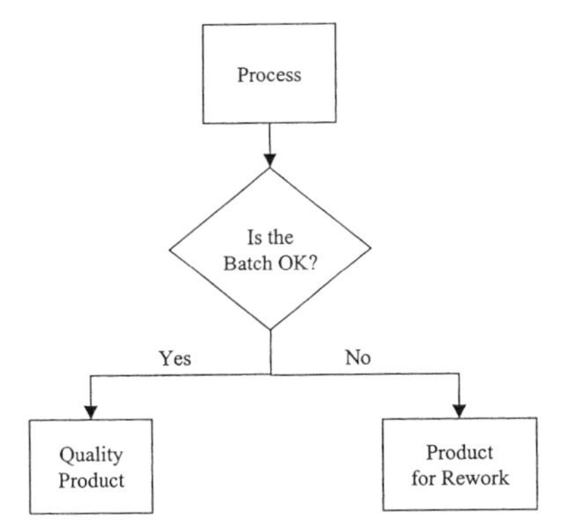

FIGURE A.10. Process with quality testing

In general, we will not know in advance whether an individual batch will pass or fail. The best we can say is that each batch has a probability, say p, of passing and hence a probability $(1 - p)$ of failing. Thus, the passing or failing of the batch can be represented by a random variable that takes a value 1 with probability p to represent passing and a value 0 with probability $(1 - p)$ to represent failing. This is a simple pass/fail probability distribution. It is a discrete distribution with a single parameter, p. The mean and standard deviation for this distribution are given by the following:

$$\mu = p, \qquad \sigma^2 = [p(1 - p)] \tag{A.17}$$

Now suppose we carry out a campaign in which we send N items through the process and each item has a probability p that it will pass the quality test. At the end of the campaign, the number of successful items in the "Quality Product" box will be a random variable with a binomial distribution. By definition, this is a discrete random variable. A typical binomial distribution is shown in figure A.11.

The equation for the binomial distribution is this:

$$P(x) = \frac{N!}{x!(N-x)!} p^x (1-p)^{N-x} \qquad \text{for } 0 \le x \le N \tag{A.18}$$

The mean and standard deviation for the binomial distribution are given by the following:

$$\mu = Np, \qquad \sigma^2 = [Np(1 - p)] \tag{A.19}$$

Thus, the pass/fail distribution is included explicitly in models, and the binomial distribution shows up as a consequence of the pass/fail distribution.

In many practical applications, the value of N is large. It can be shown that the binomial distribution is well approximated by the normal distribution. In fact, as

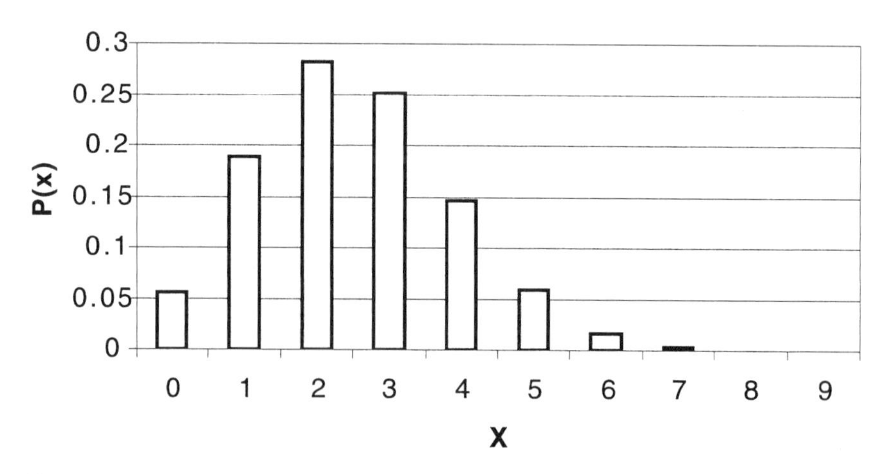

FIGURE A.11. The binomial distribution: For this graph, N = 10, and p = 0.25

$N \to \infty$ then the binomial distribution converges to a normal distribution with mean and standard deviation given by equation A.19 (Harris 1998). It might seem odd at first that a discrete distribution can converge to a continuous distribution. As $N \to \infty$, however, the binomial distribution effectively becomes continuous. A general rule of thumb is that the normal approximation is applicable when $Np > 4$ and $N(1 - p) > 4$. In many cases, p is close to 1 and $1 - p$ is small, which means that we must have a large number of items before we can use the approximation.

A.8 Bimodal distributions: Parallel process flows

In some cases, the probability distribution of a random variable is actually the sum of more than one fundamental probability distribution. This situation occurs when items follow different parallel paths through a process depending on some decisions made in the process, as was the case with the process in figure A.10. Suppose that we are measuring some aspect of the process, such as cycle time. The items that pass the quality check will have one probability distribution, and the items that fail the quality will have another probability distribution. The overall cycle time random variable, CT, can then be written as

$$CT = P * CT_{\text{Pass}} + (1 - P) * CT_{\text{Fail}} \tag{A.20}$$

where CT_{Pass} is the random variable for cycle times of items that pass the quality check and CT_{Fail} is for the items that fail the quality check. P is a random variable that represents whether the item fails of not. If p denotes the probability of an item passing the quality check, then P has a value 1 with probability p and a value 0 with probability $(1 - p)$. An example of a bimodal distribution is shown in figure A.12.

The two different distributions present in the overall distribution are evident from the two peaks in the distribution. If we find such a distribution, we do not try

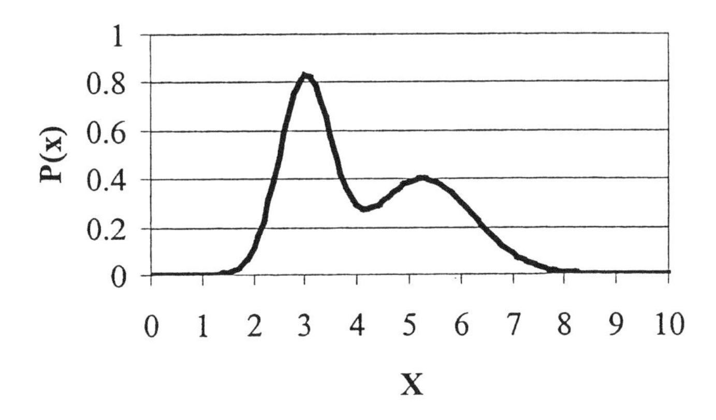

FIGURE A.12. A bimodal distribution: For this graph, the distribution is the sum of two normal distributions, one with mean 3.0 and standard deviation 0.5 and the other with mean 5.25 and standard deviation 1.0

to model it as a single distribution. Rather, we need to find what attribute of the process is leading to the two distributions; model each distribution separately; and finally, combine them to give the overall distribution. For example, in the case of cycle times, we would first identify the fact that the quality check was leading to two cycle time distributions. Then we would identify the cycle time distribution for items that pass the quality check, the distribution for the cycle times that fail the quality check, and implement them separately in the model. The decision as to which item passes or fails will be determined by a random variable with a pass/fail probability distribution. If the probability of an item passing the quality check is p, then the random variable can be modeled in ithink with a converter that uses the following equation:

$$\text{IF RANDOM}(0,1) <= p \text{ THEN 1 ELSE 0} \quad (A.21)$$

A.9 Custom distributions

Sometimes, we must model random variables where we do not have any information that can be used to identify a particular model distribution. However, we have a data set from the actual process. We need a way to fit the data to a model distribution and pick the distribution that seems to best fit the data. This gives us a best-fit theoretical distribution, but it also may lead to a choice that does not appear to make any physical sense. Consequently, we might be uncomfortable using a theoretical distribution. In such a situation, we may prefer to create a custom distribution directly from the data.

The theory behind the method is as follows (Devroye 1986). Suppose we know the Cumulative Distribution Function (CDF) for some random variable, x. Because the CDF takes on values only between 0 and 1, the CDF is like a process that maps the values of x onto the range [0,1]. This is shown in the first chart in figure A.13.

To create the random variable, we reverse this process as shown in the second chart in figure A.13. A random number is generated from a uniform distribution in the range [0,1]. Such a variable is generally referred to in the literature as U. We

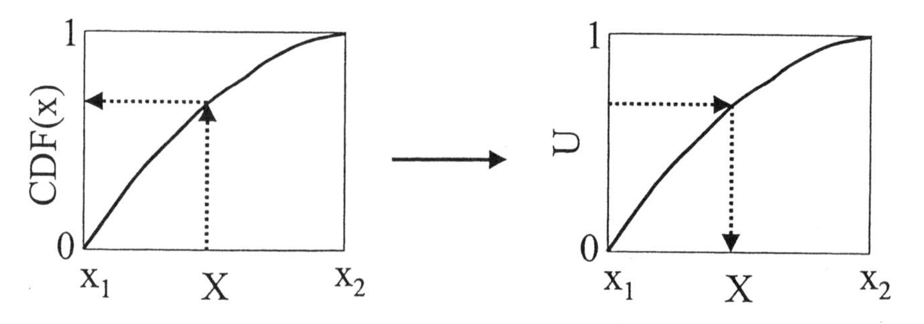

FIGURE A.13. Generating a random variable with a custom distribution

then find the value of x corresponding to a CDF value equal to the value of U generated. Now we keep generating values of U and repeating the process to generate a series of values for x. Statistical theory states that the resultant series of x values has the distribution function of the original CDF.

In a real situation, we will have a data set in the form of a series of values rather than the actual CDF. So the first thing we must do is calculate a CDF from the data we are given. To calculate the CDF, we take the data set and sort it in ascending order. Denote the random variable by v, each value in the sorted list by v_i, and M the total number of data points in the list, where $1 \leq i \leq M$. Then for each data point in the sorted list, calculate the CDF of the custom distribution at each point as (i/M). We now have a list of values (v_i, i/M). By definition, the CDF values range from 0 to 1. Now if we generate a random variable in the range (0,1), interpolate the list of CDF values to find out what value of v corresponds to this value of the CDF, then the resultant value from the v_i list[13] represents a random variable with the same distribution as the original data list. In essence, if we consider the CDF as a function that maps the values of a random variable to the region (0,1), then the procedure we are outlining here is the inverse function of the CDF.

To show how this works in practice, suppose that we had 100 values of the random variable from the process, as shown in figure A.14.

This list is now sorted and the CFD calculated. Because of the way we must implement the custom distribution in ithink, we do not calculate the CDF at each data point in the sorted list. Rather, we create a list where the values that correspond to the CDF at equal intervals between 0 and 1 are shown. For this example, we will choose 11 values of the CDF at 0, 0.1, 0.2, . . . 0.9, and 1.0. We interpolate the sorted data list to find out what value of v corresponds to values of (i/M) equal to 0, 0.1, 0.2, . . . 0.9, and 1.0. This leads to figure A.15.

We can now generate a random number on the range (0,1) and interpolate this table to find the value of V corresponding to this value of the CDF. This procedure can be implemented simply in ithink using a graphical function in a converter. A converter is defined, and the equation RANDOM(0,1) is entered as the input to the graph. Then the To Graph button is selected on the converter dialog box. The input is defined as 0 to 1 with 11 data points.[14] This gives the list 0.0, 0.1, 0.2, . . . 0.9, 1.0 as in the right hand column of figure A.15. Then the output column is defined using the left-hand column of figure A.15. Figure A.16 shows what the To Graph page in the converter looks like when the data have been entered.

Now as we run MODEL A.16.ITM, the converter output gives a random distribution equivalent to the original distribution in figure A.14.

---

13. Note that because we interpolate, we will actually end up with a value in between two values of v in the sorted list.
14. The way the graphical functions are implemented in ithink, we can only specify the input variable in equal intervals over any given range.

3.58	5.21	4.21	4.79	5.79
5.35	5.37	3.84	6.67	5.46
5.32	5.78	4.93	3.05	4.17
5.75	5.67	3.73	4.56	5.18
5.15	4.28	3.21	4.55	3.88
3.23	6.25	4.82	3.60	5.47
5.20	6.18	6.57	4.72	3.75
4.74	5.45	5.76	4.38	5.82
6.11	5.54	4.87	4.90	6.12
4.39	4.10	3.49	5.97	6.54
4.14	4.54	3.83	5.23	4.75
4.29	5.29	3.38	4.04	4.46
4.79	5.07	4.35	5.01	3.64
4.90	4.99	5.03	5.08	4.34
5.10	3.29	3.58	6.18	3.93
4.18	5.01	4.70	4.90	5.60
4.86	5.68	4.83	6.03	4.74
3.42	4.63	5.26	6.23	4.62
5.35	4.46	6.56	4.55	5.86
4.18	5.79	4.66	3.83	5.00

FIGURE A.14. Values of a random variable

A.10 The relative comparison of random variation

When we specify a variable in a model using a particular random distribution, one question of interest is this: how much random variation does this distribution create in the model versus other distributions? The motivation for this was covered in chapter 2, where we said that random variation in a process makes the behavior of a process more difficult to predict. And, of course, the larger the amount of random variation in the process inputs, the greater the difficulty in predicting behavior. So how can we measure the amount of variation in a distribution?

One way to study the amount of variation is to look at the value of σ for the distribution. Take for example, the pass/fail distribution discussed in section A.7. The equation for σ was given in equation A.16 as $\sqrt{[p(1-p)]}$. Applying simple calculus, it is easy to show that this has a maximum at $p = 0.5$. Therefore, as the probability of success drops, not only will the mean number of passing items fall, but also the amount of variability will increase, making it more difficult to predict what will happen in a particular process in the future.

A second way to study the amount of variability in a distribution is to look at the ratio of the standard deviation of a distribution to the mean of the distribution, σ/μ. This is referred to as the coefficient of variation (COV) or (less frequently) as the relative standard deviation. For a given mean, the lower the value of the COV, the less variation is being introduced into the model by the random variable.

V	CDF
3.05	0.0
3.60	0.1
4.10	0.2
4.38	0.3
4.66	0.4
4.86	0.5
5.03	0.6
5.29	0.7
5.67	0.8
6.03	0.9
6.67	1.0

FIGURE A.15. CDF values for various values of the random variable, V

For example, consider the normal distribution described in section A.4. In theory, σ and μ are independent of each other. However, many of the variables that are used to model business processes are nonnegative.[15] Most of the normal distribution is covered by the range $(\mu-3*\sigma)$ to $(\mu+3*\sigma)$, which means that $(\mu-3*\sigma)$ will be greater than 0, and hence (σ/μ) is less than 1/3.[16] Thus, for nonnegative

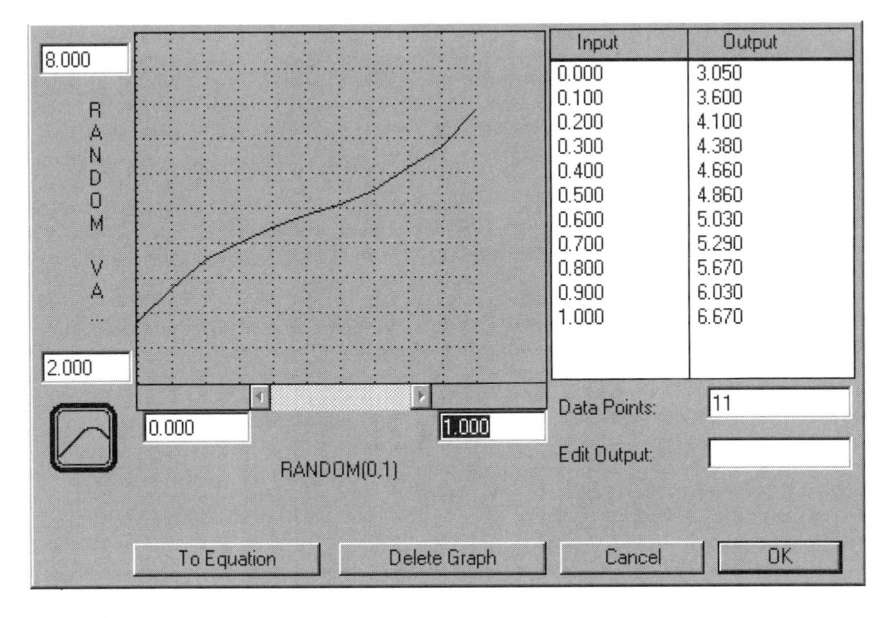

FIGURE A.16. Converter graph page configuration for a custom distribution

15. For example, cycle times, yields, and costs.
16. In theory, 99.73 percent.

variables that follow a normal distribution, the maximum COV is 1/3. In fact, a nonnegative random variable with a COV > 1/3 will tend to have a skewed distribution and so cannot be represented by a normal distribution. The zero value effectively acts as a wall, so the values bunch up against this wall. This is why random variables that are nonnegative and highly variable are generally modeled using a distribution other than the normal distribution.[17]

Compare the situation we just worked through to the case of the exponential distribution. In this case, we said that $\mu = \sigma = \lambda$. The COV $= \sigma/\mu = \lambda/\lambda = 1$! Thus, we see that the exponential distribution will introduce significantly more random variation into a model than a normal distribution for a given value of the mean, μ. This phenomenon is illustrated in chapter 4.

The information contained in this appendix provides the basis for including random variation in the models we created as we worked through this book. It puts us in a position to decide where random variation should be added to a model and what type of probability distribution will be best for a particular situation.

---

17. For example, stock prices cannot go below zero, so they are sometimes modeled using a lognormal distribution.

Appendix B

Economic Value Added

Economic Value Added (hereafter referred to an EVA) is a measure of financial performance developed by Sloan and Stern.[1] It is designed to measure how much value an organization creates for its shareholders (owners) over a period of time. It is composed of two components, profits and a capital charge. Formally, it is defined as this:

$$\text{EVA}^{\text{TM}} = \text{Net Operating Profit after Tax} - \text{Capital Charge} \tag{B.1}$$

Typically, organizations that increase the rate at which they create EVA are rewarded in the stock market with an increasing stock price. EVA reflects the two main ways that an organization creates value for its stockholders. It can increase its profitability or it can reduce the amount of money tied up in capital. An organization that uses EVA will use the measure to decide if an activity is worth doing or not. Thus, an activity will be worth doing if it increases the organization's profits more that the increase in capital that might be involved in the activity. We will now break down the components of EVA.

The Net Operating Profit After Tax (hereafter referred to as NOPAT) is given by the following:

$$\text{NOPAT} = (\text{Sales} - \text{Operating Expenses}) * (1 - \text{Tax Rate}) \tag{B.2}$$

The capital charge (hereafter referred to as CC) reflects the lost value to the shareholders because the organization has money tied up in capital that could be used elsewhere to make wealth for the shareholder. CC is given by this equation:

$$\text{CC} = \text{Cost of Capital} *\$ \text{ Tied Up In Capital} \tag{B.3}$$

In order to understand what we mean by the cost of capital, consider how a company finances the money it ties up in capital. There are basically two ways in which companies finance capital:

1. By getting a loan from the bank. The cost of capital in this case is the current interest rate. Call this interest rate I.

1. See Stern 1991.

2. By using its profits. However, these profits belong to the shareholders, so the organization is financing the capital by using shareholders' equity. When the organization uses equity to finance the capital, the shareholders have an expectation of a rate of return. This rate of return depends on the industry segment of the organization. Call this expected rate of return E.

An organization generally finances its capital needs with a mix of loans and equity. Suppose that the fraction of capital that an organization finances from equity is F. Then the Cost of Capital is defined by this equation:

$$\text{Cost of Capital} = E * F + I * (1 - F) \tag{B.4}$$

To illustrate the calculation, suppose that the current interest rate is 5 percent and the expected rate of return on shareholder equity is 15 percent. The organization finances 80 percent of its capital needs through equity and 20 percent through loans. Then the cost of capital is 15%*80%+5%*(1–80%) = 13%.

Because the current interest rate will be less than the shareholders' expected rate of return on equity, it might appear from equation B.4 that the cost of capital can be reduced by financing more of the capital by loans and not equity. However, financing too much of an organization's capital by loans is not a good idea. The shareholders know that if there is an issue with the organization, all the loans will have to be repaid before they get their money!

It is interesting to compare the EVA measure with the approach proposed by Goldratt and Cox (1992), in which they outline the theory of constraints. They argue that the three main measures that an organization should be interested in are Thruput, Operating Expenses, and Inventory—where all three of these measures are expressed in dollars. Comparing this approach to the definition of EVA presented above, it can be seen that EVA is a measure that contains all three effects together. This indicates that an organization using EVA as a measure of performance is following Goldratt and Cox's approach. However, we would contend that the EVA approach is better as an overall measure because it combines all three effects naturally into one measure. Thus, it automatically provides a way to compromise among the three effects of Thruput, Operating Expense, and Inventory.

An organization generates EVA over a period of time just like cash. When the EVA is generated over a long time period, it will generally be of interest to consider the time value of the EVA so that a net present value of the EVA can be calculated. In order to do this, future EVA values must be discounted. When this is done, the Cost of Capital is used as the discount rate.

The following example shows how EVA can be used to help make a decision about a business initiative.[2] Suppose that an organization has a sales volume of 1,000,000 units per year with a sales price of $10 per unit. The operating cost of the organization is composed of these two components:

1. a fixed cost of $1,000,000 (not including any capital charges)
2. a variable operating cost of $5 per unit

---

2. For simplicity, we will ignore any considerations of the time value of money.

The organization's tax rate is 30 percent. The organization has $2,000,000 tied up in capital and its cost of capital is 20 percent. It has been estimated that the demand for the organization's product is 1,250,000 units per year. An initiative has been proposed to increase the organization's capacity to 1,250,000 units per year. The initiative will increase the amount of capital by $2,000,000 and the fixed operating cost by $500,000. It will also increase the variable operating cost to $5.25 per unit. You have been asked to use an EVA analysis to decide if this initiative should be carried out. You proceed by calculating the current EVA and then the EVA with the initiative completed. Comparing the two EVA values will give you the answer.

Current EVA:

$$\text{NOPAT} = (1{,}000{,}000*\$10 - 1{,}000{,}000*\$5 - \$1{,}000{,}000)*(1-0.30)$$
$$= \$4{,}000{,}000*0.70 = \$2{,}800{,}000$$

$$\text{CC} = 0.2*\$2{,}000{,}000 = \$400{,}000$$

$$\text{EVA} = \text{NOPAT} - \text{CC} = \$2{,}800{,}000 - \$400{,}000 = \$2{,}400{,}000$$

EVA with initiative implemented:

$$\text{NOPAT} = (1{,}250{,}000*\$10 - 1{,}250{,}000*\$5.25 - \$1{,}500{,}000)*(1-0.30)$$
$$= \$4{,}437{,}500*0.70 = \$3{,}106{,}250$$

$$\text{CC} = 0.2*(\$2{,}000{,}000 + \$2{,}000{,}000) = \$800{,}000$$

$$\text{EVA} = \$3{,}106{,}250 - \$800{,}000 = \$2{,}306{,}250$$

The EVA has dropped. From an EVA perspective, then, you would not recommend this initiative. Notice, however, that NOPAT would increase if the change were implemented. Based on NOPAT only, we would recommend the change. So while profits would increase, the capital cost to generate these profits is too great and will not give a good enough return on the shareholder investment. The organization will have to find a way to increase the thruput without having to invest so much in capital. This is one of the advantages of EVA: it penalizes an organization for simply spending capital to improve performance!

Appendix C

Derivation of Equations 6.2, 6.3, and 6.4

Figure C.1 shows a block diagram of the repair model. Because the number of repair persons is the same or more than the number of machines, a repair person is waiting to work on the machine every time one fails. Therefore, no machines are in the queue "Machine Failed." This greatly simplifies the analysis because we do not have to worry about the number of machines in the queue.

Denote the average number of machines by N_M, the average number of machines in the working state as N_{MW} and the average of those being repaired as N_{MR}. Then we can use this equation to say the following:

$$N_M = N_{MW} + N_{MR} \tag{C.1}$$

By definition, the average cycle time of machines in the working state is *MTBF* and in the repair state is *MTTR*. Then by Little's Law, at steady state, the thruput, T_P, is given by the following equation:

$$T_P = \frac{WIP}{Cycle\,Time} = \frac{N_{MW}}{MTBF} = \frac{N_{MR}}{MTTR} \tag{C.2}$$

Combining equations C.1 and C.2 gives the following:

$$\frac{N_{MW}}{MTBF} = \frac{N_{MR}}{MTTR} = \frac{N_M - N_{MW}}{MTTR} \Rightarrow \frac{N_{MW}}{MTBF} = \frac{N_M}{MTTR} - \frac{N_{MW}}{MTTR} \tag{C.3}$$

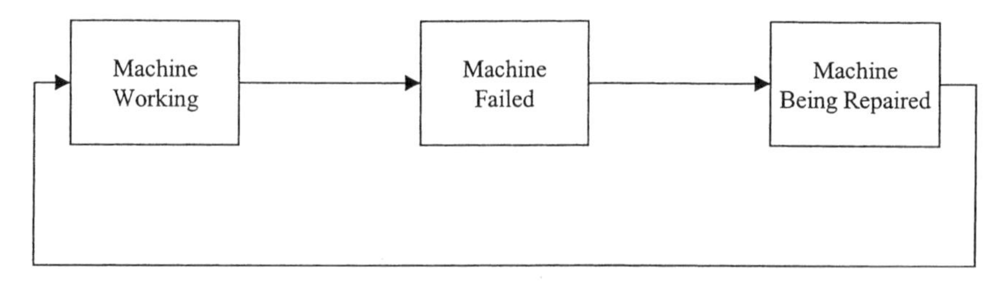

FIGURE C.1. Block diagram of machine repair process

This can be simplified to give

$$N_{MW}\left(\frac{1}{MTBF}+\frac{1}{MTTR}\right)=\frac{N_M}{MTTR}\Rightarrow N_{MW}=\frac{N_M MTBF}{MTBF+MTTR} \quad \text{(C.4)}$$

Substituting equation C.4 back into equation C.2 gives the following:

$$T_P=\frac{N_{MW}}{MTBF}=\frac{N_M}{MTBF+MTTR} \quad \text{(C.5)}$$

The repair person usage, *RU*, is given by the following:

$$RU=\frac{N_{MR}}{N_M}=\frac{N_M-N_{MW}}{N_M}=1-\frac{N_{MW}}{N_M}=1-\frac{MTBF}{MTBF+MTTR}=\frac{MTTR}{MTBF+MTTR} \quad \text{(C.6)}$$

The system availability, SA, is given by the following equation:

$$SA=\text{Probability}(N_{Wt}=N_M) \quad \text{(C.7)}$$

where N_{Wt} is the number of working machines at any time *t*. Because all the machines behave independently from one another, the probability that a particular machine is in a working state is independent of the states of the other machines. Therefore, we can use this equation to say the following:

$$SA=P(N_{Wt}=N_M)=P(\text{Machine1Working}) * P(\text{Machine2Working})\ldots \quad \text{(C.8)}$$

The probability that a machine is working is given by the following:

$$P(Machine\ Working)=\frac{MTBF}{MTBF+MTTR}=1-\frac{MTTR}{MTBF+MTTR}=1-RU \quad \text{(C.9)}$$

Therefore, we can say

$$SA=P(N_{Wt}=N_M)=(1-RU) * (1-RU) * \ldots =(1-RU)^{N_M} \quad \text{(C.10)}$$

Appendix D

Optimization Techniques for the Customer Interface Model

D.1 Optimizing the model

We are sure that we want to maximize EVA for the store problem (see appendix B). But this requires us to complicate the model with a series of cost and revenue equations. Furthermore, learning the optimization process requires that we learn another simple (but special) program. Before taking on the ultimate question, let us choose a physical objective and exercise our optimizing program. It is simpler, and it provides a way of introducing the optimization program.

Let us temporarily assume our short-term mission is to minimize the number of turned-away or unrequited customers using the model shown in figure 8.8. We will use our four control variables to accomplish this goal. The question becomes this: What are the values of the control variables that minimize the number of unrequited customers?

For this answer, we turn to another program, BERKELEY MADONNA™. This program was originally designed to allow compiling and rapid running of ithink equations. It has grown into a substantial program of its own, containing a useful optimization feature. Speed and optimization are important advantages BERKELEY MADONNA™ offers. A key disadvantage of the program is its somewhat difficult symbolic programming language. We contend that the simple iconographic programming nature of ithink allows people with different kinds of skills and backgrounds to quickly learn a common language and accurately and knowingly contribute their own expertise to the model. The ithink equations can then be easily converted by BERKELEY MADONNA™, compiled, and optimized.

We first minimize the number of customers turned away. This is a physical feature and one most favored by the store managers and salespersons. Second, we add the cost and pricing structure to the model and maximize the EVA.

D.2 Optimizing the physical model

BERKELEY MADONNA™ is found on the CD in the back of this book. The basic operating instructions for this program are provided in the associated HELP file.

To begin, we must save the equations as text from the ithink model (see figure 8.8) and open them in BERKELEY MADONNA™. Then we must modify the ithink equations to capture the function of the conveyor. The equations for the model in figure 8.8 are ready to be compiled and can be run quickly in BERKELEY MADONNA™.

Note especially the equation for STORE. This is the typical integration or summing step in ithink and BERKELEY MADONNA™. It simple says that the STORE level today is equal to the STORE level at the beginning of this time step, DT, plus the net gains, (RECEIVING—SELLING) times the time step, DT. We will always use a DT of 1.0 for these models. Another way to look at this equation is to take the STORE level at the start of the DT to the left side of this equation to form the difference, Δ STORE. Then divide both sides of this equation by DT. The result is the rate of change of STORE: simply, RECEIVING minus SELLING. This is the differential form of the equation. It is stated in the following equations in the integration or summing form.

The conveyor is not automatically translated into the language of BERKELEY MADONNA™ from ithink. Note carefully how the first two equations (SHIPPING and RECEIVING) replace the corresponding ones in the ithink equations.

The only other issue is that of blocking the stocks and preventing certain flows from becoming negative. These blocks are seen in the last four equations. The remaining equations in ithink copy directly into BERKELEY MADONNA™. Simply open the ithink equations window, go to the File menu, select Save Equations as Text, and open the equations into the edit window in BERKELEY MADONNA™. Then make the changes noted earlier until that window shows the equations listed here.

```
SHIPPED = CONVEYOR(ORDERING,TRANSIT_TIME)
      INIT_SHIPPED = 2
      TRANSIT_TIME = 7
RECEIVING = outflow(SHIPPED)
ORDERING = SELLING—round(REACTION_STORE*(STORE -
      STORE_TARGET) + REACTION_SHIPPED*(SHIPPED-
SHIPPED_TARGET))
STORE(t) = STORE(t – DT) + (RECEIVING – SELLING) * DT
      INIT STORE = 2
SELLING = IF STORE –1 – DEMAND > 0 THEN DEMAND ELSE
STORE – 1
UNREQUITED CUSTOMERS (t) = UNREQUITED CUSTOMERS (t – DT) +
      (TURNING AWAY) * DT
      INIT UNREQUITED CUSTOMERS = 0
TURNING AWAY = DEMAND – SELLING
SELLING = POISSON(2)
REACTION_SHIPPED = .05
REACTION_STORE = .3
SHIPPED_TARGET = 8
STORE_TARGET = 8
```

LIMIT STORE > = 1
LIMIT SELLING > = 0
LIMIT SHIPPED > = 0
LIMIT TURNING AWAY > = 0

Once the equations are in the BERKELEY MADONNA™ editing window, choose the Compile command, select Optimize, type in UNREQUITED_CUSTOMERS in the *minimize* window, and specify the four parameters and their approximate values as noted in the dialog box. Run the program and note how the controls change in the parameters window to optimal levels for that run.

Set up a graph that shows the UNREQUITED CUSTOMERS versus time. Run the optimization several times and note the variety of solution sets you get on subsequent runs. The variation here is due to the variation in sales, coupled with the shipping lag time. Run the graph for each set and search until you find one that gives a very low number of customers turned away yearly. On another graph, show the STORE variable and how it is changing. You should notice increases in the peak and mean store inventory for runs with very low numbers of customers turned away annually. This is a tradeoff! Try the following parameter values: R_SH = 0.043, R_ST = 0.173, SH_T = 3.48, and ST_T = 14.4.Compare them with the values given in the preceding equations figure. The result is a significantly reduction in the number of turned-away customers and an increase in the peak and mean store inventory. Note that this set of parameters gives generally desirable results in run after run. But also notice the variation between runs with the same parameters. The reason for this variation is not difficult to understand. Each run is finding a valley in the mountain range of possible UNREQUITED CUSTOMERS. Which valley is truly the lowest? From a single run, we cannot tell. But after many runs, we can become more confident that the lowest one seen is the absolute lowest.

> Learning Point: With randomness built into a model, the precise optimum solution is elusive. This elusiveness depends to some extent on where in the model the randomness and delays are located.

D.3 Disappointment with the physical criterion for optimization

What is the answer here? Neither set of parameters is "correct," because the full set of important variables is not being considered. Furthermore, we have not yet introduced a consistent way of weighing these variables. We are not considering the handling effort that must go on with various rates of sales nor the cost of space to store the peak amount of inventory. We have previously expressed concerns over the cost of making inventory and of turned-away customers. What we must do now is account for all of these costs in a systematic, commeasurable way and then determine the four parameters such that they help us meet our goals:

- minimize cost;
- maximize profits or EVA (if we know the selling price);
- maximize market share (if we are trying to gain a larger share of the market); and
- assure that we reach long-range profit goals (satisficing).

Regardless of our goals, we need a consistent set of financial considerations and definitions. Nevertheless, we have used a simple physical criterion to learn how to optimize the controls to meet a given objective.

> Learning Point: Management controls, such as inventory targets and reaction rates, used to determine ordering policies are too important to be left to the individual managers. The parameters are too important to be set without an objective criterion that encompasses the proper weighting of all the variables.

D.4 Finding the optimal financial controls

The very first step must be the definition of a proper management objective. As noted earlier, profit—the gap between revenues and costs—is involved in every business decision. One may simply maximize current profit; one may try to extend market share with the idea of controlling long-range profits; or one may set a plan into motion with a profit goal in a specified time (satisficing).

In the financial section of chapter 8, we discussed the objective of maximizing the cumulative discounted profit as a device for choosing among alternative plans. A closely related objective is to choose the EVA criterion, that is, to maximize the EVA. EVA is the after-tax profits less all direct and indirect costs. Direct costs are such expenses as labor for handling and selling. Indirect costs include the capital costs that are the opportunity cost of money spent to build the inventory and the money used to provide secure and attractive storage of the inventory. Another indirect cost is the expected present valued cost of lost customers. An optimization might be done to maximize the annual EVA or the cumulative present value of a sequence of annual EVA values.

The present value of lost customers is indeed difficult to determine. One can imagine a lost sale due to a lack of stock, causing that unrequited customer to form the habit of shopping elsewhere for the rest of his or her life. One could then examine the present value of the lost purchases during that time as a part of the cost of turning away a customer.

The DAILY EVA is a definition of the current profit rate. We will choose to maximize the ANNUAL AVERAGE EVA as our management objective and leave the present valued analysis as a homework problem. Thus, we will multiply the average DAILY EVA by 365 to get the annual value. The expected stockholder return rate is assumed to be 15 percent per year or 15/365 percent per day. The

SELLING PRICE is \$100. The unit storage cost is \$2 per widget. This is the opportunity cost of the next most profitable item in the remainder of this firm. We lose \$1,000 every time we turn a customer away (The price of the unsold widget plus the present value of a lifetime of lost sales.) The HANDLING COST is \$50 per widget. To get the storage cost, we must find the peak inventory in the STORE. This will set up the size of the space needed for the sale of widgets. The ithink model is shown in figure D.1. Go through each piece of the financial section to fully understand the individual calculations leading up to the DAILY EVA.

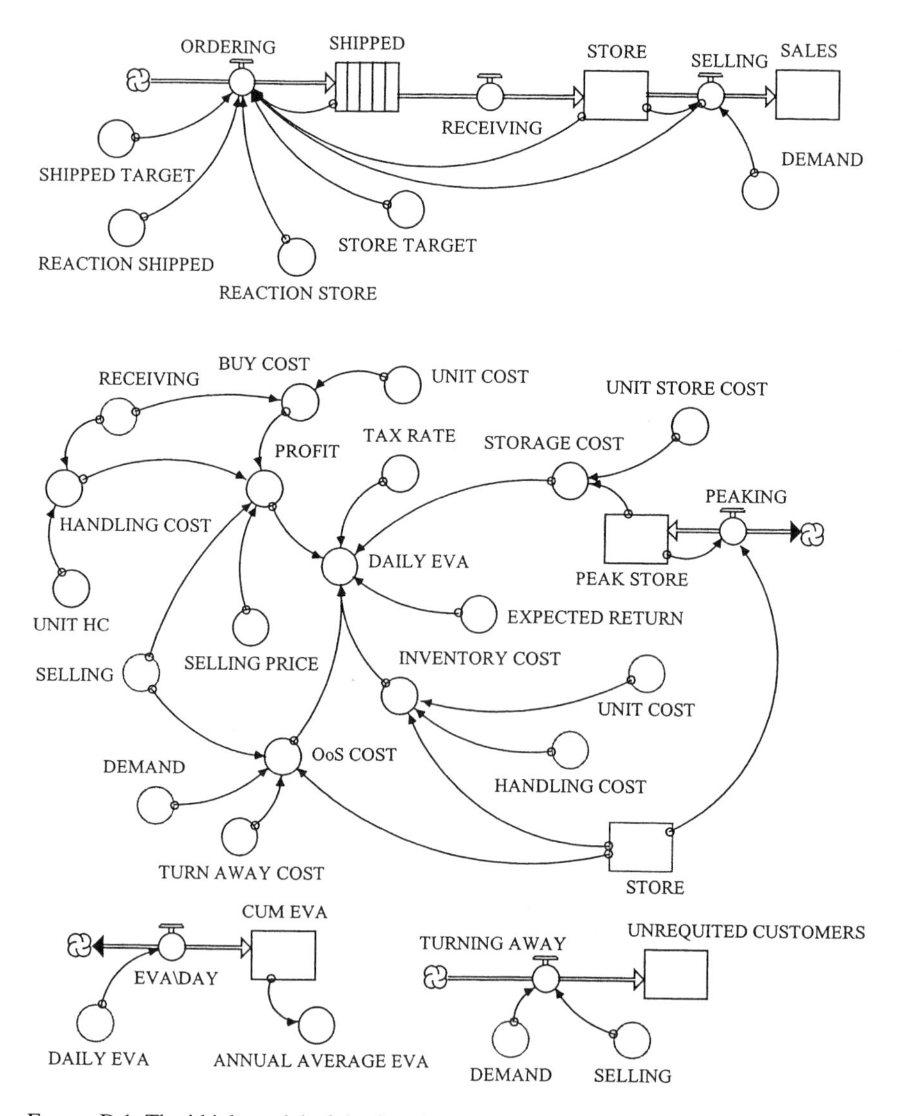

FIGURE D.1. The ithink model of the firm in which revenues, costs, and profits have been added: EVA is the economic value added; see text and appendix B for the definition

Next note the way in which the ANNUAL AVERAGE EVA is calculated. Also carefully note the way in which the number of UNREQUITED CUSTOMERS and the PEAK STORE are calculated. These are generic forms for such problems.

Here is the complete model. The upper part is the same as the model shown in figure 8.8. The central part is where all the financial calculations are made. Note how costs are separately calculated, collected, and eventually removed from after-tax revenues to form the EVA. This value is converted to the ANNUAL AVERAGE EVA in the lower section, where the number of turned-away customers is also calculated.

The equations, when transferred to BERKELEY MADONNA™ and when the conveyor language is converted, should look like this:

```
CUM_EVA(t) = CUM_EVA(t – dt) + (EVADAY) * dt
      INIT CUM_EVA = 0
EVADAY = DAILY_EVA
PEAK_STORE(t) = PEAK_STORE(t – dt) + (PEAKING) * dt
      INIT PEAK_STORE = 0
PEAKING = IF STORE > PEAK_STORE THEN STORE – PEAK_STORE
ELSE 0
SALES(t) = SALES(t – dt) + (SELLING) * dt
      INIT SALES = 0
SELLING = IF STORE –1 – DEMAND > 0 THEN DEMAND ELSE STORE -1
SHIPPED = CONVEYOR(ORDERING,TRANSIT_TIME)
      INIT_SHIPPED = 7
      RECEIVING = outflow(SHIPPED)
ORDERING = SELLING—REACTION_STORE*(STORE—
STORE_TARGET) –
      REACTION_SHIPPED*(SHIPPED – SHIPPED_TARGET)
UNREQUITED_CUSTOMERS(t) = UNREQUITED_CUSTOMERS(t – dt) +
      TURNING_AWAY*dt
      INIT_UNREQUITED_CUSTOMERS = 0
TURNING_AWAY = DEMAND – SELLING
ANNUAL_AVERAGE_EVA = CUM_EVA/(TIME+.001)*365
BUY_COST = UNIT_COST*RECEIVING
DAILY_EVA = PROFIT*(1 – TAX_RATE) -(INVENTORY_COST-
STORAGE_COST)*EXPECTED_RETURN – OoS_COST
DEMAND = POISSON(2)
EXPECTED_RETURN = .15/365
HANDLING_COST = UNIT_HC*RECEIVING
INVENTORY_COST = STORE*(UNIT_COST+HANDLING_COST)
OoS_COST = IF (STORE = 1) AND (SELLING > 0) THEN
      TURN_AWAY_COST*(DEMAND-SELLING) ELSE 0
PROFIT = SELLING_PRICE*SELLING – HANDLING_COST – BUY_COST
REACTION_SHIPPED = .017
REACTION_STORE = .099
SELLING_PRICE = 100
```

SHIPPED_TARGET = 10.4
STORAGE_COST = PEAK_STORE*UNIT_STORE_COST
STORE_TARGET = 12.2
TAX_RATE = .5/365
TURN_AWAY_COST = 1000
UNIT_COST = 20
UNIT_HC = 50
UNIT_STORE_COST = 2
LIMIT SHIPPED ≥ 0
LIMIT STORE ≥ 0
LIMIT ORDERING ≥ 0
LIMIT SELLING ≥ 0

We set the optimization option to minimize the quantity – ANNUAL AVERAGE EVA, using the four control parameters (this is equivalent to its *maximization*). The resulting typical annual average EVA is $21,000, as shown in figure D.2. The controls are shown in the preceding equations.

This parameter choice is rather critical. Run the BERKELEY MADONNA™ model optimization repeatedly and note the variation in these parameters and the resulting changes in the annual average EVA. Note that for some runs, the maximum annual average EVA remains unchanged; in contrast, the store inventory level does change. Still other optimization runs give completely different levels of profit. We are again facing a mountain range of possible peak EVA values and each run is finding another peak. Only after running a significant number of times can one be assured that the true peak has been found.

The number of turned-away customers under the optimum solution is about 25 per year out of total of about 700 customers per year. This number of turned-away

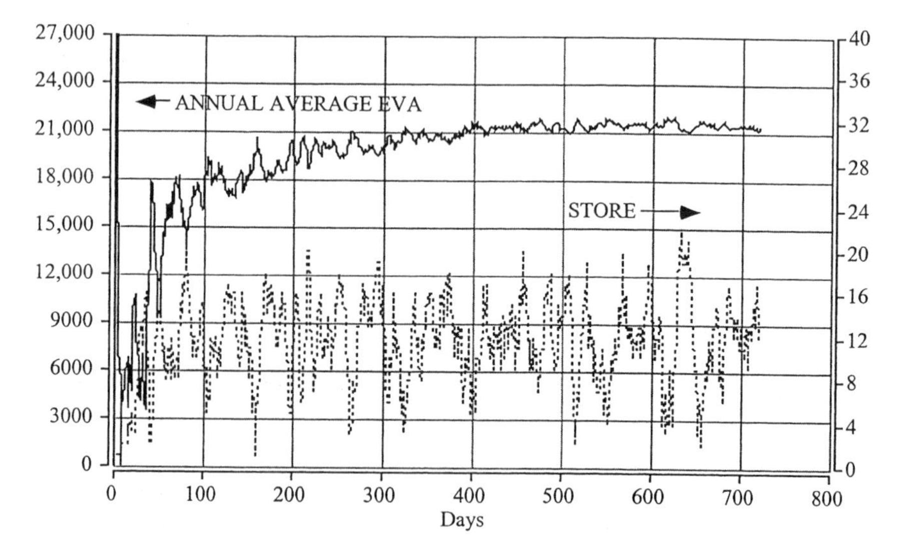

FIGURE D.2. The maximization of the ANNUAL AVERAGE EVA and the resulting variation in STORE inventory

customers may seem surprising at first glance because the assumed cost per turned-away customer is $1,000.

Associated with this set of controls is a substantial variation in the STORE inventory. The store will require space for a maximum of about 20 widgets. We can reduce this quantity only at the risk of turning more customers away or reducing profits. Continue with the following exercises:

1. Play with this model. Verify some of the points just mentioned. Vary the TURN AWAY COST and see the resulting variation in STORE and UNREQUITED CUSTOMERS. Vary the handling cost and note its effects. Did you expect to see these reactions?
2. Assume that SELLING is price sensitive. That is, assume the manager can adjust the selling price. Imagine the relationship to be SELLING PRICE = 120 – 10*DEMAND. Find the best operating parameters for the store manager, the most likely ANNUAL AVERAGE EVA, and the mean selling price for this demand curve.
3. Suppose that the transit time for shipped widgets is not fixed. Instead, it varies according to POISSON(7) days, with the transit time never less than three days. Run the optimization program and determine the best set of controls. Advise this manager on what to expect. How much different is the amount of space that must be allocated in the store?
4. Suppose the vender of these widgets offered the option of reducing the variation in shipping time to POISSON(6), where the shortest transit time is four days and the longest is 10 days, for an annual additional cost to the store of $1,500. Would you do it? Here we are addressing the cost of variation of the flows within a system. Such variations, albeit costly, may be too costly to reduce. Today, firms with single-source vendors frequently relay their current sales data to the vendors so that the vendors can adjust their production rates accordingly. Can you develop a model to incorporate this process?
5. Change the Out-of-Stock costing arrangement slightly. Have the number of unrequited customers reduce the mean in the DEMAND function slightly, in proportion to the number of turned-away customers.
6. Instead of maximizing the AVERAGE ANNUAL EVA, accumulate the present value of the DAILY EVA and maximize this sum. Use the EXPECTED RETURN rate to find the present value of each future DAILY EVA. Maximize the sum of these accumulated present-valued terms.

Learning Point: Many different physical elements are part of the operation of this simple system. To single out the optimization of one of them will likely make conditions worse for some of the others. For example, minimizing the number of turned-away customers greatly increased the onsite inventory need. An objective function is required. To use such a function, all of the physical parameters must be monetized to allow them to be sensibly combined. Then the objective can be optimized.

Learning Point: A model is never completely representing reality. It is always unfinished, always needing modification to meet some newly recognized special feature of the real world. At some point, however, incorporating these additional features ceases to add to the skill with which you meet the objective.

Appendix E

System Requirements for the CD-ROM

E.1 ithink 8® demo installation instructions

To install the ithink 8® demo software:

1. For Windows, double-click the file ithink 8 demo.exe. For Macintosh, double-click the file ithink_8_demo_installer.hqx.
2. Follow the instructions in the installer wizard.

E.2 ithink 8® demo system requirements

Windows minimum:

Pentium-class processor
Microsoft Windows® 95/98/NT/2000/XP
16 MB RAM
35 MB of hard-disk space required
VGA display of at least 256 colors

Windows recommended:

233 MHz Pentium (or better)
Microsoft Windows® 95/98/NT/2000/XP
64 MB RAM or more
70 MB hard-disk space
16-bit color (or better)
Soundblaster-compatible sound card
QuickTime 4.0

Macintosh minimum:

PowerPC
System software 7.1 through 9 (Classic mode in OS X)
16 MB RAM (4 MB available to the program)
35 MB of hard-disk space required
256 colors
Quicktime

Macintosh recommended:

120 MHz PowerPC (or better)
MacOS 8.1 (or higher)
32 MB RAM
70 MB hard disk space
Thousands of colors
Quicktime 4.x
IE 4.x (or later)

E.3 Berkeley Madonna system requirements

Windows:

Berkeley Madonna for Windows runs on personal computers with Intel x86 or compatible processors under Microsoft Windows® 95, Windows® NT 4.0, and later versions of these operating systems. If you want to use the Flowchart Editor flowchart.html, ensure that JRE 1.1.8 is installed on your system. *Important: Berkeley Madonna's Flowchart Editor does not work with newer versions of the JRE; you must install version 1.1.8 in order for it to work.* You can download JRE 1.1.8 from Sun Microsystems, http://java.sun.com/products/archive/jdk/1.1.8_010/jre/index.html.

Macintosh:

Berkeley Madonna for Macintosh runs on an Apple Power Macintosh and 100% compatible computers with a PowerPC CPU (601, 604, G3, etc.). Older machines with 68K CPUs are not supported. If you want to use the Flowchart Editor flowchart.html, ensure that MRJ 2.1 (Macintosh runtime for Java™) or later is installed on your system. You can download the latest version of MRJ from Apple Computer, http://www.apple.com/java.

Bibliography

Allen, T.F.H., and T.B. Starr. 1982. "Hierarchy as a Context for Mathematical Modeling." In *Hierarchy.* Chicago: University of Chicago Press.

Allen, T.F.H., and E.P. Wyleto. 1983. "A Hierarchical Model for Complexity of Plant Communities." *Journal of Theoretical Biology* 101: 529–40.

Anderson, P.W. 1995. *Science* 267.

Arthur, B. 1990. "Positive Feedback in the Economy." *Scientific American* (February): 92–99.

Bequette, B.W. 1998. *Process Dynamics Modeling, Analysis, and Simulation.* Upper Saddle River, N.J.: Prentice Hall.

Botkin, First Name. 1977. *Title.* New York: John Wiley.

Breyfogle, F.W. 1999. *Implementing Six Sigma: Smarter Solutions Using Statistical Methods.* New York: Wiley-InterScience.

Conquest, R. 1993. "History, Humanity and Truth." Lecture broadcast on National Public Radio, September 23.

Deming, W.E. 1986. *Out of the Crisis.* Cambridge, Mass.: Center for Advanced Study, Massachusetts Institute of Technology Press.

Devroye, L. 1986. *Non-Uniform Random Variate Generation.* New York: Springer-Verlag.

Dodson, B., and D. Nolan. 1995. *Reliability Engineering Bible.* Houston: Quality Publishing.

Garvin, D.A. 1998. "Building a Learning Organization." in *The Harvard Business Review on Knowledge Management,* Boston: Harvard Business School Press.

Gleick, J. 1998. Chaos: Making a New Science. New York: Penguin.

Goldratt, E.M., and J. Cox. 1992. *The Goal: A Process of Ongoing Improvement.* Great Barrington, N.Y.: North River Press.

Hammer, M. 1996. "Beyond Reengineering." New York: Harper Business.

Hannon, B., and M. Ruth. 1997. *Modeling Dynamic Biological Systems.* New York: Springer-Verlag.

Harrington, H. 1991. *Business Process Improvement.* New York: McGraw-Hill.

Harris, J.W., and H. Stocker. 1998. *Handbook of Mathematics and Computational Science.* New York: Springer-Verlag.

Hopp, W.J., and M.L. Spearman. 1996. *Factory Physics—Foundations of Manufacturing Management.* New York: McGraw-Hill.

Hoseyni-Nasab, M., and S. Andradottir. 1998. Time Segmentation Parallel Simulation of Tandem Queues with Manufacturing Blocking. Proceedings of the 1998 Winter Simulation Conference, Dec. 13–16, Washington, D.C.

Johnson, A.R. 1993. Spatiotemporal Hierarchies in Ecological Theory and Modeling. Paper presented at the Second International Conference on Integrating Geographic Inform. Systems and Environmental Modeling, Breckenridge, Colorado.

Johnson, H.T. 2001. "Manage by Means, not Results." *The Systems Thinker* 12 (6): 1–5.

Joshi, Y.V. 2000. Information Visibility and Its Effect on Supply Chain Dynamics. Master's thesis, Massachusetts Institute of Technology.

Kaplan, R.S. and D.P. Norton. 1996. *The Balanced Scorecard—Translating Strategy into Action.* Boston: Harvard Business School Press.

Kaplan, R.S., and D.P. Norton. 1996. "Linking the Balanced Scorecard to Strategy." *California Management Review* 39 (1): 53–79.

Lee, H.A., V.P. Padmanabhan, and S. Whang. 1995. "The Paralyzing Curse of the Bullwhip Effect in a Supply Chain." March 1, Stanford University Press.

———. 1997. "The Bullwhip Effect in Supply Chains." *Sloan Management Review* (spring): 53–77.

Lee, Hau L., P. Padmanabhan, and S. Whang. 1997. "Information Distortion in a Supply Chain: The Bullwhip Effect." *Management Science* 43 (4): 546–58.

Levy, F. 1965. "Adaptation in the Production Process." *Management Science* 11 (6).

Lind, R.C., K.J. Arrow, G.R. Corey, P. Dasgupta, A.K. Sen, T. Stauffer, J.E. Stiglitz, J.A. Stockfisch, and R. Wilson. 1982. *Discounting for Time and Risk in Energy Policy.* Washington, D.C.: Resources for the Future.

Marek, R.P., D.A. Elkins, and D.R. Smith. 2001. Understanding the Fundamentals of Kanban and CONWIP Pull Systems Using Simulation. Proceedings of the 2001 Winter Simulation Conference.

McConnell, J. 1997. *Metamorphosis.* Brisbane, Australia: Wysowl.

Myers, R. 1990. Classical and Modern Regression with Applications. 2d ed. Belmont, Calif.: Duxbury Press.

Nadler, G., and W. Smith. 1963. "Manufacturing Progress Functions." *International Journal of Production Research* 2 (2).

Nahmias S. 1997. *Production and Operations Analysis.* Homewood, IL: Times Mirror Higher Education Group.

O'Connor, P. 1991. *Practical Reliability Engineering.* 3d ed. New York: John Wiley & Sons.

Pagels, H. 1988. *The Dreams of Reason.* New York: Simon & Schuster.

Ross, S. 1970. *Applied Probability Models with Optimization Applications.* San Francisco: Holden-Day.

Rozanov, Y.A. 1977. *Probability Theory: A Concise Course.* New York: Dover.

Ruth, M., and B. Hannon. 1997. *Modeling Dynamic Economic Systems.* New York: Springer-Verlag.

Schneiderman, A. 1986. "Optimum Quality Costs and Zero Defects: Are They Contradictory Concepts?" *Quality Progess* (November): 28.

———. 1988. "Setting Quality Goals." *Quality Progress* (April): 51.

Scholtes, P. 1998. *The Leader's Handbook.* New York: McGraw-Hill.

Senge, P. 1990. *The Fifth Discipline.* New York: Doubleday/Currency.

Shewart, W. 1980. *Economic Control of Quality of Manufactured Product.* Milwaukee: ASQ Quality Press.

———. 1986. *Statistical Method from the Viewpoint of Quality Control.* New York: Dover Publications.

Stata, R. 1989. "Organizational Learning—The Key to Management Innovation." *Sloan Management Review* 30 (3): 63–73.

Steedman, I. 1970. “Some Improvement Curve Theory.” *The International Journal of Production Research* 8 (3): 189–205.

Sterman, J. 2000. *Business Dynamics.* New York: Irwin McGraw-Hill.

Stern, J. M., J.S. Shiely, and I. Ross. 1991. *The EVA Challenge—Implementing Value Added Change in an Organization.* New York: John Wiley & Sons.

Stommel, H. 1963. “Varieties of Oceanographic Experience.” *Science* 130: 573–76.

Suresh, S., and W. Whitt. 1990a. “Arranging Queues in Series: A Simulation Experiment.” *Management Science* 36: 1080–91.

———. 1990b. “The Heavy-Traffic Bottleneck Phenomenon in Open Queuing Networks.” Oper. Res. Letters 9: 355–62.

Swaminathan, J., S. Smith, and N. Sadeh-Koniecpol. 1997. “Modeling Supply Chain Dynamics: A Multiagent Approach.” *Decision Sciences* 3 (April): 607–632.

Taylor, W.A. 1991. *Optimization and Variation Reduction in Quality.* New York: McGraw-Hill.

Tsiotras, G.D., and H. Badr. 1989. “Modeling the Blocking Phenomenon of Service Facilities.” *Mathl. Comput. Modeling* 12 (6): 641–49.

Wadsworth, H.M. 1997. *Handbook of Statistical Methods for Engineers and Scientists.* New York: McGraw-Hill.

Weber, R.R. 1979. “The Interchangeability of Tandem ./M/1 Queues in Series.” *J. Appl. Prob.* 16 (9): 690–95.

Wheeler, D.J., and D.S. Chambers. 1992. Understanding Statistical Process Control. Knoxville, Tenn.: SPC Press.

Whitt, W. 1994. “Towards Better Parametric-Decomposition Approximations for Open Queuing Networks.” *Ann. Oper. Res.* 48: 221–48.

Winston, W.L. 1993. *Operations Research—Applications and Algorithms.* 3d ed. Washington, D.C.: International Thompson Publishing.

Wright, T.P. 1936. “Factors Affecting the Cost of Airplanes. *Journal of Aeronautical Science* 3 (3): 122–28.

Index